Agroforestry as a Nature-Based Solution

Agroforestry as a Nature-Based Solution:

A Holistic Approach to Food Security and Climate Change

Edited by

Amit Kumar, Gopal Shukla, Ram Gopal and Dinesha S.

Cambridge Scholars Publishing

Agroforestry as a Nature-Based Solution:
A Holistic Approach to Food Security and Climate Change

Edited by Amit Kumar, Gopal Shukla, Ram Gopal and Dinesha S.

This book first published 2025

Cambridge Scholars Publishing

Lady Stephenson Library, Newcastle upon Tyne, NE6 2PA, UK

British Library Cataloguing in Publication Data
A catalogue record for this book is available from the British Library

ISBN: 978-1-0364-4384-9
ISBN (Ebook): 978-1-0364-4385-6

Contents

CONTRIBUTORS

Sonam Rajput
Wildlife Institute of India, Dehradun-

Deepshikha Nirala
Forest Research Institute Deemed to be University, Dehradun-248001

Karuna Gusain
G.B. Pant National Institute for Himalayan Environment, Kosi-Katarmal, Almora-263643

Vivek Chauhan
G.B. Pant National Institute for Himalayan Environment, Kosi-Katarmal, Almora-263643

Gargi Navpute
Forest Research Institute, Dehradun-248006

Suchitra Surbhi
Forest Research Institute, Dehradun-248006

Santan Barthwal
Forest Research Institute, Dehradun-248006

Hukum Singh
Forest Research Institute, Dehradun-248006

Sarita Bisht
Forest Research Institute, Dehradun-248006

Navjot Singh Kaler
College of Horticulture and Forestry, Neri-Hamirpur, Himachal Pradesh, India-177001

Kamal Sharma
College of Horticulture and Forestry, Neri-Hamirpur, Himachal Pradesh, India-177001

Kritika
College of Horticulture and Forestry, Neri-Hamirpur, Himachal Pradesh, India-177001

Shiwangee
College of Horticulture and Forestry, Neri-Hamirpur, Himachal Pradesh, India-177001

Amit Kumar
CSIR-Institute of Himalayan Bioresource Technology Palampur-176061

Akanksha Bisht
Sam Higginbottam University of Agriculture, Technology and Sciences, Prayagraj, (U.P.)

Praveen Kumar Singh
Faculty of Agricultural Sciences, Mahayogi Gorakhnath University Gorakhpur, (U.P.)

K. Dujeshwer
North-Eastern Hill University, Tura Campus, Meghalaya-794002

Aradhana
Gurukul Kangri (Deemed to be) University, Haridwar, India

Neha
Department of Horticulture & Food Processing, Government of Uttarakhand

Nidhi
Department of Animal Husbandry, Fisheries & Dairy Development, Government of Uttarakhand

Mukesh Kumar
Gurukula Kangri (Deemed to be University), Haridwar, India

Rakesh Negi
Gurukula Kangri (Deemed to be University), Haridwar, India

Dinesha S.
North-Eastern Hill University, Tura Campus, Tura-794002, Meghalaya, India

Shridevi B. Telib
Sri Krishnadevaraya College of Horticultural Sciences, Anantapur-515002, Andhra Pradesh

Naresh Singh
College of Forestry, Ranichauri, Tehri Garhwal, Uttarakhand

R.S. Bali
College of Forestry, Ranichauri, Tehri Garhwal, Uttarakhand

Priyanka Thakur
College of Forestry, Ranichauri, Tehri Garhwal, Uttarakhand

Pankaj Lal
College of Forestry, Ranichauri, Tehri Garhwal, Uttarakhand

Ram Gopal
North Eastern Hill University, Tura Campus, Meghalaya

Gopal Shukla
North Eastern Hill University, Tura Campus, Meghalaya

Khulakpam Apshahana
North Eastern Hill University, Tura Campus, Meghalaya

Khwairakpam Rolince
North Eastern Hill University, Tura Campus, Meghalaya

Varsha Kumari
KVK, Jehanabad, Bihar Agricultural University, Sabour, Bhagalpur, Bihar, India-813210

Muneshwar Prasad
KVK, Jehanabad, Bihar Agricultural University, Sabour, Bhagalpur, Bihar, India-813210

Abhay Kumar
KVK, Jehanabad, Bihar Agricultural University, Sabour, Bhagalpur, Bihar, India-813210

Dinesh Mahto
KVK, Jehanabad, Bihar Agricultural University, Sabour, Bhagalpur, Bihar, India-813210

Rikesh Kumar
Faculty of Forestry, Birsa Agricultural University, Ranchi, Jharkhand, India-834006.

Tanu Shree Lakra
Institute of Forest Productivity, Ranchi, Jharkhand, India-835303

Rashmi Priyadarshi
KVK, Purnea, Bihar Agricultural University, Sabour, Bhagalpur, Bihar, India-813210.

ABBREVIATIONS

IDRC	International Development Research Centre
NTFP	Non-timber forest products
ICRAF	International Centre for Research in Agroforestry
CSA	Climate-Smart Agroforestry
FAO	Food and Agriculture Organization
IPM	Integrated Pest Management
SDG	Sustainable Development Goal
UNFCCC	United Nations Framework Convention on Climate Change
SOC	Soil Organic Carbon
AFS	Agroforestry Systems
ITLS	Integrated Tree-livestock systems
INM	Integrated nutrient management
IWM	Integrated water management
BBF	Broad Bed and Furrow
PMKSY	Pradhan Mantri Krishi Sinchayee Yojana
IWMI	International Water Management Institute
WOTR	Watershed Organization Trust
BD	Bulk density

FC	Field capacity
BNF	Biological nitrogen fixation
IKS	Indian Knowledge System
IK	Indigenous knowledge
ITK	Indian Traditional Knowledge
PAFERN	Philippine Agroforestry Education and Research Network
TEK	Traditional ecological knowledge
MBDA	Meghalaya Basin Development Authority
FRA	Forest right act
AFSI	Agroforestry Systems inventory
MAP	Medicinal and aromatic plant
NMMP	National Mission on Medicinal Plants
MbAF	Millet based Agroforestry Systems
LDN	Land Degradation Neutrality
CAFRI	Central Agroforestry Research Institute
ICFRE	Indian Council of Forestry Research and Education
IIMR	Indian Institute of Millet Research
MSP	Minimum Support Price
MIIRA	Millet International Initiative for Research and Awareness
GHG	Greenhouse gas
APEDA	Agricultural and Processed Food Products Export Development Authority

PREFACE

Agroforestry, the integration of trees and shrubs into agricultural landscapes, stands as a beacon of hope in addressing some of the most pressing challenges of our time—food security and climate change. This book, "Agroforestry as a Nature-Based Solution: A Holistic Approach towards Food Security and Climate Change," delves into the multifaceted benefits and potential of agroforestry systems.

In an era where conventional agriculture often falls short of meeting the rising demands for food while concurrently degrading the environment, agroforestry offers a sustainable alternative. By mimicking natural ecosystems, agroforestry enhances biodiversity, improves soil health, and sequesters carbon, thereby mitigating climate change. This holistic approach not only bolsters food production but also strengthens the resilience of agricultural systems against climate variability and extreme weather events.

The inspiration for this book stemmed from witnessing the transformative impact of agroforestry practices in various regions across the globe. From smallholder farmers in tropical countries to large-scale agricultural enterprises in temperate zones, the integration of trees into farming systems has shown remarkable results. It is this potential to harmonize agricultural productivity with ecological integrity that we seek to highlight and promote.

Throughout this book, we explore a range of topics crucial to understanding and implementing agroforestry systems. We begin with the foundational principles of agroforestry and its historical context, setting the stage for a deeper exploration of its benefits. Subsequent chapters delve into specific agroforestry practices, such as alley cropping, silvopasture, and forest farming, providing practical insights and case studies from around the world. We also address the socioeconomic aspects, emphasizing the role of agroforestry in enhancing livelihoods and promoting social equity.

A significant portion of the book is dedicated to discussing agroforestry's role in climate change mitigation and adaptation. Through detailed analysis and empirical evidence, we illustrate how agroforestry can contribute to reducing greenhouse gas emissions, enhancing carbon sequestration, and building resilient agricultural landscapes.

This book is intended for a diverse audience, including researchers, practitioners, policymakers, and students. By presenting a comprehensive overview of agroforestry's potential as a nature-based solution, we aim to inspire and equip readers with the knowledge and tools necessary to implement and advocate for agroforestry practices in their respective contexts.

As we face the dual challenges of ensuring food security and combating climate change, it is imperative that we embrace innovative and sustainable solutions. Agroforestry, with its myriad benefits, offers a path forward. We hope this book will serve as a valuable resource and a catalyst for action, encouraging a broader adoption of agroforestry systems worldwide.

Sincerely,
Amit Kumar
Gopal Shukla
Ram Gopal
Dinesha S

INTRODUCTION

The global landscape of agriculture and forestry is undergoing profound changes driven by the escalating challenges of food insecurity, environmental degradation, and climate change. As the world population continues to grow, the pressure on our agricultural systems to produce more food is intensifying, leading to unsustainable practices that degrade soils, reduce biodiversity, and contribute to greenhouse gas emissions. In this critical juncture, the concept of agroforestry presents a beacon of hope—a nature-based solution that harmonizes agricultural productivity with environmental sustainability.

Agroforestry, the practice of integrating trees and shrubs into agricultural landscapes, offers a multifaceted approach to land management that enhances biodiversity, improves soil health, and sequesters carbon, thereby playing a pivotal role in mitigating climate change. Unlike conventional monocropping systems, agroforestry promotes a symbiotic relationship between trees and crops, creating resilient ecosystems capable of withstanding environmental stresses. This book, "Agroforestry as a Nature-Based Solution: A Holistic Approach towards Food Security and Climate Change," seeks to illuminate the diverse benefits of agroforestry systems and their potential to transform agricultural practices. By leveraging the natural synergies between forestry and agriculture, agroforestry provides a sustainable pathway to address some of the most pressing issues facing our planet today.

Agroforestry systems are uniquely positioned to offer solutions to multiple global challenges simultaneously. Trees in agricultural landscapes contribute to carbon sequestration, reducing atmospheric carbon dioxide levels and mitigating climate change. They enhance soil fertility through nutrient cycling, reduce erosion, and improve water retention, leading to more stable and productive farming systems. Furthermore, agroforestry can diversify farm incomes by providing additional products such as fruits, nuts, timber, and non-timber forest products. This book is structured to provide a comprehensive understanding of agroforestry and its applications. It begins with an exploration of the fundamental principles and historical development of agroforestry, setting the stage for a deeper dive into specific practices and their ecological, economic, and social impacts.

CHAPTER ONE

INTRODUCTION TO AGROFORESTRY: NATURE'S BLUEPRINT FOR SUSTAINABLE ENVIRONMENT

SONAM RAJPUT
WILDLIFE INSTITUTE OF INDIA, DEHRADUN

Abstract

Agroforestry involves the growing of trees with crops and sometimes with animals in interacting combinations for a variety of objectives. Agroforestry is a sustainable land use practice and system that can be implemented worldwide in any land cover type. Agroforestry is the potential choice for increasing the tree cover outside the notified forest areas to reduce the pressure on forest land. Agroforestry practices can provide multiple ecosystem services such as carbon sequestration, biodiversity conservation, maintenance of soil fertility, prevention of soil erosion, improvement of water infiltration, reduction of wind speed, etc. This chapter aims to cover the concept of agroforestry, the origin of agroforestry, and systems of agroforestry. Agroforestry has received much more attention from researchers, policymakers, and other academicians for its perceived ability to contribute significantly to economic growth, poverty, environmental quality, and employment opportunities.

1.1 Agroforestry as an overview

A short definition of agroforestry is 'farming with trees'. Agroforestry includes the integration of trees on farmland and the use of crops and livestock in woodlands.

OR

A more detailed definition is “the practice of deliberately integrating woody vegetation (tree or shrub) with crop and animal systems to benefit from the resulting ecological and economic interaction”.

- ✓ Agroforestry is the collective name for land use systems involving trees combined with crops and animals on the same unit of land.
- ✓ Combines production of multiple outputs with the protection of resource base.
- ✓ Places emphasis on the use of multiple Indigenous trees and shrubs
- ✓ Suitable for input conditions and fragile environment
- ✓ Involves the interplay of sociocultural values more than in most other land use systems
- ✓ Structurally and functionally more complex than monoculture

Some of the definitions given by the authors are as follows:

According to Bene et al. (1977), agroforestry is a sustainable land management system for land that increases overall production, combines crops, forest plants, tree crops, and animals simultaneously, and applies management practices that are compatible with the cultural pattern of a local population.

Nair (1979) defines agroforestry as a land use system that integrates trees, crops, and animals in a way that is scientifically sound, ecologically desirable, practically feasible, and socially acceptable to that farmer.

The cycle of an agroforestry system is always more than one year. Even the simplest agroforestry system is more complex, ecologically (structurally and functionally) and economically than a mono-cropping system.

Lundgren and Raintree (1982) defined, agroforestry as a collective name for land use systems and technologies, where woody perennials (trees, shrubs, palm bamboo, etc.) are deliberately used in the same piece of land management units as crops and animals in some form of spatial arrangement or temporal sequence. In agroforestry systems, there are both ecological and economic interactions between the different components.

King and Chandler (1978) – "Agroforestry is a sustainable land management system which increases the overall yield of the land, combines the production of crops (including tree and crops) and forest plants and animals sequentially, on the same unit of the land and applies management practices that are compatible with the cultural practices of the local population. Agroforestry is intentionally designed land use management combining trees and shrubs with crops and livestock.

1.1.2 Need of Agroforestry

Agroforestry offers a joined-up way of thinking about rural land use that addresses the negative environmental effects of intensive agriculture and also addresses climate change.

There are clear environmental benefits from agroforestry, relative to agriculture alone or forestry alone, including increased biodiversity, reduced run-off, increased carbon sequestration, and reduced water pollution. The most appropriate form of agroforestry will depend on the individual farm situation but there are forms of agroforestry for arable farmers, livestock farmers, horticulturalists, foresters, and householders.

1.1.3 Roots of Agroforestry

Trees have been used in cropping systems since the beginning of agriculture. Throughout the world, at one period or another, in its history, it has been the practice to cultivate tree species and crops in intimate combination. As a part of the process, human beings cut down trees, cleared the debris by burning, and sowed crops in the ash-enriched soil. Thus was born slash-and-burn agriculture, a primary forerunner of the present day agroforestry and a practice that may have originated in the Neolithic period, around 7000 BC.

In the 1960s and early 1970s, there was increasing concern for the forested lands of the tropics. It was recognized that they were under severe pressure. Some thought that commercial exploitation was the problem some believed that fuel wood needs were the culprit, while some thought that shifting cultivation was the root cause. The president of the International Development Research Centre (IDRC), located in Ottawa, Canada, to study the problem. Bene assembled a small team in Canada called the advisory committee, and recruited experts in the various continents, to prepare studies pertinent to their area. The culmination of these various activities, including extensive travel by Bene, 1977.

1.2 Criteria of Agroforestry

Four key criteria characterize agroforestry

Intentional

The combination of trees, crops, and livestock is intentionally designed, established, and managed to work together and yield multiple products and benefits, rather than as individual elements rich may occur together but are managed separately. Agroforestry is neither monoculture farming, nor is it a mixture of monocultures.

Intensive

Agroforestry practices are created and intensively managed to maintain their productive and protective functions, and often involve cultural operations such as cultivation, fertilization, irrigation, pruning, and thinning.

Integrated

Components are structurally and functionally combined into an integrated management unit tailored to meet the objectives of the landowner. Integration may be horizontal or vertical, above or below ground, or simultaneous. Integration of multiple crops utilizes more of the productive capacity for economic production with resource conservation.

Interactive

Agroforestry actively manipulates and utilizes the interactions among components to yield multiple harvestable products, while concurrently providing numerous conservation and ecological benefits.

1.3 Attributes of Agroforestry

1.3.1 Productivity

Agroforestry systems aim to maintain or increase production commodities as well as productivity. Agroforestry can improve productivity in many different ways. These include tree products, improved yields of associated crops, reduction of cropping system inputs, and increased labor efficiency.

1.3.2 Sustainability

By conserving the production potential of the resource base, mainly through the beneficial effects of woody perennials on soils, agroforestry can achieve and indefinitely maintain conservation and fertility goals.

1.3.3 Adoptability

The word “adopt” here means “accept” and it may be distinguished from another commonly used word adapt, which implies “modify” or change”. Agroforestry can increase adaptive capacity, and help cope with risk or uncertainty.

1.4 Potential of Agroforestry

The different aspects in which agroforestry can help in enhancing the productivity of our lands to meet the demand of the ever-growing human and livestock population are as follows:

Meeting the demand for food and fodder

- Enhanced food production from crops associated with trees
- Food for man from trees such as fruits, nuts, and cereal substitutes
- Fodder for meeting rural needs

Water conservation

- Improvement of soil-moisture retention in rainfed croplands and pastures through improved soil structure and micro-climate effects of trees
- Regulation of stream flow, reducing flood hazards, and a more even supply of water through reduction of run-off and improvement of interjection and storage in infiltration galleries.
- Improvement in drainage from waterlogged or saline soil by trees with high water requirements.

Fuelwood and energy

- Fuel wood for direct combustion
- Ethanol produced from fermentation of high-carbohydrate fruits
- Oils, latex, and other combustible saps and resins.

Shelter from trees

- Shade trees for people, livestock, and shade-loving crops
- Building materials for shelter construction
- Windbreak and shelter-belts for protection of settlements, crops lands, pastures, and roadways

Raw material for industries

- Tannins, essential oils and medicinal ingredients
- Fiber for weaving
- Woods for agricultural implements and various crafts
- Raw material for the pulp and paper industry

Cash benefits

- Direct cash benefits from the sale of tree products
- Indirect cash benefits from increased productivity

Increased yield and maximized production

- Utilizing available solar radiations throughout the year and thus enhancing overall productivity.
- Many leguminous tree species fix nitrogen from the atmosphere and return much more in leaf fall than they take from the soil.

- Leaves of tree species could be used as green manure and help the farmer to increase soil productivity at optimum levels over a long period of time.

Diversified products

- Several trees, shrubs, herbs, and climbers yield a substantial quantity of food materials that are used by the rural poor and particularly by tribal.
- By adopting agroforestry one can get diversified products i.e. fuel, fodder, fruits timber, etc.
- Agroforestry aims to maximize the production of biomass of trees and agricultural crops.

Provides employment opportunity

- Agroforestry systems increase the employment opportunities.
- Agroforestry systems can enhance farmer's income.

Environmental benefits

- Reduction of pressure on forest
- More efficient recycling of nutrients by deep-rooted trees on the site
- Better protection of ecological systems
- Reduction of surface run-off, nutrient leaching, and soil erosion through the impending effect of tree roots and stems on these processes
- Improvement of microclimate, such as lowering of soil surface temperature and reduction of evaporation of soil moisture through a combination of mulching and shading
- Increment in soil nutrients through addition and decomposition of litter-fall

- Improvement of soil moisture through the constant addition of organic matter from decomposed litter.

Economic benefits

- Increase in the output of food, fuel wood, fodder, fertilizer, and timber,
- Reduction in incidence of total crop failure, which is common to single-cropping or monoculture systems.
- Increase in levels of farm income due to improved and sustained productivity.

Social benefits

- Improvement in rural living standards from sustained employment and higher income
- Improvement in nutrition and health due to increased quality and diversity of food outputs.
- Stabilization and improvement of communities through the elimination of the need to shift sites of farm activities.

1.5 Constraints in Agroforestry

- The interference of trees decreases the crop yield which is lower than the monocropping
- The felling of trees causes damage to the arable crop
- Some of the trees serve as hosts to pests that harm the main crop
- Agroforestry systems require more for its management
- Longer gestation periods for trees delay the return to the farmer
- Competition for moisture between trees and arable crops is maximum when the trees have no deep tap root system
- Farmers give more weightage to field crops compared to tree crop

- The tree canopy absorbs maximum light and causes competition for light
- Certain tree species produce chemical exudation which affects the growth of agricultural crops
- Lack of demonstration sites
- The expense of additional management
- Lack of financial assistance
- Competition between trees, crops, and animals

1.6 Agroforestry systems and practices

The system refers to a group of physical components and assemblage of objects connected or related in such a manner to form and or act as a unit e.g., an ecosystem that consists of a living organism and its non-living environment with which they are inseparably interrelated.

In land-use system refers to a type of land use specific to an area and described according to its biotechnical composition and arrangement, and level of technical management of socio-economic features, e.g., rice production system, and plantation crop system.

Sub-system: It refers to a part of the system, with more or less restricted role, content, and complexity than the system itself. A sub-system produces a defined 'basic needs' as its major output so that there can be a food sub-system, an energy production sub-system, and a cash sub-system.

Practices: agroforestry practices denote specific land management operations of any nature, carried out on a farm or other management unit. Such practices are involved in the constitution and maintenance of an agroforestry system, e.g., cropping boundary of trees and shrubs, shelterbelts and windbreaks, etc.

Technology: This refers to an innovation or improvement, usually through scientific, intervention, to either modify an existing system or practice or develop a new one.

1.6.1 Characteristics of Agroforestry

- Land-use practices are deliberately integrated into the whole farm.
- Has two or more outputs
- Contain complex interactions among components suited to particular environments and human needs.
- The "cycle" of an agroforestry practice is always more than one year.
- Competition and management are critical.
- The farmer, is often judged successful or not by the bottom line "Does it pay?"
- Even the simplest agroforestry practice is more complex, ecologically (in terms of structure and function) and economically than monocropping.

1.6.2 Classification of Agroforestry

After the conceptualization and definition of agroforestry, evolving some broad-based and widely acceptable schemes for the classification of agroforestry systems was a major issue in the process of understanding, evaluating, and improving existing agroforestry systems and designing new ones.

1.6.3 Need for Classification

provide a practical framework for the synthesis and analysis of information about existing systems and the development of new and promising ones.

- Focus and emphasis on strategies for the development of improved systems
- It indicates how the system is managed
- It offers flexibility for regrouping the information
- We usually understood and readily handled

1.6.4 Criteria of agroforestry systems classification

Agroforestry system can be categorized according to these set of criteria

Structural basis: refers to the composition of the components, including spatial arrangement of the woody component, vertical stratification of all the components, and temporal arrangement of the different components. Based on structure agroforestry systems can be divided into two categories.

A) Nature of components

B) Arrangement of components

1.6.5 A) Nature of components

Based on the nature of components agroforestry systems can be classified into the following categories.

Agrisilviculture systems

Silvopastoral system

Agrosilvopastoral system

Other system

Agrisilviculture/Silviagriculture/Agrosilviculture: This system involves the conscious and deliberate use of land or the concurrent production of agricultural crops including trees, crops, and forest crops. Based on the nature of the components this system can be grouped into various forms:

i. Improved fallow species in shifting cultivation

ii. Taungya system

iii. Multispecies tree garden

iv. Alley cropping

v. Multipurpose trees and shrubs on farmlands

vi. Crops combinations with plantation crops

vii. Agroforestry for fuelwood production

viii. Shelter belts

ix. Windbreaks

x. Soil conservation hedges etc

i. Improved fallow species in shifting cultivation

Shifting cultivation: Shifting cultivation refers to growing crops on the same plot for a few growing seasons until it is exhausted of soil nutrients, then temporarily abandons the plot of land for several years until it regenerates. Shifting cultivation is called by different names in different parts of the world.

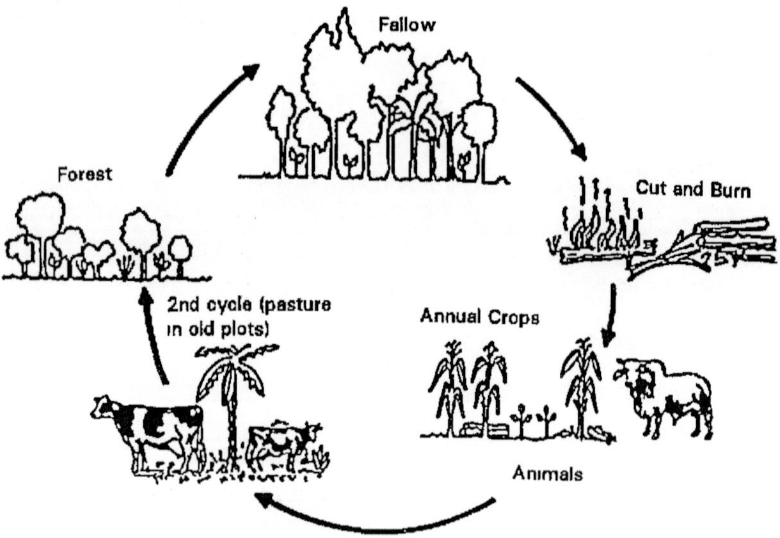

Shifting cultivation

It is generally known as 'slash and burn' and 'bush fallow' agriculture. It is known as Jhum or Jum in the hilly states of Northeast India, as Podu, Dabi, Koman or Bringa in Orissa, as Kumari in Western Ghats, as Watra in southeast Rajasthan, as Penda, Bear or Dahia and Deppa or Kumari in the Bastar district of Madhya Pradesh.

ii. Taungya system: The Taungya (taung = hill, ya = cultivation) is a Burmese word coined in Burma in 1850. The system was introduced to India by Brandis in 1890 and the first taungya plantations were raised in 1896 in North Bengal. It is still practiced in the states of Kerala, West Bengal, Orissa, Karnataka, and the north-eastern hill region.

The Taungya system is a modified form of shifting cultivation in which the labor is permitted to raise agri-crops in the area but only side by side with the forest species planted by it. The practice consists of land preparation, tree planting, growing agricultural crops for 1-3 years until shade becomes too dense, and then moving on to repeat the cycle in a different area.

Types of Taungya systems

Departmental Taungya: In this system, crops and tree species are raised by the forest department by employing a number of laborers on daily wages.

Leased Taungya: The plantation land is given on lease to the person who offers the highest money for raised crops for a specified number of and ensures the care of the tree plantation.

Village Taungya: In this system, crops are raised by the people who have settled down in a village inside the forest for this purpose. Usually, each family has about 0.8 to 1.7 ha of land to raise trees and cultivate crops for 3 or 4 years.

Advantages of the Taungya system

- Low-cost method of forest plantation establishment
- Problems of unemployment are solved
- Helps towards maximum utilization of the site
- Artificial regeneration of the forest is obtained cheaply
- Provision of food crops from forest land and
- Weed, climber growth, etc. is eliminated.
- In every case, highly remunerative to the departments

The disadvantage of the Taungya system

- Danger of epidemics
- Legal problems created
- Loss of soil fertility and exposure to soil
- It is a form of exploitation of human labour
- The susceptibility of land to accelerated erosion increases

iii. Multi-species tree Gardens: In this system, various kinds of tree species are grown mixed. The major function of this system is the production of food, and wood products for home consumption and sale.

iv. Alley cropping (Hedge row intercropping): In this system, trees managed as hedgerows, are grown in wide rows and the crop is planted in the interspace or alley between the tree rows. The position and spacing of hedgerows and crop plants in an alley cropping system depend on plant species, climate, slope, soil conditions, and the space required for the movement of people. The spacing used in fields is usually 4 to 8 meters between rows and 25 cm to 2 m between trees within rows. The closer spacing is generally used in humid areas and the wider spacing in sub-humid or semi-arid regions.

Species for the alley cropping: *Gliricidia sepium, Flemingia macrophylla, Leucaena, Calliandra calothyrsus, Erythrina subumbrans, Albizia saman, Pithecellobium dulce, Paraserianthes falcataria, Acacia spp., Paraserianthes falcataria, and Cajanus cajan.*

***v.* Multipurpose trees and shrubs on farmlands:** In this agroforestry system, various multipurpose tree species are scattered haphazardly. Examples of multipurpose trees employed in agroforestry are *Leucaena leucocephala, Acacia albida, Cassia siamea, Casuarina equisetifolia, Azadirachta indica, Acacia senegal, Cocos nucifera, etc.*

vi. Crop combinations with plantation: perennial trees and shrubs such as coffee, tea, coconut, and cocoa are combined into intercropping systems in numerous ways including an integrated multistory mixture of plantation crops. Cacao (*Theobroma cacao*) is grown under the shade of coconut areca nut, and *Dipterocarpus macrocarpa*.

vii. Agroforestry for fuelwood production: In this system, various multipurpose fuelwood species are inter-planted on or ground agricultural lands. Tree species commonly used as fuel wood are *Acacia nilotica, Albizia lebbek, Cassia siamea, Casuarina equisetifolia, Dalbergia sissoo, Prosopis juliflora, Eucalyptus tereticornis*, etc.

a. **Shelterbelt: A** Shelterbelt is a wide belt of trees, shrubs, and grasses, planted in rows that goes right across the land right-angle to the direction of the prevailing winds to deflect air current, reduce wind velocity, and give general protection to cultivated areas against wind erosion and desiccating effects of the hot winds in lee-wards side. The width of the shelterbelt depends upon local climate conditions, wind velocity, and soil type.

   The choice of species to be raised in the shelterbelt depends on the area's climate, soil, and topography. It is better to grow local species that may serve the object in view, as they can be quickly established.

   Tree species: *Acacia arabica, Acacia senegal, Albizzia lebbek, Azadirachta indica, Dalbergia sissoo, Parkinsonia aculeata, Prosopis cineraria, Prosopis juliflora, Ponogamia pinnata, etc.*

   Shrubsspecies: *Calotropis procera, Crotolaria burhia, Calligonum polygonoides, Clerodendron phlomoides, Cassia auriculata, Dodonaea viscosa, Jatropha curcas*

   Grassesspecies: *Cenchrus barbatus, Saccharum spontaneum, Saccharum munja, Panicum turgidum, Panicum antidotale.*

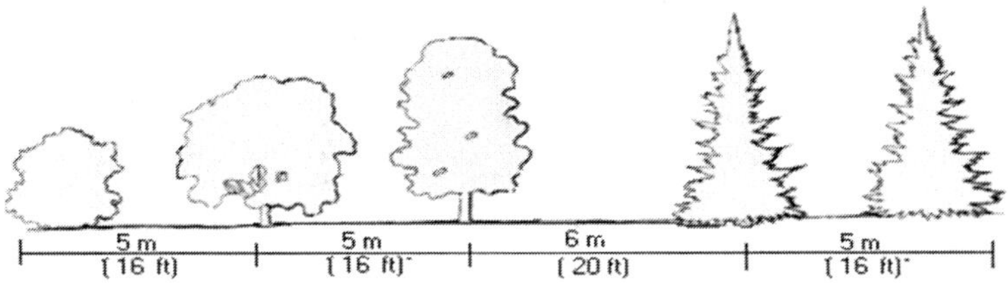

Shelterbelt

Purpose of shelterbelt

- To deflect air currents
- To reduce the velocity of winds
- To provide general protection to the downwind areas against the effects of wind erosion
- To protect the leeward areas from the desiccating effects of hot wind
- To provide fuel, fodder timber, etc
- To protect field crops/livestock from cold/hot wind.
- To prevent soil erosion.
- To reduce evaporation from farmlands
- To improve the microclimate
- Fencing and boundary demarcation.

1.6.6 B) Based on the Arrangement of components

Arrangement of components refers to the plant components of the system even in an agroforestry system involving animals the management of such animals according to a definite plan such as a rotational grazing scheme is in consideration more of the plan than an animal.

i) Spatial arrangement of plants in agroforestry mixture

- Mixed dense, e.g. home gardens
- Mixed sparse, e.g. most systems of trees in pastures
- Zonal micro-zonal, microzonal

ii) Temporal arrangement of plants in agroforestry

Coincident: when two component woody and nonwoody components occupy the land together as coffee under shade trees and pasture under shade trees.

Concomitant: when two components woody or nonwoody stay together for some part of life as in taungya.

Intermittent: when annual crops are grown with perennial crops such as paddy with coconut.

Interpolated: when different components occupy space during different times as in home gardens.

Functional basis: refers to the major function usually furnished by the woody components.

Socioeconomic basis: refers to the level of inputs of management or scale of management and commercial goals.

Ecological basis: refers to the environmental condition and ecological suitability of systems, based on the assumption that certain types of systems can be more appropriate for certain ecological conditions.

1.6.7 Agroforestry Systems of India

In general, the common agroforestry systems being practiced in different agro-ecological regions of India are- agri-silviculture, Boundary plantation, block plantation, energy plantation (trees+ crops during initial years), alley cropping (hedges + crops), agri-horticulture (fruit tree + crops), Agri-silvi-horticulture (tees+ fruit trees+ crops), agri-silvi-pasture (trees + crops + pasture or animal), Silvi-polyculture (tree + vegetables), horti-pasture (fruit trees + pasture or animals), horti-olericulture (fruit tree + vegetables), silvi-pasture (trees + pasture or animals), forage forestry (forage trees + pasture), shelter belts (trees + crops), windbreak (trees + crops), live fence (shrub and under trees on boundary), silvi-sericulture (trees + sericulture), horti-apiculture (fruits tree + honey or bee), aqua-forestry (trees + fishes), and homestead (multiple combinations of timber or fruit or fodder trees, vegetables and animals.

Conclusion

Agroforestry is bound to play a major role in the future, not only for its importance of food and livelihood security but also for its role in combating environmental challenges. This explains the concept the agroforestry and the history of agroforestry. Why we need agroforestry to increase the farmer's welfare. And how to help agroforestry increase employment, reduce poverty, economic and environmental aspects. Agroforestry will be a key issue in providing a solution to global warming, and climate change and enhancing the productivity of the land converting degraded and marginal lands into productive areas. The major focus of research in the coming years will be on developing agroforestry technologies for critical areas like aid and semi-arid zones and other fragile ecosystems such as the Himalayan region and the Coastal eco-system to sustain these areas for higher productivity and natural resource management.

References

Rana A. 2012. Introductory Agroforestry. Dr. Yaswant Singh Parmar University of Horticulture and Forestry Solan, Himachal Pradesh.

Bene J.G., Beall H.W and Cote A. 1977. Trees, Food, and People. IDRC, Ottawa, Canada. Blanford, H.R. 1958. Highlights of one hundred years of forestry in Burma. Empire Forestry Review 37(1): 33-42.

Jeetram, Joshi P.C., Sag S., Pant P. D. and Joshi H.C. Social Forestry and Agroforestry. Department of Forestry and Environment Science School of Earth and Environment Science. Uttarakhand Open University.

King K. F. S. and Chandler L.T. 1978. The Wasted Lands. ICRAF, Nairobi, Kenya.

Lundgren B.O. and Raintree J.B. 1982. Sustained agroforestry. In: Nestel, B. (ed.). Agricultural Research for Development: Potentials and Challenges in Asia, pp. 37-49. ISNAR, The Hague, The Netherlands.

M. O. Adedire. Principles of Agroforestry. Lecture Note. Department of Forestry and Wildlife Management University of Agriculture, Abokuta.

Nair, P.K.R. 1979. Intensive Multiple Cropping with Coconuts in India. Verlag Paul Parey, Berlin/Hamburg, Germany.

Torralba, M., Fagerholm, N., Burgess, P.J., Moreno, G., Plieninger, T. (2016). Do European agroforestry systems enhance biodiversity and ecosystem services? A meta-analysis. Agriculture, Ecosystems and Environment 230: 150-161.

CHAPTER TWO

ECOSYSTEM SUSTAINABILITY UNDER AGROFORESTRY AND THEIR CLIMATE MITIGATION POTENTIAL

AMIT KUMAR[1], SARITA BISHT[2], NARENDRA KUMAR[3] AND HUKUM SINGH[2]
[1]CSIR-INSTITUTE OF HIMALAYAN BIORESOURCE TECHNOLOGY, PALAMPUR-176061
[2]FOREST RESEARCH INSTITUTE DEHRADUN-248006
[3]COLLEGE OF AGRICULTURE, KRYDEMKULAI, CENTRAL AGRICULTURE UNIVERSITY IMPHAL

Abstract

Agroforestry emerges as a transformative paradigm for sustainable land use, offering a holistic approach that reconciles ecological integrity, economic prosperity, and social equity. By harnessing the synergies between trees, crops, and livestock, agroforestry embodies the ethos of sustainability, fostering resilient landscapes that nourish both people and the planet. As we confront the pressing challenges of climate change, biodiversity loss, and food insecurity, agroforestry stands as a beacon of hope, offering a path toward a more sustainable and regenerative future. Ecosystem resilience in agroforestry systems arises from their diverse species composition, functional redundancy, resource use efficiency, ecosystem services provisioning, and adaptive management practices. By harnessing the synergies between trees, crops, and livestock, agroforestry embodies the principles of resilience, offering a sustainable approach to land use management that enhances the capacity of agricultural landscapes to withstand and recover from disturbances, while supporting the well-being of both people and the environment.

Introduction

Agroforestry stands as a beacon of hope in the realm of sustainable land use, offering a multifaceted approach that harmonizes ecological resilience, economic viability, and social equity. At its core, agroforestry represents a departure from conventional monoculture systems, embracing diversity as a cornerstone principle (Anand et al., 2022). By integrating trees with crops and/or livestock, agroforestry systems create dynamic landscapes that mimic natural ecosystems, thereby enhancing their capacity to deliver a multitude of ecosystem services while mitigating environmental degradation. One of the paramount virtues of agroforestry lies in its ability to promote soil health and fertility. Trees in agroforestry systems contribute organic matter to the soil through leaf litter and root exudates, fostering soil structure, moisture retention, and nutrient cycling (Kumar et al., 2024). This enrichment of soil biology and chemistry translates into improved crop yields, reduced erosion, and enhanced resilience to drought and floods, thereby safeguarding the foundation of agricultural productivity (Bisht et al., 2022). Furthermore, agroforestry serves as a bastion of biodiversity conservation, providing habitat and refuge for a diverse array of flora and fauna. The presence of trees within agricultural landscapes creates niches for native species, fostering ecological balance and resilience. By enhancing habitat connectivity and promoting genetic diversity, agroforestry mitigates the risk of species loss and ecosystem degradation, thereby bolstering the resilience of both natural and agricultural ecosystems (Kaushal et al., 2021).

From an economic perspective, agroforestry offers a pathway towards sustainable livelihoods and economic empowerment for rural communities. By diversifying income streams and reducing dependence on a single crop, agroforestry buffers farmers against market fluctuations and climate variability, enhancing their resilience and adaptive capacity. Additionally, the integration of high-value tree species into agroforestry systems opens up avenues for value-added products, such as timber, fruits, nuts, and medicinal plants, thereby expanding economic opportunities and fostering local entrepreneurship (Dasila et al., 2023). Beyond its ecological and economic benefits, agroforestry holds profound social significance, fostering community cohesion, cultural identity, and food sovereignty. Agroforestry systems often reflect traditional land management practices rooted in indigenous knowledge and wisdom, honoring the intergenerational transmission of ecological wisdom and cultural heritage. By revitalizing traditional agroforestry practices and fostering community stewardship of natural resources, agroforestry promotes social cohesion, cultural resilience, and equitable access to land and resources (Kumar et al., 2020a).

➢ Types of agroforestry systems and their characteristics in different agroclimatic regions

Agroforestry systems vary in their suitability and characteristics across different agroclimatic regions, reflecting the diverse environmental conditions, land use, and socio-economic contexts found around the world. Here, we explore how various types of agroforestry systems manifest in different agroclimatic regions and their distinctive characteristics:

Tropical Regions:

Agroforestry Homegardens:

In tropical regions, agroforestry home gardens thrive, featuring a rich diversity of tree species intercropped with vegetables, fruits, and medicinal plants. These systems are highly intricate, often reflecting traditional knowledge and cultural practices. The dense canopy of trees provides shade, while diverse plant species contribute to soil fertility and biodiversity.

Silvopasture:

Silvopasture systems integrate trees with pasture or forage crops, providing shade and fodder for livestock in tropical regions. In humid tropical areas, fast-growing leguminous trees like Leucaena or Gliricidia are common choices, enhancing forage quality and soil fertility. Livestock grazing occurs under the canopy, promoting symbiotic relationships between trees, crops, and animals.

Subtropical Regions:

Alley Cropping:

Alley cropping systems are well-suited to subtropical regions, where a diverse range of tree and crop species can thrive. Trees like nitrogen-fixing Acacias or Leucaena are planted in rows with crops like maize or soybeans grown in between. The trees provide shade, wind protection, and nitrogen fixation, enhancing soil fertility and crop yields.

Windbreaks and Shelterbelts:

Windbreaks play a crucial role in protecting crops from strong winds and soil erosion in subtropical regions. Fast-growing species like Eucalyptus or Casuarina are commonly planted along field boundaries to create sheltered microclimates. These windbreaks also provide habitat for beneficial insects and birds, enhancing biodiversity in agricultural landscapes.

Temperate Regions:

Forest Farming:

Forest farming thrives in temperate regions, where hardwood species like oak, maple, and chestnut provide the canopy for cultivating specialty crops such as mushrooms, berries, and medicinal herbs. The dappled light and moist microclimate of the forest floor create ideal conditions for crop growth, while the trees contribute to ecosystem stability and biodiversity.

Riparian Agroforestry:

Riparian agroforestry is particularly important in temperate regions, where rivers and streams are vulnerable to erosion and pollution. Planting native riparian species like willow, alder, and poplar along watercourses helps stabilize banks, filter water, and provide habitat for aquatic and terrestrial species. These riparian buffers enhance water quality, biodiversity, and landscape resilience.

Arid and Semi-arid Regions:

Agroforestry Homegardens:

In arid and semi-arid regions, agroforestry home gardens serve as oases of productivity and resilience. Drought-tolerant trees like moringa, neem, and jujube are intercropped with drought-resistant crops and vegetables, utilizing water-efficient irrigation techniques like drip irrigation. The shade and mulch provided by trees help conserve soil moisture and protect crops from heat stress.

Alley Cropping:

Alley cropping systems are adapted to arid and semi-arid regions, where water availability is limited. Deep-rooted trees like mesquite or acacia are planted in rows with crops like sorghum or millet grown in between. The trees help improve soil structure, moisture retention, and nutrient cycling, enabling crops to withstand drought conditions and produce reliable yields.

Therefore, agroforestry systems exhibit diverse characteristics across different agroclimatic regions, reflecting the unique environmental conditions and socio-economic contexts of each region. By selecting and adapting appropriate agroforestry practices to local conditions, farmers can enhance the resilience, productivity, and sustainability of their agricultural landscapes, contributing to food security, biodiversity conservation, and climate change adaptation.

➢ Ecosystem Services Provided by Agroforestry

Agroforestry systems are renowned for their capacity to deliver a wide array of ecosystem services, which play a vital role in supporting human well-being, environmental sustainability, and socio-economic development. These services encompass provisioning, regulating, supporting, and cultural functions, each contributing to the resilience and functionality of agroforestry landscapes (Kumar et al., 2020b). Here, we delve into the diverse ecosystem services provided by agroforestry:

Provisioning Services:

Food Production: Agroforestry systems contribute to food security and nutrition by producing a diverse range of crops, fruits, nuts, and animal products. The integration of trees with crops and livestock increases agricultural productivity and resilience, providing a continuous supply of nutritious food throughout the year.

Timber and Non-timber Forest Products: Agroforestry systems produce timber, fuelwood, fibers, resins, and medicinal plants, serving as valuable sources of raw materials for construction, fuel, crafts, and traditional medicine. Trees grown in agroforestry systems can be selectively harvested without compromising ecosystem integrity, providing sustainable livelihoods for rural communities.

Regulating Services:

Soil Erosion Control: The presence of trees in agroforestry systems helps stabilize soils, reduce erosion, and mitigate the impacts of rainfall and wind. Tree roots bind soil particles, while tree canopies intercept rainfall, reducing the velocity of runoff and soil loss. Agroforestry mitigates soil erosion, preserving soil fertility and preventing sedimentation of water bodies (Kumar et al., 2020).

Water Regulation: Agroforestry systems play a crucial role in regulating water quantity and quality by enhancing water infiltration, retention, and recharge. Trees improve soil structure, increase organic matter content, and promote groundwater recharge, contributing to water availability and resilience to drought and floods. Agroforestry also filters pollutants and reduces non-point source pollution, improving water quality.

Supporting Services:

Biodiversity Conservation: Agroforestry systems enhance biodiversity by providing habitat and food sources for a diverse range of plant and animal species. The structural complexity and spatial heterogeneity of agroforestry landscapes create niches for native flora and fauna, promoting species richness, genetic diversity, and ecosystem resilience.

Nutrient Cycling: Agroforestry systems enhance nutrient cycling and soil fertility through the deposition of leaf litter, root exudates, and organic matter. Trees fix atmospheric nitrogen, cycle nutrients from deeper soil layers, and facilitate nutrient uptake by crops and soil microorganisms. Agroforestry promotes sustainable nutrient management, reducing the need for external inputs and enhancing agricultural productivity (Kumar et al., 2021).

Cultural Services:

Recreation and Aesthetics: Agroforestry landscapes provide aesthetic value and recreational opportunities for local communities and visitors. The diverse array of trees, crops, and wildlife creates visually appealing and culturally significant landscapes that contribute to cultural identity, spiritual well-being, and ecotourism.

Traditional Knowledge and Cultural Heritage: Agroforestry systems embody traditional knowledge and cultural practices rooted in indigenous wisdom and ancestral stewardship of the land. These systems preserve cultural heritage, strengthen community cohesion, and foster intergenerational transmission of ecological wisdom, storytelling, and traditional practices (Kumar et al., 2022).

Agroforestry systems offer a multitude of ecosystem services that are essential for sustaining human livelihoods, maintaining ecological integrity, and fostering cultural resilience. By harnessing the synergies between trees, crops, and livestock, agroforestry contributes to food security, biodiversity conservation, climate change mitigation, and cultural preservation, embodying the principles of sustainability and resilience in agricultural landscapes (Kumar et al., 2021c&e).

➢ Ecosystem Sustainability and Resilience in Agroforestry Systems

Agroforestry systems exemplify a holistic approach to land use management that prioritizes ecosystem sustainability, balancing ecological integrity, economic viability, and social equity. These systems are characterized by their ability to harness ecological processes, enhance biodiversity, and promote resilience in agricultural landscapes. Here, we delve into the principles and practices that underpin ecosystem sustainability in agroforestry:

Enhanced Biodiversity:

Agroforestry systems are renowned for their capacity to enhance biodiversity by providing habitat and ecological niches for a diverse array of plant and animal species. The structural complexity and spatial heterogeneity of agroforestry landscapes create opportunities for species richness, genetic diversity, and ecosystem resilience. Trees, crops, and livestock coexist in symbiotic relationships, fostering ecological interactions and promoting the coexistence of multiple species. Agroforestry systems serve as ecological corridors, facilitating the movement and dispersal of wildlife across fragmented landscapes, thereby contributing to landscape connectivity and biodiversity conservation (Kumar et al., 2021a).

Improved Soil Health:

One of the cornerstones of ecosystem sustainability in agroforestry is the enhancement of soil health and fertility. Trees in agroforestry systems play a crucial role in improving soil structure, nutrient cycling, and organic matter accumulation. Tree roots penetrate deep into the soil, breaking up compacted layers and facilitating water infiltration and nutrient uptake. The deposition of leaf litter, root exudates, and organic matter enriches the soil, enhancing microbial activity, soil aggregation, and nutrient availability. Agroforestry systems mitigate soil erosion, conserve soil moisture, and promote soil carbon sequestration, thereby preserving soil fertility and supporting sustainable agriculture (Kumar et al., 2018).

Water Resource Management:

Agroforestry systems contribute to water resource management by enhancing water infiltration, retention, and recharge. Trees improve soil structure, increase water-holding capacity, and reduce surface runoff, mitigating the risk of soil erosion and water loss. Agroforestry systems act as natural buffers against floods and droughts, moderating hydrological cycles and reducing the vulnerability of agricultural landscapes to extreme weather events. By promoting groundwater recharge and regulating streamflow, agroforestry systems enhance water availability for irrigation, domestic use, and ecosystem functioning, thereby supporting both human livelihoods and ecological integrity (Anand et al., 2022).

Carbon Sequestration:

Agroforestry systems play a vital role in climate change mitigation by sequestering carbon in trees, vegetation, and soils. Trees act as carbon sinks, absorbing carbon dioxide from the atmosphere and storing it in biomass and organic matter. Agroforestry enhances carbon sequestration by increasing vegetation cover, promoting tree growth, and fostering soil carbon accumulation. The integration of trees with crops and livestock diversifies carbon pools, enhancing the resilience and stability of carbon stocks. Agroforestry systems contribute to climate resilience by reducing greenhouse gas emissions, enhancing ecosystem resilience, and fostering adaptation to climate variability and change (Singh et al., 2018).

Ecological Resilience:

Ecosystem resilience is a key aspect of agroforestry systems, reflecting their ability to adapt, recover, and maintain functionality in the face of disturbances, environmental changes, and external pressures. Agroforestry systems exhibit resilience at multiple levels, from the ecological dynamics within the system to the socio-economic dynamics of the surrounding landscape. Here, we explore the characteristics and mechanisms that underpin ecosystem resilience in agroforestry (Tewari et al., 2021).

Diverse Species Composition:

Agroforestry systems are characterized by their diverse species composition, encompassing a mix of trees, crops, and livestock that interact synergistically to enhance resilience. The presence of multiple species within agroforestry landscapes increases biological diversity, creating redundancies and buffers against environmental stressors. Diverse species composition provides ecological insurance against pests, diseases, and climatic extremes, ensuring the continued functioning and productivity of agroforestry systems under changing conditions (Kumar et al., 2021d).

Functional Redundancy:

Agroforestry systems exhibit functional redundancy, whereby multiple species perform similar ecological functions, ensuring the continuity of ecosystem services even in the absence of specific species or components. For example, different tree species may contribute to soil fertility, water regulation, or pest control, providing overlapping benefits that enhance system resilience. Functional redundancy enhances the stability and adaptive capacity of agroforestry systems, enabling them to maintain functionality and productivity in the face of disturbances or perturbations (Verma et al., 2023).

Resource Use Efficiency:

Agroforestry systems optimize resource use efficiency by capturing and utilizing resources more effectively than monoculture systems. The spatial and temporal complementarity between trees, crops, and livestock allows for efficient utilization of sunlight, water, nutrients, and space. Trees with deep root systems can access nutrients and water from deeper soil layers, complementing the nutrient and moisture uptake of shallow-rooted crops. Livestock grazing in agroforestry systems can recycle nutrients and organic matter, contributing to soil fertility and ecosystem productivity. Resource use efficiency enhances the resilience of agroforestry systems by maximizing productivity while minimizing inputs and waste (Singh et al., 2024a).

Ecosystem Services Provisioning:

Agroforestry systems provide a diverse array of ecosystem services that contribute to resilience at multiple scales. These services include soil erosion control, water regulation, carbon sequestration, biodiversity conservation, and climate regulation, among others. The multifunctionality of agroforestry landscapes ensures the provision of essential services even in the face of disturbances or changes in environmental conditions (Kumar et al., 2020). For example, windbreaks in agroforestry systems mitigate soil erosion and protect crops from wind damage, enhancing landscape resilience to extreme weather events.

Adaptive Management Practices:

Agroforestry systems are often managed using adaptive management practices that emphasize flexibility, learning, and innovation in response to changing conditions. Farmers in agroforestry systems employ diverse agroecological techniques, such as intercropping, agroforestry rotations, and mixed-species plantations, to optimize productivity and resilience. Adaptive management allows farmers to experiment with new crops, varieties, and management practices, adapting to local conditions and evolving challenges. This iterative process of learning and adaptation enhances the resilience of agroforestry systems, enabling them to thrive in dynamic and uncertain environments (Pandey et al., 2018).

➢ **Adaptive strategies of agroforestry to mitigate the impact of Climate change**

Agroforestry systems employ a variety of adaptive strategies to mitigate the impacts of climate change, enhancing the resilience and sustainability of agricultural landscapes. These adaptive strategies leverage the synergies between trees, crops, and livestock to mitigate climate-related risks, conserve resources, and maintain productivity under changing climatic conditions. Here are some key adaptive strategies employed by agroforestry systems:

Diversification of Species and Varieties:

Agroforestry systems promote species diversification, incorporating a mix of trees, crops, and livestock species that are resilient to climate variability and extremes. By diversifying species and varieties, agroforestry systems reduce the vulnerability of agricultural production to specific pests, diseases, and climatic stresses. Farmers select and cultivate a range of tree species with different growth habits, phenology, and tolerance to drought, pests, and diseases, ensuring the continuity of ecosystem services and agricultural productivity under changing climatic conditions (Kumar et al., 2022).

Enhanced Water Management:

Agroforestry systems employ water management techniques that enhance water availability, retention, and efficiency, mitigating the impacts of drought and water scarcity. Trees in agroforestry systems contribute to soil moisture conservation through their deep root systems, shading effects, and transpiration rates. Agroforestry practices such as contour planting, alley cropping, and riparian buffers optimize water use, reduce runoff, and enhance groundwater recharge, improving the resilience of agricultural landscapes to drought and water stress (Kumar et al., 2021a).

Improved Soil Health and Fertility:

Agroforestry systems prioritize soil health and fertility, implementing practices that enhance soil organic matter, nutrient cycling, and structure. Trees contribute organic matter and nutrients to the soil through leaf litter, root exudates, and nitrogen fixation, improving soil fertility and resilience to erosion and degradation. Agroforestry practices such as agroforestry rotations, cover cropping, and intercropping enhance soil carbon sequestration, nutrient retention, and microbial activity, fostering soil health and productivity under changing climatic conditions (Singh et al., 2020).

Carbon Sequestration and Climate Regulation:

Agroforestry systems play a crucial role in climate change mitigation by sequestering carbon in trees, vegetation, and soils. Trees act as carbon sinks, absorbing and storing atmospheric carbon dioxide through photosynthesis and biomass accumulation. Agroforestry practices such as agroforestry afforestation, reforestation, and silvopasture enhance carbon sequestration, reducing greenhouse gas emissions and mitigating climate change. By enhancing carbon sinks and reducing emissions from agriculture and land use, agroforestry contributes to climate regulation and resilience at both local and global scales (Kumar et al., 2022).

Integrated Pest and Disease Management:

Agroforestry systems employ integrated pest and disease management strategies that minimize reliance on synthetic inputs and enhance ecological resilience. Trees in agroforestry systems provide habitat and food sources for beneficial insects, birds, and microorganisms that contribute to pest and disease control. Agroforestry practices such as polyculture, mixed-species plantations, and biological control enhance biological diversity and natural enemy populations, reducing pest and disease pressure and increasing resilience to outbreaks and infestations.

Adaptive Management and Knowledge Exchange:

Agroforestry systems embrace adaptive management approaches that prioritize flexibility, learning, and innovation in response to changing climatic conditions. Farmers in agroforestry systems engage in knowledge exchange, participatory research, and community-based adaptation initiatives to share experiences, strategies, and best practices for climate resilience. Adaptive management allows farmers to experiment with new crops, varieties, and management practices, adapting to local conditions and evolving challenges while enhancing the resilience and sustainability of agroforestry systems.

Microclimate Modification:

Agroforestry systems modify microclimates through the strategic arrangement of trees, crops, and other components. The presence of trees in agroforestry systems provides shading, windbreak, and evapotranspiration, creating microclimatic conditions that buffer against temperature extremes, wind damage, and moisture loss. By moderating temperature fluctuations and reducing heat stress, agroforestry systems enhance the resilience of crops and livestock to climate variability and extremes (Singh et al., 2018).

Crop Diversity and Resilience:

Agroforestry systems promote crop diversity and resilience by integrating multiple crop species, varieties, and genotypes within the same landscape. Diverse cropping systems increase genetic diversity, phenotypic variation, and adaptability to changing environmental conditions. Farmers select and cultivate a mix of crop species with different growth habits, maturity periods, and tolerance to pests, diseases, and climatic stresses. Agroforestry practices such as intercropping, polyculture, and relay cropping optimize resource use, reduce competition, and enhance resilience to climate variability and change (Tewari et al., 2024).

Livestock Integration and Resilience:

Agroforestry systems integrate livestock with trees and crops, leveraging synergies between different components to enhance resilience and productivity. Livestock grazing in agroforestry systems promotes nutrient cycling, soil fertility, and pasture quality, enhancing resilience to drought and forage scarcity. Agroforestry practices such as silvopasture, agroforestry grazing systems, and integrated crop-livestock-forestry systems optimize livestock management, reduce environmental impacts, and increase resilience to climate-related risks.

Water Harvesting and Conservation:

Agroforestry systems employ water harvesting and conservation techniques that capture, store, and utilize rainwater more effectively, mitigating the impacts of water scarcity and drought. Techniques such as contour planting, swales, and terraces capture runoff, reduce soil erosion, and enhance water infiltration and retention. Agroforestry practices such as alley cropping, contour hedgerows, and agroforestry buffers optimize water use efficiency, reduce irrigation demand, and increase resilience to water stress and variability (Singh et al., 2022).

Adaptive Tree Management:

Agroforestry systems adapt tree management practices to enhance resilience and productivity under changing climatic conditions. Farmers select and cultivate tree species that are well-suited to local climatic conditions, including drought tolerance, pest resistance, and fast growth. Agroforestry practices such as agroforestry thinning, coppicing, and pruning optimize tree growth, biomass production, and ecosystem services provisioning, increasing resilience to climate variability and change (Singh et al., 2024b).

Community-Based Adaptation and Social Resilience:

Agroforestry systems foster community-based adaptation initiatives that enhance social resilience and adaptive capacity at the local level. Farmers collaborate in agroforestry networks, cooperatives, and knowledge exchange platforms to share experiences, strategies, and best practices for climate resilience. Community-based adaptation initiatives prioritize local knowledge, traditional practices, and participatory decision-making, empowering farmers to cope with climate-related risks and uncertainties while enhancing the resilience and sustainability of agroforestry systems (Verma et al., 2022).

Policy and Institutional Support

Policy and institutional support are crucial for promoting and scaling up agroforestry practices, facilitating their adoption, and maximizing their contributions to sustainable development, climate change mitigation, and resilience building. Effective policies and institutions provide the necessary enabling environment, resources, incentives, and capacity-building opportunities to farmers, researchers, extension agents, and other stakeholders involved in agroforestry. Here are some key aspects of policy and institutional support for agroforestry:

- **Policy Frameworks and Strategies:**

Governments develop policy frameworks and strategies that recognize the importance of agroforestry in achieving national development goals, such as food security, poverty alleviation, biodiversity conservation, and climate change mitigation. These policies provide a clear vision, direction, and guidance for integrating agroforestry into national agricultural, forestry, environmental, and climate change policies, plans, and programs.

Legal and Regulatory Frameworks:

Legal and regulatory frameworks establish the rights, responsibilities, and incentives for implementing agroforestry practices, including land tenure, property rights, environmental regulations, and forest management policies. Governments enact laws, regulations, and land-use planning instruments that promote agroforestry, protect natural resources, and ensure the sustainable management of trees, crops, and land within agroforestry systems.

Financial Incentives and Support Mechanisms:

Governments provide financial incentives, subsidies, grants, loans, and other forms of financial support to farmers, communities, and organizations engaged in agroforestry activities. These financial mechanisms help offset the costs of establishing, maintaining, and managing agroforestry systems, making them more economically viable and attractive to farmers. Financial support also encourages investment in agroforestry research, extension, training, and infrastructure development.

Extension and Advisory Services:

Extension and advisory services play a critical role in disseminating information, knowledge, and best practices on agroforestry to farmers and communities. Governments, research institutions, NGOs, and extension agencies provide technical assistance, training, capacity building, and farmer-to-farmer exchange programs to promote agroforestry adoption, innovation, and entrepreneurship. Extension services help farmers make informed decisions, overcome barriers, and maximize the benefits of agroforestry practices.

Research and Innovation:

Governments fund research, innovation, and technology transfer initiatives that generate scientific knowledge, tools, and technologies to support agroforestry development. Research institutions, universities, and agricultural research centers conduct applied research, experiments, trials, and demonstrations to improve agroforestry practices, species selection, management techniques, and value chains. Innovation hubs, incubators, and technology parks facilitate the commercialization and uptake of agroforestry innovations by farmers and entrepreneurs (Tewari et al., 2022).

Market Access and Value Chains:

Governments support the development of market access and value chains for agroforestry products, linking farmers to domestic and international markets, buyers, processors, retailers, and consumers. Policies, infrastructure investments, market incentives, quality standards, certification schemes, and market information systems help farmers market their agroforestry products, obtain fair prices, and capture value-added opportunities along the supply chain.

Partnerships and Multi-stakeholder Collaboration:

Governments foster partnerships and multi-stakeholder collaboration among public sector agencies, private sector actors, civil society organizations, research institutions, farmers' associations, and community groups to promote agroforestry development. Collaborative platforms, networks, alliances, and partnerships facilitate knowledge sharing, resource mobilization, policy dialogue, and coordinated action on agroforestry issues at local, national, regional, and global levels.

➢ Future Directions and Recommendations

As we look towards the future of agroforestry, several key directions and recommendations emerge to further enhance its contribution to sustainable development, climate change mitigation, and resilience building. These directions and recommendations encompass policy, research, capacity building, and stakeholder engagement to unlock the full potential of agroforestry. Here are some future directions and recommendations for advancing agroforestry:

Policy Integration and Coherence:

Governments should prioritize the integration of agroforestry into national policies, strategies, and plans related to agriculture, forestry, environment, climate change, and rural development. Coherent policy frameworks that recognize the multifunctional benefits of agroforestry and provide incentives for its adoption are essential for scaling up agroforestry practices and mainstreaming them across sectors.

Enhanced Research and Innovation:

Increased investment in research, innovation, and technology transfer is crucial for advancing agroforestry science, practices, and systems. Research priorities should focus on improving agroforestry models, species selection, management techniques, and value chain development. Interdisciplinary research approaches that integrate ecological, social, and economic dimensions are needed to address complex challenges and opportunities in agroforestry.

Climate-Smart Agroforestry:

Agroforestry systems should be designed and managed to enhance climate resilience, carbon sequestration, and adaptation to climate change. Climate-smart agroforestry practices that optimize resource use efficiency, conserve water, protect biodiversity, and mitigate greenhouse gas emissions should be promoted. Agroforestry can play a critical role in climate change adaptation and mitigation efforts, contributing to global climate goals and targets (Tewari et al., 2021)

Scaling Up and Scaling Out:

Efforts to scale up agroforestry should focus on expanding its adoption, impact, and reach across diverse agroecological zones, landscapes, and farming systems. Scaling out involves disseminating successful agroforestry practices, models, and innovations to new regions, communities, and stakeholders through extension, capacity building, and knowledge sharing. Scaling up involves integrating agroforestry into large-scale development programs, initiatives, and investments.

Market Access and Value Addition:

Enhancing market access and value addition for agroforestry products is essential for improving the livelihoods and incomes of smallholder farmers and rural communities. Governments, private sector actors, and development partners should support the development of market infrastructure, value chains, certification schemes, and market information systems for agroforestry products. Investment in processing, storage, and marketing facilities can unlock the economic potential of agroforestry.

Capacity Building and Extension Services:

Strengthening capacity building and extension services is crucial for empowering farmers, researchers, extension agents, and other stakeholders involved in agroforestry. Training programs, workshops, field demonstrations, and knowledge exchange platforms should be provided to build technical skills, knowledge, and awareness of agroforestry practices and benefits. Extension services should be decentralized, participatory, and tailored to local contexts to ensure relevance and effectiveness.

Policy Dialogue and Stakeholder Engagement:

Promoting policy dialogue and multi-stakeholder engagement is essential for creating an enabling environment for agroforestry. Governments, civil society organizations, research institutions, farmers' associations, and private sector actors should collaborate to advocate for supportive policies, mobilize resources, and address barriers to agroforestry adoption. Platforms for dialogue, networking, and collaboration should be established at local, national, and regional levels to facilitate knowledge sharing, learning, and collective action.

Conclusion

Agroforestry systems employ a diverse array of adaptive strategies to mitigate the impacts of climate change, enhancing the resilience, productivity, and sustainability of agricultural landscapes. By integrating trees, crops, and livestock in synergistic ways, agroforestry offers a multifaceted approach to climate resilience that addresses ecological, social, and economic dimensions, supporting the well-being of both people and the planet in a changing climate. Policy and institutional support are essential for mainstreaming agroforestry into agricultural and forestry policies, programs, and practices, enabling its widespread adoption and scaling up its impact on sustainable development, climate change mitigation, and resilience building. Governments, international organizations, development agencies, and other stakeholders have a crucial role to play in creating an enabling environment that empowers farmers, promotes innovation, and fosters collaboration for realizing the full potential of agroforestry in addressing global challenges and achieving sustainable development goals.

References

Anand, R., Chaturvedi, S., Kumar, A., Tewari, S., Sah, V.K., Dhyani, V.C., Kaushal, R. 2022. Rainfall partitioning in important multipurpose tree species of Himalayan foothills. Tropical Ecology, 1, 1-5

Bisht, S., Kumar, A., Kumar, N., Singh, H. and Kumar, P. (2022). Biofuel Production by Using Biomass and Its Application. In book: Renewable Energy and Green Technology in CRC Press.

Dasila, H., Joshi, D., Verma, S., Maithani, D., Rawat, S.K., Kumar, A., Suyal, N., Kumar N., and Suyal, D.C. (2023). Hazardous waste: impact and disposal strategies. Advanced Microbial Techniques in Agriculture, Environment, and Health Management. Edition: 1, Publisher: Elsevier

Kaushal R, Tewari S., Thapliyal SD., Kumar, A., Roy, T., Islam, S., Lepcha, STS., Durai, J. 2021. Build-up of Labile, and non-labile carbon fractions under fourteen-year-old bamboo plantations in the Himalayan foothills. Heliyon 7(3):e07850

Kumar, A., Dwivedi, G. K., Tewari, S., Sah, V. K., Singh, H., Kumar, P., Kumar, N., Kaushal, R. 2020a. Soil organic carbon pools under *Terminalia chebula* Retz. based agroforestry system in the Himalayan foothills, India. Current Science (00113891), 118(7).

Kumar, A., Dwivedi, G.K., Tewari, S., Paul, J., Anand, R., Kumar, N., Kumar, P., Singh, H. and Kaushal, R. 2020b. Carbon Mineralization and Inorganic Nitrogen Pools under *Terminalia chebula* Retz.-Based Agroforestry System in Himalayan Foothills, India. Forest Science. 66(5), 634-643

Kumar, A., Kumar, P., Singh, H., and Kumar, N., 2021a. Impact of plant functional traits on near-saturated hydraulic conductivity of soil under different forests of Kempty watershed in Garhwal Himalaya. Indian Journal of Soil Conservation, 49(1), pp.38-44.

Kumar, A., Kumar, P., Singh, H., Kumar, N., 2021b. Adaptation and mitigation potential of roadside trees with bio-extraction of heavy metals under vehicle emissions and their impact on physiological traits during seasonal regimes. Urban Forestry & Urban Greening, 58, p.126900.

Kumar, A., Kumar, P., Singh, H., Kumar, N., 2021c. Modulation of Plant Functional Traits Under Essential Plant Nutrients During Seasonal Regime in Natural Forests of Garhwal Himalayas. Plant and Soil 465(1):1-16

Kumar, A., Kumar, P., Singh, H., Kumar, N., 2021d. Relationship of physiological plant functional traits with soil carbon stock in the temperate forest of Garhwal Himalaya. Current Science 120(7): 1064-1068.

Kumar, A., Singh, H., Kumari, G., Bisht, S., Malik, A., Kumar, N., Singh, M., Raturi, A., Barthwal, S., Thakur, A. and Kaushal, R., 2022. Adaptive resilience of roadside trees to vehicular emissions via leaf enzymatic, physiological, and anatomical trait modulations. Environmental Pollution, p.120191.

Kumar, A., Tewari, S., Singh, H., Kumar, P., Bisht, S., 2021e. Biomass accumulation and carbon stock in different agroforestry systems prevalent in the Himalayan foothills. Current Science 120(6), 1083-1088 IF: 1.16

Kumar, A., Tewari, S., Singh, Indra., Panday, R., Kumar, D. and Anand, R. 2018. Effect of nutrient sources on growth, yield, and quality of turmeric under harad (*Terminalia chebula*) based agroforestry system Indian J. of Agroforestry, 20(2). 1-6

Kumar, N., Jeena, N., Kumar, A., Khwairakpam, R., Singh, H. 2021. Comparative response of rice cultivars to elevated air temperature in Bhabar region of Indian Himalaya: Status on Yield attributes. Heliyon 7(6):e07474

Kumar, N., Kumar, A., Jeena, N., Singh, R., and Singh, H. (2020). Factors Influencing Soil Ecosystem and Agricultural Productivity at Higher Altitudes. Microbiological Advancements for Higher Altitude Agro-Ecosystems & Sustainability. Springer. p. 55- 71.

Kumar, N., Singh, H., Giri, K., Kumar, A., Joshi, A., Yadav, S., Singh, R., Bisht, S., Kumari, R., Jeena, N., Khairakpam, R., and Mishra, G. (2024). Physiological and molecular insights into the allelopathic effects on agroecosystems under changing environmental conditions. Physiology and Molecular Biology of Plants. 248. 119255.

Kumar, N., Verma, S., Kumar, A., Dashila, H., Suyal, D.C., Kumari, G., Rawat, S., Jeena, N., Singh, M., Kumar, A., (2022). Molecular insights into stress-responsive genes in the mitigation of environmental stresses. Unravelling Plant-Microbe Synergy, Edition: 1, Publisher: Elsevier

Pandey R., Dubey S., Tewari L., Kumar A., Kumar D., Sharma N. and Tewari S.K. 2018. Performance of soybean under poplar-based agroforestry system in Terai region. Indian J. of Agroforestry, 20(1), 31-34.

Singh H, Yadav M, Kumar N, Kumar A, Kumar M. 2020. Assessing adaptation and mitigation potential of roadside trees under the influence of vehicular emissions: A case study of *Grevillea robusta* and *Mangifera indica* planted in an urban city of India. PLoS ONE 15(1): e0227380.

Singh, H., Bandyopadhyay, S., Kumar, A., Singh, M., Malik, A., Kumari, G., Raturi, A., Yadav, S.K., Kumar, M., Barthwal, S. and Thakur, A., 2022. Understanding the physiological and biophysical response of urban roadside plantations for assessing adaptation and mitigation mechanisms toward vehicular emissions. Urban Climate, 44, p.101183.

Singh, H., Kumar, N. and Kumar, A. (2024a). Enhancing Resource Use Efficiency in Crops Through Plant Functional Traits. Plant Functional Traits for Improving Productivity, Edition: 1, Publisher, Springer, 97-117

Singh, I., Rawat, P., Kumar, A. and Bhatt, P., 2018. Soil physical-biochemical properties under different agroforestry systems in the Terai region of the Garhwal Himalayas. Journal of Pharmacognosy and Phytochemistry, 7(5), pp.2813-2821.

Singh, M., Singh, H., Kumar, A., Kumar, N., Kumar, M., Barthwal, S., and Thakur, A. (2024b). Soil nitrogen availability determines the CO_2 fertilization effect on tree species (*Neolamarckiacadamba*): growth and physiological evidence. Environmental Sustainability. 24, 119246.

Tewari, S., Tewari, L., Kumar, A., Bisht, S, Kaushal, R. and Kumar, A. (2024). Germination Behaviour of Teak Genotypes and Effect of Different Treatments on Seed Germination Characteristics in Himalayan Foothills, India. Indian Forestry, 150 (3): 244-254

Tewari, S., Tewari, L., Kumar, A., Kaushal, R. 2021. Performance of *Glycine max* L. varieties under *Tectona grandis* L. and *Dalbergia sissoo* Roxb. based agroforestry system Indian J. of Agroforestry, 20(2) 1-6

Tewari, S., Tewari, L., Dubey, A., Kumar, A., Kumar, N., Kaushal, R. 2022. Use of the RAPD marker to determine the genetic diversity of various *Dalbergia sissoo* Roxb. (Shisham) genotypes and their evolutionary relationship in Indian subcontinents. Vegetos,

Verma, R., Dwivedi, G.K., Singh, J.P., Singh, A.P., Tamta, N., Kumar, A. (2023). Characterization of synthesized zinc oxide nanoparticles and their effect on growth, productivity, and zinc use efficiency of wheat and field pea in the Indian Himalayan foothills. Current Science. 124,

Verma, T., Bhardwaj, D.R., Sharma, U., Sharma, P., Kumar, D., Kumar, A. and Kumar, A. (2023). Agroforestry systems in the mid-hills of the north-western Himalaya: A sustainable pathway to improved soil health and climate resilience. Journal of Environmental Management. 348, 119264.

CHAPTER THREE

THE ROLE OF TREES IN AGROECOSYSTEMS FOR ENHANCING RESILIENCE AND BIODIVERSITY

DEEPSHIKHA NIRALA
FOREST RESEARCH INSTITUTE DEEMED TO BE UNIVERSITY, DEHRADUN

Abstract

Agroecosystems, which cover a variety of agricultural practices and landscapes, are important to global food security and a sustainable environment. Integrating trees into these systems, commonly referred to as agroforestry offers various benefits that enhance biodiversity and resilience. Producing enough food for an expanding global population while protecting natural capital presents a significant challenge to humanity. Tree-based ecosystem services are acknowledged as important components of more sustainable agroecosystems, but the strategic management of tree properties for ecosystem service provision is poorly understood. This chapter explores the role of trees in agroecosystems providing a comprehensive understanding of how these trees contribute to biodiversity conservation, ecological stability, economic, and social benefits by enhancing resilience and biodiversity, they contribute to the stability and productivity of agricultural landscapes.

3.1 Introduction

Agroforestry is referred to as a land use management system where shrubs or trees are grown among pasturelands or crops. This diversification of the farming system helps a lot in initiating agroecological succession, which is very similar to that of natural ecosystems. Moreover, trees also help produce a wide range of marketable and useful substances such as medicines, nuts, fruits, wood products, etc. This intentional combination of forestry and agriculture has multiple benefits, from increased biodiversity to reduced erosion and improved soil structure and health.

An agroecosystem is created when human manipulation and alteration of an ecosystem take place for the purpose of establishing agricultural production. The process of understanding agroecosystem sustainability is based on two kinds of ecosystems: traditional and modern agroforestry ecosystems. Generally, "agroforestry is defined as a collective name for land-use systems involving trees combined with crops and/or animals on the same unit of land". The tree-based agroecosystems, commonly known as agroforestry systems, manage the same piece of land which aims to create more diverse, productive, profitable, healthy, and sustainable land-use systems by integrating the best characteristics of forestry and agriculture.

Agroforestry, the inclusion of woody perennials within farming systems, has been practiced as a traditional land use and livelihood option since time immemorial. It was being practiced on agricultural lands and other rural areas for a variety of reasons like, bio-fertilizer trees to enrich the soil and help in land regeneration enabling food security, fuel- ---wood and fodder trees, fruit trees as cash crops, medicinal trees, providing non-timber forest products, timber, shelters, etc. It has been observed that farmers usually prefer multipurpose tree species.

It is clear from the definition that trees and agricultural crops are planted on the same land management unit and have some sort of arrangement that can be spatial or temporal, and there is interaction between trees and agricultural crops. This leads to the investigation of tree-crop interfaces within the system. According to the definition of ICRAF, Agroforestry systems are not simply systems where trees and crops or animals give useful products to farmers but systems where trees and crops (and or animals) interact with each other.

Agroforestry has considerable potential, not as the only way to improve agriculture production, but as one important way to enhance and maintain overall productivity of the small upland farm, the agricultural unit that is becoming more prevent in many parts of the world.

Ecosystem resilience is the ability of an ecosystem to absorb change and return to the same equilibrium state after a temporary disturbance. Ecosystems with higher species diversity tend to be more resilient. If an ecosystem has a diverse community of organisms, they are not all likely to be affected by a disturbance in the same way. So, if one species is nearly going to be extinct, functionally a similar species can take its place, ensuring the function of the ecosystem as a whole.

Benefits of Trees in Agroforestry System

Agroforestry can occur at a variety of spatial scales (e.g., field or woodlot, farm, watershed) in different ecosystems and cultures. The use of trees in agroforestry systems can improve livelihoods through enhanced health and nutrition of soil, increased economic growth for the farmers, and strengthened environmental resilience and sustainability of the ecosystem. Trees have different uses, including cultural benefits, and they typically range over various uses. Furthermore, the agroforestry system is also used as land boundary markers to confer land use rights. Some of the benefits of trees in agroforestry are as follows:

- Trees help to bring up nutrients and water from deep underground.
- Trees help to create and regulate microclimates.
- Trees help to eliminate carbon from the atmosphere.
- Trees build soil carbon and soil organic matter.
- Trees help to provide a framework for both below and above-ground diversity to flourish.
- Trees help to innovate diversified farm enterprises and enhance the economic growth of farmers.
- Trees provide shelter and fodder for all livestock.
- Trees help in making agricultural landscapes more and more resilient.

3.2 Historical Context

3.2.1 Traditional Agroforestry Practices

- **Alley Cropping:** The practice of planting trees in rows with crops grown in between the rows to maximize the use of space and resources.
- **Silvopasture:** The practice of integrating trees with pastures and livestock systems for providing shade and forage for animals, and improving their health.
- **Agroforestry Homegardens:** Combination of various crops, trees, and sometimes livestock in a small land unit, managed garden space to meet household needs and improve food security.
- **Windbreaks and shelterbelts:** Windbreaks and shelterbelts both are designed to protect agricultural crops from high wind damage, reduce soil erosion, and improve microclimatic conditions.
- **Riparian Buffers:** Trees and shrubs are planted along waterways to enhance water quality by filtering runoff, reducing erosion, and providing habitat for wildlife.
- **Woodlots:** Use of forest and timber species for firewood, fodder, and land reclamation.

3.2.2 Evolution of Agroforestry Practices

The shift from traditional to more modern forms of agroforestry began with the initiation of the scientific study of these systems. At the start of the 20th century, scientists started to take an interest in the advantages of combining trees with crops or livestock. Agroforestry became more structured and recognized as a distinct discipline in the end. With the advancement of technology, different agroforestry systems have been created with the aim of achieving certain goals like soil preservation, carbon sequestration, and increasing agricultural productivity. For example, introducing fast-growing tree species led to the development of various agroforestry systems designed for rapid biomass production. Table 3.2 highlights the most popular agroforestry practices along with their key attributes. These systems are related to the specific locality or the region where they practice:

Table 3.2 Agroforestry practices and their main characteristics

Agroforestry practice	Brief description (of arrangement of components)	Major groups of components	Agroecological adaptability
Agrisilvicultural systems (crops - including shrub/vine/tree crops - and trees)			
(1) Improved fallow	Woody species planted and left to grow during the 'fallow phase'	w: fast-growing preferably leguminous h: common agricultural crops	In shifting cultivation areas
(2) Taungya	Combined stand of woody and agricultural species during early stages of establishment of plantations	w: usually plantation forestry spp. h: common agricultural crops	All ecological regions (where taungya is practiced); several improvements possible
(3) Alley cropping (hedge-row intercropping)	Woody species in hedges; agricultural species in alleys in between hedges; microzonal or strip arrangement	w: fast-growing, leguminous, that coppice vigorously h: common agricultural crops	Subhumid to humid areas with high human population pressure and fragile (productive but easily degradable) soils
(4) Multilayer tree gardens	Multispecies, multilayer dense plant associations with no organized planting arrangements	w: different woody components of varying form and growth habits h: usually absent; shade tolerant ones sometimes present	Areas with fertile soils, good availability of labour, and high human population pressure
(5) Multipurpose trees on crop lands	Trees scattered haphazardly or according to some systematic patterns on bunds, terraces or plot/field boundaries	w: multipurpose trees and other fruit trees h: common agricultural crops	In all ecological regions esp. in subsistence farming; also commonly integrated with animals
(6) Plantation crop combinations	(i) Integrated multistorey (mixed, dense) mixtures of plantation crops (ii) Mixtures of plantation crops in alternate or other regular arrangement (iii) Shade trees for plantation crops; shade trees scattered (iv) Intercropping with agricultural crops	w: plantation crops like coffee, cacao, coconut, etc. and fruit trees, esp. in (i); fuelwood/fodder spp., esp in (iii) h: usually present in (iv), and to some extent in (i); shade-tolerant species	In humid lowlands or tropical humid/subhumid highlands (depending on the plantation crops concerned); usually in smallholder subsistence system
(7) Homegardens	Intimate, multistorey combination of various trees and crops around homesteads	w: fruit trees predominate; also other woody species, vines, etc. h: shade tolerant agricultural species	In all ecological regions, esp. in areas of high population density
(8) Trees in soil conservation and reclamation	Trees on bunds, terraces, raisers, etc. with or without grass strips; trees for soil reclamation	w: multipurpose and/or fruit trees h: common agricultural species	In sloping areas, esp. in highlands, reclamation of degraded, acid, alkali soils, and sand-dune stabilization
(9) Shelterbelts and windbreaks, live hedges	Trees around farmland/plots	w: combination of tall-growing spreading types h: agricultural crops of the locality	In wind-prone areas
(10) Fuelwood production	Interplanting firewood species on or around agricultural lands	w: firewood species h: agricultural crops of the locality	In all ecological regions
Silvopastoral systems (trees + pasture and/or animals)			
(11) Trees on rangeland or pastures	Trees scattered irregularly or arranged according to some systematic pattern	w: multipurpose; of fodder value f: present a: present	Extensive grazing areas
(12) Protein banks	Production of protein-rich tree fodder on farm/rangelands for cut-and-carry fodder production	w: leguminous fodder trees h: present f: present	Usually in areas with high person: land ratio
(13) Plantation crops with pastures and animals	Example: cattle under coconuts in south-east Asia and the south Pacific	w: plantation crops f: present a: present	In areas with less pressure on plantation crop lands
Agrosilvopastoral systems (trees + crops + pasture/animals)			
(14) Homegardens involving animals	Intimate, multistorey combination of various trees and crops, and animals, around homesteads	w: fruit trees predominate; also other woody species a: present	In all ecological regions with high density of human population
(15) Multipurpose woody hedgerows	Woody hedges for browse, mulch, green manure, soil conservation, etc.	w: fast-growing and coppicing fodder shrubs and trees h: (similar to alley cropping and soil conservation)	Humid to subhumid areas with hilly and sloping terrain
(16) Apiculture with trees	Trees for honey production	w: honey producing (other components may be present)	Depending on the feasibility of apiculture
(17) Aquaforestry	Trees lining fish ponds, tree leaves being used as 'forage' for fish	w: trees and shrubs preferred by fish (other components may be present)	Lowlands
(18) Multipurpose woodlots	For various purposes (wood, fodder, soil protection, soil reclamation, etc.)	w: multipurpose species; special location-specific species (other components may be present)	Various

Note: w = woody; h = herbaceous; f = fodder for grazing; and a = animals.
Source: Nair (1991).

3.3 Ecological Functions of Trees in Agroecosystems

In agroecosystems, trees play an important role by promoting environmental sustainability, economic viability, and social well-being. There are a number of benefits to planting trees in agricultural landscapes that raise farming systems' resilience and production. The specific functions of trees in agroecosystems are as follows:

3.3.1. Soil Health and Fertility

Trees contribute significantly to soil health through various mechanisms. Their roots help bind the soil, reducing erosion caused by wind and water. This is particularly important in areas prone to soil degradation. Furthermore, trees enhance soil fertility by adding organic matter i.e. leaf litter and root exudates (Figure 1). This organic matter improves soil structure, soil moisture retention, and nutrient availability, through creating a healthy and resilient agroecosystem.

- **Nutrient Cycling:** Enhancing nutrient cycling by drawing up nutrients from deeper soil layers through their deep root systems. Leaf litter and organic matter from trees decompose, adding nutrients to the topsoil and improving soil fertility.
- **Soil Structure:** Tree roots help to improve soil structure by preventing soil compaction and promoting the formation of soil aggregates. This increases soil aeration, water infiltration, and moisture retention into the soil.
- **Erosion Control:** The root systems of trees stabilize the soil, reducing erosion by water and wind. This is very important in hill and heavy rainfall areas.

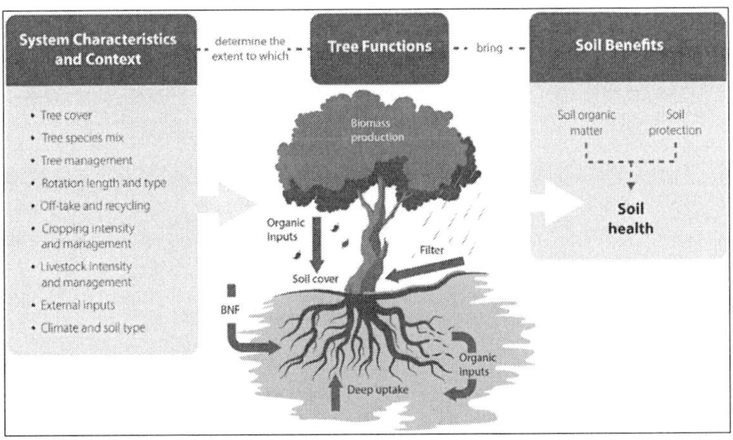

Figure 1: The influence on soil functions and soil health benefits provided by agroforestry trees
(Source: *Barrios et.al.,* 2018)

3.3.2. Microclimate Regulation

- **Temperature Moderation:** Trees provide shade, reducing air and soil temperatures in the agricultural fields. This helps in creating a more favorable microclimate for agricultural crops and livestock.

- **Windbreaks:** Trees act as windbreaks, protecting crops from wind damage, reducing evapotranspiration, and preventing the drying out of soils.

3.3.3. Water Management

Tree roots can access deeper water sources, making them crucial during periods of drought. Trees also facilitate the infiltration of water into the soil, reducing surface runoff and increasing groundwater recharge. This natural water management system helps maintain a stable water supply for crops and other plants, thereby enhancing the resilience of the entire agroecosystem.

- **Water Infiltration:** The presence of trees enhances water infiltration into the soil, by reducing stormwater in several ways and removing macro and micropollutants (Figure 2).

- **Moisture Retention:** Tree canopies reduce the impact of raindrops on the soil, minimizing surface crusting and increasing soil moisture retention.

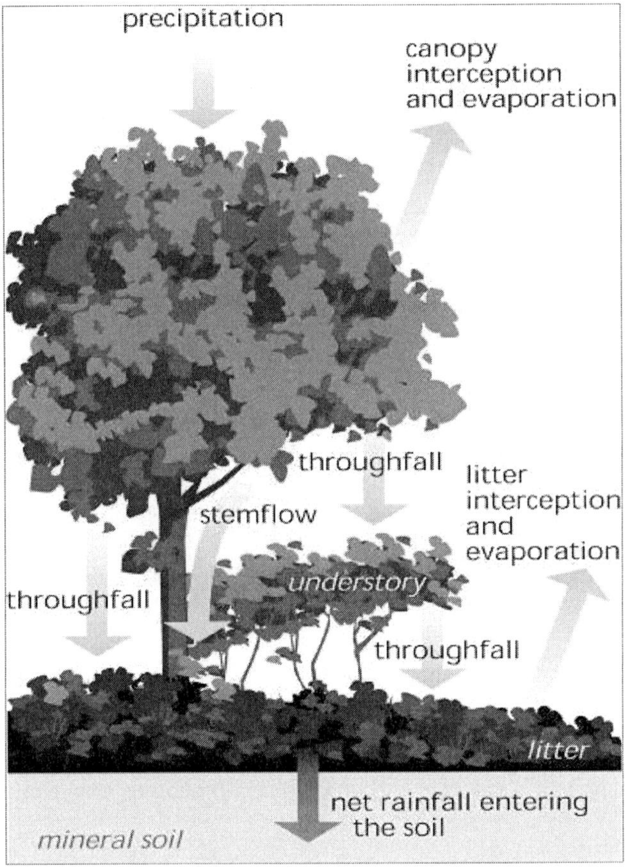

Figure 2: Canopy Interception and Infiltration

3.3.4 Economic Benefits

- **Diverse Income Streams:** In agroecosystems, trees can provide additional sources of income through the production of multiple products such as fruits, nuts, timber, fuelwood, and non-timber forest products (NTFP).

- **Improved Crop Yields:** By enhancing soil fertility, water availability, and pest control, trees can improve the yields and quality of agricultural crops.

3.3.5 Social and Cultural Benefits

- **Livelihood Diversification**
  Beyond ecological benefits, trees provide sustainable livelihoods to rural communities and offer a variety of products (i.e. fruits, nuts, timber, fuelwood, and medicinal plants) that contribute to livelihood diversification for farmers and local communities. The diversification of income sources reduces the economic risk which is associated with reliance on a monocrop and enhances the overall resilience of farming communities. It offers employment opportunities and harvesting these resources supports income and nutrition for local populations.

- **Cultural and Aesthetic Values**
  Trees often hold significant cultural and aesthetic value for local communities. They can be an integral part of traditional agricultural practices, spiritual beliefs, and local heritage. Conserving and incorporating trees into agroecosystems can also support cultural identity and community well-being, helping to create a deeper connection between people and their environment.

3.4 Biodiversity Enhancement

In agroecosystems, biodiversity performs a variety of ecological services beyond the production of food, including recycling of nutrients, regulation of microclimate and local hydrological processes, suppression of undesirable organisms, and detoxification of noxious chemicals. Therefore, trees contribute to the conservation of biodiversity and ecological functions nourishing selected ecosystem services assessed across dominant land uses in agricultural landscapes representing different levels of tree cover:

- **Habitat Provision:** Trees provide habitats for a wide range of species, including birds, insects, and other wildlife, promoting biodiversity within agricultural landscapes. In agroecosystems, they create vertical and horizontal stratification, which supports diverse species at different levels of the ecosystem. The canopy, understory, and forest floor each offer unique habitats that promote biodiversity. This increased habitat complexity supports various organisms, including birds, insects, mammals, and microorganisms, contributing to a balanced and healthy ecosystem.

- **Germplasm Preservation:** Agroecosystem helps to preserve the genetic diversity of sensitive species. By maintaining diverse tree species, we safeguard valuable genetic resources for future generations.

- **Pest Control:** Trees can support populations of beneficial insects and birds that act as natural pest control agents, reducing the need for chemical pesticides.

- **Habitat Creation:** In agroecosystems, they create vertical and horizontal strata, which support diverse species at different levels of the ecosystem. The canopy, understory, and forest floor each offer unique habitats that promote biodiversity. This increased habitat complexity supports various organisms, including birds, insects, mammals, and microorganisms, contributing to a balanced and healthy ecosystem.

- **Creating Connectivity:** Trees in agroecosystems create corridors between habitat remnants. These corridors allow movement for area-sensitive floral and faunal species, maintaining ecological connectivity.

- **Ecosystem Services:** Trees offer additional benefits such as erosion control and water recharge, indirectly contributing to biodiversity conservation.

3.5 Resilience of Biodiversity

In Tropical Agroforestry System, resilience is the capacity of an ecosystem to recover from its previous disturbance, including maintaining its essential characteristics such as taxonomic composition, structures, ecosystem functions, and process rates. Applying this concept to agroforestry systems can suggest two things: (1) that the systems have features that enable to regaining of their biodiversity back to levels found in the previous land use system, before the disturbance of land conversion, and (2) that features of managing system are adequately benign to avoid irreversible changes during its lifespan.

3.5 Carbon Sequestration and Climate Change Mitigation

In recent years, agroecosystem has emerged as a vital means to address the challenges of climate resilience, a combination of agriculture and forestry to produce synergetic effects. Trees are essential for climate regulation in agroecosystems. They provide shade and decrease wind speeds, leading to a more consistent microclimate. This stability can lessen the impact of extreme weather conditions like heatwaves, droughts, and intense rainfall, which are becoming more common due to climate change. The presence of trees helps to balance temperature changes and retain soil moisture, thereby improving the resilience of crops.

- **Carbon Storage:** Trees absorb carbon dioxide from the atmosphere and store it as biomass, helping to mitigate climate change by reducing greenhouse gas concentrations.

- **Climate Resilience:** By improving soil health, water retention, and microclimate conditions, trees enhance the resilience of agroecosystems to climate variability and extreme weather events. The shedding of leaves, bark, and other organic material from the trees adds nutrient-rich organic matter to the soil. This improves soil fertility, making it a healthier medium for crops to grow. Studies have shown that, compared to conventional agricultural systems, agroforestry can improve soil quality by as much as 40%. The trees in these systems can also act as a buffer against soil salinity, a growing concern under changing climatic conditions.

3.6 Challenges and Considerations

The opportunities for agroforestry are exciting, but not without challenges. Adoption of agroecosystems has been surprisingly low, considering the well-documented benefits. Barriers have included the expense of establishment, the landowner having lack of experience with trees, and the time and knowledge required for management.

- **Management Complexity:** Integrating trees into agricultural systems requires careful planning and management to balance the needs of both trees and crops.

- **Economic Trade-offs:** Farmers may face short-term economic trade-offs, such as reduced crop yields while trees are established.

- **Policy Support:** Effective agroforestry implementation often requires supportive policies, extension services, and financial incentives to encourage farmers to adopt these practices.

  Since many of the useful outcomes from agroforestry trees are less tangible or longer-term, it may be difficult for landowners to predict its future yield. For agroforestry systems that produce edible products such as fruits and nuts, the logistics of harvest can be challenging. For agroforestry systems to be economically competitive, mechanization may be required for larger plantings.

3.7 Conclusion

This chapter portrays that the 'Trees' are an integral part of the sustainability and productivity of agroecosystems. They provide ecological, economic, and social benefits that enhance the resilience and functionality of agricultural landscapes. This chapter also explains the challenges facing agroecosystems and how tree-based land management practices contribute to biodiversity conservation and ecosystem services including agricultural production.

References

Barrios E., Coe R., Place F., Gudeta W.S. and Sinclair F. Nurturing Soil Life through Agroforestry: The Roles of Trees in the Ecological Intensification of Agriculture "Biological Approaches to Regenerative Soil Systems" 265-274pp.

Barrios E., Valencia V., Jonsson M., Brauman A., Hairiah K., Mortimer P.E. and Okubo S. 2018. Contribution of trees to the conservation of biodiversity and ecosystem services in agricultural landscapes, *International Journal of Biodiversity Science, Ecosystem Services & Management*, 14(1): 1-16, DOI: 10.1080/21513732.2017.1399167

Current D.A., Brooks K.N., Ffolliott P.F. and Keefe M. 2008. Moving agroforestry into the mainstream. *Agroforestry Systems* 75:1–3.

Faulkner, P.E.; Owooh, B.; Idassi, J. Assessment of the Adoption of Agroforestry Technologies by Limited-Resource Farmers in North Carolina. J. Ext. 2014, 52, Article 5. Available online: http://www.joe.org/joe/2014october/rb7.php (accessed on 16 October 2015).

Garbach K., Milder J.C., Montenegro M., Karp D.S., and DeClerck F.A.J., 2014. Biodiversity and Ecosystem Services in Agroecosystems *Encyclopedia of Agriculture and Food Systems* 2: 21-39.

In Agroforestry—The Future of Global Land Use; Nair, P.K.R., Garrity, D., Eds.; Springer: Dordrecht, The Netherlands, 2012; pp. 247–283.

Jacobson, M.; Kar, S. Extent of Agroforestry Extension Programs in the United States. J. Ext. 2013, 51, Article 4.

Nair, P.K.R. 1993. Classification of agroforestry systems In An Introduction to Agroforestry Kluwer Academic Publishers London International Centre for Research in Agroforestry (ICRAF).

Norgrove L. and Beck J. 2016. Biodiversity Function and Resilience in Tropical Agroforestry

Systems Including Shifting Cultivation, *Current Forestry Reports* 2:62–80.

Pancholi R., Yadav R., Gupta H., Vasure N., Choudhary S., Singh M.N. and Rastogi M. 2023. The Role of Agroforestry Systems in Enhancing Climate Resilience and Sustainability- A Review, *International Journal of Environment and Climate Change* 13 (11): 4342-4353.

Quinn, C.E.; Quinn, J.E.; Halfacre, A.C. Digging Deeper: A Case Study of Farmer Conceptualization of Ecosystem Services in the American South. Environ. Manag. 2015, 56, 802–813.

Strong, N.; Jacobson, M.G. A case for consumer-driven extension programming: Agroforestry adoption potential in Pennsylvania. Agrofor. Syst. 2006, 68, 43–52.

Thevathasan, N.V.; Gordon, A.M.; Bradley, R.; Cogliastro, A.; Folkard, P.; Grant, R.; Kort, J.; Liggins, L.; Njenga, F.; Olivier, A.; et al. Agroforestry Research and Development in Canada: The Way Forward.

Trozzo, K.E.; Munsell, J.F.; Chamberlain, J.L. Landowner interest in multifunctional agroforestry Riparian buffers. In Agroforestry Systems; Springer: Dordrecht, The Netherlands, 2014; Volume 88, pp. 619–629.

Valdivia C., Barbieri C. and Gold M.A. 2012. Between Forestry and Farming: Policy and Environmental Implications of the Barriers to Agroforestry Adoption. Can. J. Agric. Econ. 60: 155–175.

Wilson M.H. and Lovell S.T. 2016. Review Agroforestry—The Next Step in Sustainable and Resilient Agriculture *Sustainability* 8:1-15.

Chapter Four

An Introduction to Climate-Smart Agroforestry for Sustainable Food Production

Karuna Gusain[1], Mithilesh Singh[1], Vivek Chauhan[1], Gargi Navpute[2], Sarita Bisht[2] and Hukum Singh[2]

[1]G.B. Pant National Institute for Himalayan Environment, Kosi-Katarmal, Almora-263643

[2]Forest Research Institute, Dehradun-248006

Abstract

Climate change poses significant challenges to traditional agroforestry systems globally, impacting agricultural productivity, biodiversity, and livelihoods. In response, Climate-Smart Agroforestry (CSA) has emerged as a transformative approach, integrating innovative practices to enhance the resilience and sustainability of agroecosystems. This paper explores the principles, benefits, and challenges of CSA, drawing on global and Indian contexts. It examines the impacts of climate change on agroforestry systems, potential shifts in agroforestry zones, and implications for different regions. By highlighting the benefits of CSA, including increased resilience, biodiversity conservation, and enhanced livelihoods, the paper underscores its potential as a holistic strategy for sustainable land management. The imperative for collaborative efforts between researchers, policymakers, and local communities is emphasized, along with the need for context-specific approaches to foster widespread adoption. In the Indian context, where agriculture plays a pivotal role in the economy, CSA offers a promising pathway to address climate change impacts and promote sustainable agriculture. The paper concludes by advocating for the integration of CSA into broader climate change agendas, envisioning a future where agroforestry contributes to global climate goals while ensuring food security and livelihoods for farming communities.

Introduction

Agroforestry encompasses diverse land-use systems and technologies characterized by the intentional integration of woody perennials such as trees, shrubs, palms, or bamboos with crops, grasses, and/or animals on the same parcel of land, organized in specific spatial and temporal arrangements (Choudhury and Jansen, 1999).

In the face of escalating climate change impacts, including a projected 60% increase in world food production demand by 2050, there is a critical need for adaptive strategies in agriculture (McKenzie and Williams, 2015). However, the developing world is grappling with a concerning shortfall in domestic cereal production, expected to widen from approximately 100 million tons in 1997 to around 190 million tons in 2020 (Rosegrant et al., 2001; Verchot et al., 2007). Traditional approaches, such as the introduction of new crop varieties and increased fertilizer use, are encountering limitations in many regions globally (Huang, Pray, and Rozelle, 2002; Balemi and Negisho, 2012).

Against this backdrop, agroforestry emerges as a promising solution for building climate-resilient agricultural systems, integrating trees and shrubs into conventional agricultural landscapes (Luedeling et al., 2014). These systems offer ecological, economic, and social benefits, promoting crop diversification and providing multiple sources of income for farmers. Agroforestry not only enhances biodiversity and soil fertility but also contributes to increased resilience in the face of climate change. Successful implementation requires a holistic approach, considering local contexts and engaging communities in the process of adaptation, as highlighted by Mimura et al. (2015). In this context, agroforestry stands out as a beacon of hope, offering a sustainable and resilient alternative to address the challenges posed by climate change.

Agroforestry systems vary considerably from landscape to landscape, country to country, and region to region, depending on human needs and capabilities and on the prevailing environmental, cultural, and socio-economic conditions. Generally speaking, three main types of agroforestry systems can be identified depending on the components associated with the trees: agrosilvicultural systems, silvopastoral systems, and agrosilvopastoral systems.

Although Agroforestry integrates trees and shrubs with agricultural crops and livestock, has long been celebrated for its ecological, economic, and social benefits. Traditional agroforestry systems face several challenges and vulnerabilities, both globally and in the specific context of India.

- Land Use Changes and Deforestation: Globally, and particularly in developing countries like India, there is a persistent challenge of land use changes and deforestation. The conversion of forested areas into agricultural land poses a threat to traditional agroforestry systems (FAO, 2018). In India, this is exacerbated by population pressure and the need for agricultural expansion (Kumar et al., 2019).

- Climate Change Impacts: Climate change introduces uncertainties, including altered rainfall patterns, increased temperatures, and extreme weather events, affecting the viability of traditional agroforestry practices (Dawoe et al., 2016). In India, changing climatic conditions contribute to the vulnerability of agroforestry systems, impacting crop yields and the overall resilience of the system (Chaturvedi et al., 2012).

- Lack of Policy Support: A global challenge for traditional agroforestry lies in insufficient policy support and recognition. Policies often prioritize conventional agriculture, neglecting the potential of agroforestry to contribute to sustainable development (Torquebiau et al., 2019). In India, policy frameworks need to integrate agroforestry more comprehensively to unlock its potential for addressing food security and environmental sustainability (Joshi et al., 2017).

- Socioeconomic Pressures: In both global and Indian contexts, socioeconomic pressures on land, such as population growth and changing market dynamics, contribute to the challenges faced by traditional agroforestry systems (Kumar et al., 2019; Garrity et al., 2010). These pressures often lead to the abandonment of traditional practices in favor of more intensive, short-term agricultural strategies.

- Limited Access to Resources: Resource constraints, including limited access to credit, quality planting material, and technical knowledge, hinder the adoption and sustainability of traditional agroforestry systems (Garrity et al., 2010). In India, smallholder farmers often lack the necessary support to invest in agroforestry practices (Mishra et al., 2018).

- Biodiversity Loss: Globally, the loss of biodiversity is a critical issue, impacting the ecological balance of agroforestry systems (Chaplin-Kramer et al., 2015). In India, where agroforestry is practiced across diverse agroecological zones, biodiversity loss can affect ecosystem services and the resilience of traditional systems (Chaturvedi et al., 2012).

- Knowledge Gaps and Technological Barriers: Both globally and in India, there are knowledge gaps and technological barriers that hinder the optimization of traditional agroforestry practices. Limited access to information about best practices and the application of modern technologies constrains the efficiency and adaptability of agroforestry systems (Chaturvedi et al., 2012; Garrity et al., 2010).

Addressing these challenges requires a holistic approach, involving policymakers, researchers, and local communities to enhance the resilience and sustainability of traditional agroforestry systems. Global and national efforts should prioritize policies that recognize and support the multifunctional benefits of agroforestry in promoting environmental conservation and sustainable livelihoods.

Climate-Smart Agroforestry (CSA) represents an innovative approach that addresses the challenges faced by traditional agroforestry systems. By integrating climate resilience into agricultural practices, CSA aims to enhance the adaptive capacity of agroforestry in the face of changing climatic conditions. Unlike traditional systems, CSA incorporates climate-smart technologies and practices such as improved water management, agroecological diversification, and precision farming techniques. Through these innovations, CSA not only sustains the ecological and economic benefits of agroforestry but also mitigates the impacts of climate change on crop yields and overall system resilience. By promoting a holistic and adaptive approach, CSA offers a pathway to overcome the vulnerabilities associated with traditional agroforestry, ensuring sustainable food production, environmental conservation, and improved livelihoods for farmers in a changing world.

Principles of Climate-Smart Agroforestry

Climate-Smart Agroforestry (CSA) integrates climate resilience into traditional agroforestry practices, addressing the challenges posed by climate change. The key principles and characteristics of CSA, both in a global and Indian context, encompass the following:

- Adaptation to Climate Change:

Global Context: CSA prioritizes practices that enhance the adaptive capacity of agroforestry systems in the face of changing climate conditions (FAO, 2010).

Indian Context: In India, where climate variability is pronounced, CSA focuses on adapting agroforestry systems to ensure sustained productivity and ecosystem health (Jat et al., 2019).

- Mitigation of Greenhouse Gas Emissions:

Global Context: CSA emphasizes practices that reduce greenhouse gas emissions, such as minimizing deforestation, adopting tree-based systems, and promoting efficient use of fertilizers (Smith et al., 2014).

Indian Context: In India, where agriculture is a significant contributor to emissions, CSA in agroforestry systems contributes to India's commitment to mitigating climate change (Kumar et al., 2019).

- Biodiversity Conservation:

Global Context: CSA recognizes the importance of biodiversity in maintaining ecosystem services and resilience, promoting diverse agroforestry systems (Sayer et al., 2013).

Indian Context: In India, CSA encourages the integration of native tree species in agroforestry to enhance biodiversity, supporting ecological balance and sustainable production (Saxena et al., 2018).

- Enhanced Resilience and Productivity:

Global Context: CSA focuses on building resilience in agroforestry systems by diversifying crops, improving water management, and adopting climate-smart technologies (Minang et al., 2014).

Indian Context: In India, where smallholder farmers face climatic uncertainties, CSA enhances resilience by promoting agroforestry practices that ensure reliable and sustainable yields (Joshi et al., 2017).

- Community Engagement and Livelihoods:

Global Context: CSA recognizes the importance of involving local communities in decision-making processes and ensuring that agroforestry practices contribute to improved livelihoods (Van Noordwijk et al., 2014).

Indian Context: In India, CSA aligns with the government's focus on doubling farmers' income by 2022 by integrating agroforestry for diversified income sources (NITI Aayog, 2018).

- Integrated Landscape Management:

Global Context: CSA promotes an integrated approach to landscape management, recognizing the interconnectedness of agriculture, forestry, and water resources (Minang et al., 2014).

Indian Context: Integrated landscape planning in India incorporates CSA principles, ensuring harmonized land use for sustainable agriculture and conservation (Kala et al., 2017).

Table 1. Climate Smart agroforestry practices globally.

Climate-Smart Agroforestry Practice	Description	Geographic Suitability	Benefits
Alley Cropping	Planting rows of trees within crops	Global, varies by tree species	Improved soil fertility, reduced erosion, increased biodiversity
Windbreaks	Planting trees to protect crops from wind	Worldwide	Reduced wind erosion, improved microclimate, enhanced water use efficiency
Agroforestry Systems	Integrating trees with crops or livestock	Various climates	Increased yields, enhanced biodiversity, improved resilience

Silvopasture	Combining trees, forage, and livestock	Global, varies by tree and forage species	Improved animal welfare, increased productivity, carbon sequestration
Perennial Crops in Agroforestry	Incorporating perennial crops with trees	Suitable for various climates	Enhanced biodiversity, reduced soil erosion, increased carbon sequestration
Conservation Buffer Strips	Planting trees along water bodies or field edges	Global, especially in areas prone to erosion	Improved water quality, reduced runoff, habitat for beneficial organisms
Forest Garden	Mimicking natural ecosystems with diverse plantings	Various climates	Enhanced biodiversity, increased resilience, sustainable food production

These principles and characteristics collectively define Climate-Smart Agroforestry, offering a comprehensive framework to address the challenges posed by climate change while promoting sustainable and resilient agroecosystems.

Climate Change Impacts on Agroforestry Systems

Climate change poses significant challenges to agroforestry systems, impacting the delicate balance of tree-crop interactions. Changes in temperature, precipitation, and the frequency of extreme weather events can have multifaceted effects on the productivity, biodiversity, and resilience of agroforestry. Here, we delve into these impacts, drawing on key studies and research findings:

- Temperature Changes:

Effects: Altered temperature regimes influence the phenological patterns of both trees and crops, affecting flowering, fruiting, and overall growth. Higher temperatures can lead to mismatches in phenological stages, disrupting synchronized interactions between trees and crops (Dai et al., 2018).

- Changes in Precipitation Patterns:

Effects: Shifts in precipitation patterns, including changes in rainfall intensity and distribution, can impact water availability for both trees and crops. Water stress or excess moisture can disrupt nutrient uptake, root growth, and overall plant health (Verma et al., 2018).

- Extreme Weather Events:

Effects: Extreme weather events, such as storms, floods, and droughts, can cause direct physical damage to both trees and crops. Additionally, these events disrupt the ecological balance within agroforestry systems, leading to losses in productivity and increased vulnerability (Fisher et al., 2018).

- Shifts in Pests and Diseases:

Effects: Climate change alters the distribution and abundance of pests and diseases, affecting both trees and crops. Increased temperatures and changing precipitation patterns create favorable conditions for the spread of pests and diseases, adding stress to agroforestry systems (Bebber et al., 2013).

- Impact on Biodiversity and Ecosystem Services:

Effects: Climate change can disrupt the delicate balance of biodiversity within agroforestry systems. Shifts in temperature and precipitation patterns may affect the availability of suitable habitats for both beneficial and harmful species, altering ecosystem services such as pollination and pest control (Isaacs et al., 2017).

In conclusion, the impacts of climate change on agroforestry are intricate and interconnected. It is imperative to recognize these challenges to develop adaptive strategies that enhance the resilience of tree-crop interactions.

Table 2. Impact of climate change on agroforestry.

Climate Change Impact	Description	Global Consequences
Temperature Increase	Warmer temperatures can affect plant growth, alter phenology, and impact water availability.	Changes in crop and tree growth patterns, altered species distribution, increased risk of heat stress.
Altered Precipitation Patterns	Changes in precipitation can affect water availability, leading to drought or increased flooding.	Water stress, reduced crop yields, and changes in suitable tree species.
Increased Frequency of Extreme Weather Events	More frequent and intense events such as storms and hurricanes.	Increased risk of wind damage, flooding, and soil erosion.
Shifts in Growing Seasons	Changes in temperature and precipitation patterns can alter the timing of planting and harvesting.	Disruptions in traditional agricultural calendars, affecting crop and tree development.
Pests and Diseases	Climate change can influence the distribution and prevalence of pests and diseases.	Increased risk of new pest and disease outbreaks, impacting both crops and trees.
Sea-Level Rise	Coastal agroforestry systems may face salinity intrusion and loss of productive land.	Threats to coastal agroforestry, affecting both crops and trees.
Changes in Carbon Sequestration	Altered conditions may impact the ability of trees to sequester carbon.	Variable effectiveness of agroforestry systems in mitigating climate change.
Shifts in Ecosystem Services	Changes in climate can affect the ecosystem services provided by agroforestry systems.	Altered water regulation, nutrient cycling, and biodiversity support.

Potential Shifts in Agroforestry Zones and Implications for Different Regions

Climate change is expected to bring about shifts in agroforestry zones globally, impacting ecosystems, agricultural practices, and livelihoods (Van Noordwijk et al., 2013). Examining these potential shifts provides insights into the implications for different regions, encompassing environmental, socio-economic, and adaptive considerations.

- Temperature and Precipitation Changes:

Implications: Regions experiencing increased temperatures and altered precipitation patterns may witness shifts in agroforestry zones. Traditional agroforestry belts might migrate polewards or to higher elevations, affecting the suitability of certain tree-crop combinations and altering ecosystem dynamics.

Example: In temperate regions, such as parts of Europe or North America, warming temperatures may expand the range of certain tree species, potentially transforming existing agroforestry systems (Dai et al., 2018).

- Water Scarcity and Drought-Prone Areas:

Implications: Agroforestry zones in regions facing water scarcity and frequent droughts may contract. Water-demanding tree species may face challenges, leading to changes in tree-crop interactions and the need for more resilient, drought-tolerant species.

Example: Areas in sub-Saharan Africa or parts of Australia may experience shifts in agroforestry suitability, necessitating the adoption of climate-resilient tree-crop combinations (Verma et al., 2018).

- Increased CO_2 Levels and Carbon Sequestration:

Implications: Elevated atmospheric CO_2 levels can impact tree growth and carbon sequestration rates. Agroforestry systems may need adjustments to maximize carbon sequestration potential, influencing choices of tree species and planting configurations.

Example: Regions with high carbon sequestration potential, such as tropical rainforests, may need adaptive strategies to maintain or enhance their role as carbon sinks (Fisher et al., 2018).

- Pest and Disease Dynamics:

Implications: Changes in temperature and precipitation patterns can influence the distribution and abundance of pests and diseases. Agroforestry zones may need to adapt to evolving pest pressures, requiring the introduction of pest-resistant tree varieties and integrated pest management practices.

Example: Regions in Asia, facing the threat of new pest invasions due to climate change, may need to reevaluate agroforestry strategies to minimize pest-related risks (Bebber et al., 2013).

- Socio-Economic Considerations:

Implications: Shifts in agroforestry zones have socio-economic consequences, affecting the livelihoods of farmers and communities dependent on agroforestry. Adaptation strategies must consider the economic implications of changing tree-crop combinations.

Example: Smallholder farmers in Latin America relying on specific agroforestry practices may face challenges if shifting agroforestry zones necessitate changes in their traditional practices.

- Ecosystem Services and Biodiversity:

Implications: Changes in agroforestry zones can impact ecosystem services provided by these systems. Biodiversity considerations become crucial as certain tree species may no longer be viable in their existing regions, affecting pollination services and overall ecosystem health.

Example: Southeast Asian regions, known for their diverse agroforestry practices, may see alterations in the availability of ecosystem services, necessitating conservation efforts (Isaacs et al., 2017).

- Adaptation Strategies:

Implications: Regions experiencing shifts in agroforestry zones will need to implement adaptive strategies, including selecting resilient tree species, adjusting planting calendars, and incorporating climate-smart technologies.

Example: Sub-Saharan Africa, with its diverse agroecological zones, may need to invest in research and extension services to facilitate the adoption of adapted agroforestry practices (Montagnini & Nair, 2004).

Benefits of Climate-Smart Agroforestry (CSA)

Climate-Smart Agroforestry (CSA) stands as a transformative approach to sustainable land management, strategically combining the principles of agroforestry with climate resilience. In response to the challenges posed by a changing climate, CSA integrates innovative practices that enhance agricultural productivity, mitigate greenhouse gas emissions, and improve the adaptive capacity of farming systems. By fostering tree-crop interactions, CSA not only ensures food security but also promotes biodiversity, carbon sequestration, and resilient livelihoods for farmers. This holistic approach signifies a forward-thinking paradigm that seeks to harmonize agricultural productivity with environmental stewardship in the face of climate uncertainty.

Some benefits of CSA are:

- Increased Soil Health and Water Conservation: Climate-smart agroforestry employs practices such as composting, soil management techniques, and efficient water harvesting and retention systems, contributing to improved soil health and water conservation on farms (Pretty et al., 2006).

- Promotion of Greenhouse Gas Emission Reduction: CSA actively promotes practices that reduce greenhouse gas emissions by avoiding encroachment into forested areas, adopting tree-based systems with additional planted trees, and promoting more efficient fertilizer use to mitigate methane emissions (Smith et al., 2008).

- Diversification for Food Security and Livelihoods: The diversification of crops through Climate-Smart Agroforestry not only addresses food security but also creates alternative income opportunities and sustainable economic livelihoods for farmers (Remans et al., 2014).

- Enhanced Climate Change Resilience: Climate-Smart Agroforestry helps farmers mitigate vulnerability to crop losses in the face of climate change-induced weather impacts by fostering resilient agricultural practices (Lin, Huang, & Hsu, 2017).

- Integrated Planning for Synergies: Integrated planning of land, agriculture, forests, fisheries, and water at local, watershed and regional scales ensure synergies are properly captured, promoting sustainable development (Pretty et al., 2011).

- Carbon Storage and Soil Fertility Improvement: Activities promoted by Climate-Smart Agroforestry, such as combining animal husbandry and trees with food production, contribute to increased carbon storage and improved soil fertility (Garrity et al., 2010).
- Emission Reduction from Agriculture: Climate-smart agroforestry helps in reducing various emissions from agriculture, including nitrous oxide from fertilizer application, livestock emissions, and methane from rice cultivation (Smith et al., 2008).
- Exploration of Carbon Finance: The exploration of carbon finance as a lever to promote sustainable agricultural practices provides smallholder farmers with direct benefits and contributes to environmental conservation (Minang et al., 2014).
- Biodiversity Conservation: Climate-smart agroforestry contributes to biodiversity conservation by creating agroecosystems that support diverse flora and fauna (Jose, 2009).
- Community Resilience: These systems enhance community resilience by fostering local self-sufficiency and reducing dependence on external resources (Garrity et al., 2010).

Table 3. Benefits of CSA.

Benefit	Description
Enhanced Resilience	Agroforestry systems are more resilient to extreme weather events, such as droughts and floods, due to diversified and integrated land use.
Increased Biodiversity	The combination of trees, crops, and sometimes livestock in agroforestry systems promotes biodiversity, providing habitats for various species.
Improved Soil Health	Agroforestry practices contribute to better soil structure, increased organic matter, and improved nutrient cycling, leading to healthier soils.

Carbon Sequestration	Trees in agroforestry systems sequester carbon, helping to mitigate climate change by offsetting greenhouse gas emissions.
Water Conservation	Agroforestry practices can improve water-use efficiency, reduce runoff, and contribute to better water conservation in the landscape.
Sustainable Crop Yields	Combining trees with crops can enhance overall productivity by providing shade, windbreaks, and nutrient cycling, leading to more stable and sustainable yields.
Income Diversification	Agroforestry systems often provide multiple products (timber, fruits, nuts, etc.), diversifying income sources for farmers and increasing economic stability.
Erosion Control	Tree cover in agroforestry systems helps prevent soil erosion, protecting against the loss of fertile topsoil during heavy rainfall or wind events.
Adaptation to Climate Change	The diversity and flexibility inherent in agroforestry make it an effective strategy for adapting to changing climate conditions.
Improved Water Quality	Trees in agroforestry systems act as buffer strips, filtering runoff and enhancing water quality by reducing the impact of agrochemicals.
Community and Social Benefits	Agroforestry practices often contribute to community development by fostering social cohesion, knowledge sharing, and sustainable resource management.

Case Studies

Global Case Studies

EverGreen Agriculture in Malawi:

Evergreen Agriculture is an agricultural approach that integrates trees with crop and livestock production systems to create more sustainable and resilient farming systems. In Malawi, where agriculture is a crucial component of the economy and livelihoods, EverGreen Agriculture has gained attention for its potential to address environmental and food security challenges (Garrity et al., 2010).

Key features of Evergreen Agriculture in Malawi include:

- Agroforestry Systems: Evergreen Agriculture promotes the integration of trees on farms through agroforestry systems (Garrity et al., 2010). This involves planting trees alongside crops and livestock to enhance overall productivity and ecosystem services.

- Soil Fertility Improvement: Certain tree species in agroforestry systems have the ability to fix nitrogen, improve soil structure, and increase nutrient availability (Sanchez & Jama, 2002). This helps in maintaining or enhancing soil fertility, which is critical for sustained agricultural productivity.

- Water Conservation: Trees play a crucial role in water conservation. Their deep roots help prevent soil erosion, improve water infiltration, and contribute to overall water management. This is particularly important in regions with erratic rainfall patterns, such as Malawi (Garrity et al., 2010).

- Climate Change Adaptation: EverGreen Agriculture contributes to climate change adaptation by increasing the resilience of farming systems. Trees provide shade, which helps in mitigating the impacts of extreme temperatures. Additionally, they can act as windbreaks, reducing the risk of crop damage during storms (Garrity et al., 2010).

- Diversification of Income Sources: Agroforestry systems diversify income sources for farmers. Trees can provide additional products such as fruits, nuts, and timber, contributing to household income and food security (Garrity et al., 2010).

- Biodiversity Conservation: Integrating trees into agricultural landscapes enhances biodiversity by providing habitat for various species. This can have positive effects on pest control and overall ecosystem health (Garrity et al., 2010).

Agroforestry in Brazil's Cerrado Region:

Description: Agroforestry systems in Brazil's Cerrado region represent a sustainable land-use approach that involves the integration of crops with native tree species (Gomide et al., 2007). In this system, farmers strategically combine agricultural activities with the cultivation of trees, often native to the region. This integration is designed to achieve multiple objectives, including the enhancement of biodiversity within agricultural landscapes (Smith et al., 2010), the improvement of soil structure through the contributions of tree roots and organic matter (Franzel et al., 2004), and the provision of economic benefits to farmers through diversified and sustainable yields (Garrity et al., 2010).

Taungya System in Southeast Asia:

Description

The Taungya system, practiced in countries like Myanmar and Thailand, is a unique agroforestry approach that integrates tree plantations with agriculture. In the initial years, farmers cultivate crops beneath the newly established tree plantations. Over time, as the trees mature, the agricultural component is gradually phased out. The Taungya system plays a pivotal role in reforestation efforts, providing a sustainable land-use model that combines economic benefits with environmental conservation (Montagnini, 2000).

Key Features:

- Phased Integration: The Taungya system involves a phased approach, with the initial cultivation of crops underneath young tree plantations, gradually transitioning to a predominant focus on tree growth.

- Reforestation: The system actively contributes to reforestation, helping combat deforestation and promoting the restoration of tree cover in the region.

- Carbon Sequestration: The integration of trees with agriculture enhances carbon sequestration, playing a crucial role in mitigating climate change (Montagnini & Nair, 2004).

- Biodiversity Enhancement: The combination of trees and crops promotes biodiversity, providing habitats for various species and supporting ecosystem health.

- Sustainable Land Use: The Taungya system represents a sustainable land-use practice, balancing agricultural productivity with long-term environmental benefits.

Agroforestry in Costa Rica's Coffee Plantations:

Description: Costa Rican coffee plantations exemplify a sustainable approach through shade-grown agroforestry systems. In this practice, farmers integrate shade trees within the coffee plantations, not only enhancing coffee quality but also promoting biodiversity conservation and mitigating the impacts of climate change (Philpott & Dietsch, 2003). This agroforestry system is a testament to the harmonious coexistence of agricultural production and environmental stewardship.

Key Features:

- Shade-Grown Coffee: Costa Rican coffee plantations prioritize shade-grown coffee, utilizing shade trees to provide cover for coffee plants. This helps regulate temperature, reduce evaporation, and improve overall coffee quality.

- Biodiversity Conservation: The integration of shade trees contributes to the conservation of biodiversity by providing habitats for various bird species, insects, and other wildlife, fostering a balanced ecosystem.

- Climate Change Mitigation: Shade trees assist in mitigating the impacts of climate change by regulating microclimates, reducing temperature extremes, and enhancing the resilience of coffee plants to environmental stresses.

- Sustainable Agriculture: This agroforestry system exemplifies sustainable agriculture, emphasizing the long-term health of both the coffee crops and the surrounding environment.

China's Sloping Land Conversion Program:

Description: China's innovative program focuses on the conversion of sloping lands into agroforestry systems. This initiative involves integrating fruit and timber trees into agricultural landscapes, resulting in reduced soil erosion, enhanced biodiversity, and economic benefits for farmers (Zhang et al., 2007). The program showcases a holistic approach to sustainable land use, emphasizing the integration of environmental conservation and agricultural productivity.

Key Features:

- Sloping Lands Conversion: The program specifically targets sloping lands for conversion, addressing the challenges of soil erosion common in such areas and contributing to landscape stability.

- Agroforestry Integration: Fruit and timber trees are strategically integrated into the agricultural landscape, promoting a diverse land use system that combines the benefits of both forestry and agriculture.

- Soil Erosion Reduction: By introducing agroforestry systems, the initiative effectively reduces soil erosion, preserving fertile topsoil and maintaining the long-term productivity of the land.

- Biodiversity Enhancement: The integration of trees into the landscape enhances biodiversity by providing habitat for various species, and promoting ecological balance.

- Economic Benefits: The program provides economic benefits to farmers through the cultivation of fruit and timber trees, diversifying income sources, and contributing to the financial sustainability of agricultural practices.

Indian Case Studies:

Jal Shakti Abhiyan in Rajasthan

Description: The Jal Shakti Abhiyan in Rajasthan, India, launched by the Ministry of Jal Shakti, actively promotes agroforestry to enhance water conservation and combat desertification. As part of the initiative, farmers are encouraged to plant drought-resistant trees alongside traditional crops, contributing to sustainable water resource management and improving land resilience (Ministry of Jal Shakti, Government of India, 2019).

Key Features:

- Drought-Resistant Species: Emphasis on planting trees such as *Prosopis juliflora* and *Acacia senegal* that are well-adapted to arid conditions.

- Water Harvesting Techniques: Adoption of water harvesting techniques integrated with agroforestry practices to maximize water availability.

Social Forestry in Himachal Pradesh:

Description: Himachal Pradesh's Social Forestry Program, led by the Himachal Pradesh Forest Department, involves planting fruit-bearing and timber trees on community lands. This community-based initiative aims to enhance forest cover, provide livelihoods, and improve climate resilience through sustainable forest management practices (Himachal Pradesh Forest Department, 2021).

Key Features:

- Community-Led Reforestation: Active involvement of local communities in the planning, implementation, and maintenance of social forestry projects.

- Economic Diversification: Planting fruit-bearing trees to provide economic opportunities for local communities, enhancing overall livelihoods.

Agroforestry in Andhra Pradesh's Rayalaseema Region:

Description: In Andhra Pradesh's Rayalaseema region, farmers practice agroforestry by integrating horticultural crops with native trees. Led by B. M. Kumar et al. (2005), this approach aims to improve water retention, enhance soil fertility, and diversify income sources for farmers.

Key Features:

- Water Management: Planting native trees with deep root systems to improve water retention in the soil.

- Income Diversification: Integrating horticultural crops such as mangoes and guavas for additional income sources.

Village Watershed Development in Maharashtra:

Description: Watershed development projects in Maharashtra, undertaken by the Watershed Organization Trust, incorporate agroforestry components. By planting trees such as *Acacia* and *Eucalyptus*, these projects aim to enhance water availability, prevent soil erosion, and support rural livelihoods.

Key Features:

- Species Selection for Erosion Control: Planting specific trees like Acacia and Eucalyptus to prevent soil erosion and improve soil structure.
- Community Livelihood Support: Contributing to rural livelihoods through sustainable agroforestry practices.

Van Dhan Yojana in Madhya Pradesh

Description: Madhya Pradesh's Van Dhan Yojana, as outlined by the Ministry of Tribal Affairs, encourages tribal communities to practice agroforestry. This initiative focuses on cultivating medicinal and aromatic plants alongside traditional crops, promoting sustainable resource utilization and economic empowerment for tribal communities (Ministry of Tribal Affairs, Government of India, 2020).

Key Features:

- Tribal Empowerment: Focusing on tribal communities and encouraging their active involvement in agroforestry practices.
- Medicinal Plant Cultivation: Integrating cultivation of medicinal and aromatic plants for economic, environmental, and health-related benefits.

Adaptation of CSA

Different agroecological zones require a nuanced approach that considers the unique characteristics and challenges of each zone.

- Arid and Semi-Aid Zones: Given Rajasthan's predominantly arid and semi-arid climate, the choice of drought-resistant tree species becomes crucial. Species with deep roots and high water-use efficiency would be suitable.

- Water Harvesting Techniques: Adaptations may involve incorporating location-specific water harvesting techniques such as check dams or contour trenches to maximize rainwater capture.

- Mountainous Regions: In mountainous agroecological zones, adaptations may include selecting tree species that are well-suited to the altitude and slope conditions. Contour planting can be employed to prevent soil erosion on hilly terrains.

- Semi-Aid Zones: In semi-arid regions, emphasis should be placed on water-efficient horticultural crops and native trees that can thrive in limited water conditions.

- Soil Amendments: Depending on soil types in different zones, adaptations may involve incorporating specific soil amendments to enhance fertility.

- Variable Agroecological Zones: Maharashtra encompasses diverse agroecological zones. Adaptations should be made based on specific zones, such as coastal areas, plateaus, or plains, taking into account variations in rainfall and soil types.

- Species Selection: Species selection may vary based on local ecological conditions, ensuring they are well-suited to prevent erosion and contribute to water availability.

- Tribal and Forested Regions: In tribal and forested regions, adaptations may involve incorporating indigenous knowledge and traditional agroforestry practices that are well-suited to the local ecosystems.

- Biodiversity Considerations: Recognizing the existing biodiversity in different regions and adapting agroforestry practices to complement and enhance local biodiversity.

Cross-Cutting Adaptation Strategies

- Participatory Approaches: In all agroecological zones, involving local communities in decision-making and implementation ensures practices align with local needs and knowledge.

- Research and Extension Services: Continuous research and extension services are crucial for adapting practices to changing environmental conditions and disseminating updated information to farmers.

- Policy Support: Government policies should be flexible to accommodate regional variations and provide incentives that encourage the adoption of agroforestry practices suitable for each agroecological zone.

- Climate-Resilient Agriculture: Given the ongoing climate change, incorporating climate-resilient agricultural practices within agroforestry systems can help mitigate risks associated with extreme weather events.

In summary, successful adaptation of agroforestry practices across diverse agroecological zones necessitates a context-specific approach, considering climate, soil, and socio-economic factors unique to each region. It also underscores the importance of local knowledge, community engagement, and continuous learning for sustainable and effective agroforestry implementation.

Integrating Technology, Innovation, and Policy Making

Technology and innovation, alongside effective policy-making, collectively propel the evolution of climate-smart agroforestry practices, fostering a resilient and sustainable agricultural landscape. Precision agriculture, driven by technologies such as remote sensing and GIS, empowers farmers with precise insights into their agroforestry systems. This data-driven approach aids in optimal resource allocation, strategic planning, and informed management decisions. Mobile applications and digital tools further facilitate real-time information exchange, equipping farmers with timely updates on weather conditions, soil health, and crop performance.

Innovative practices within agroforestry, including climate-resilient tree varieties and improved cultivation techniques, benefit from advancements in technology. Smart irrigation systems and sensor-based monitoring, enabled by technological solutions, enhance water management efficiency. Additionally, the integration of blockchain technology in agroforestry supply chains promotes transparency and traceability, aligning with sustainable practices.

Effective policy-making is crucial to creating an enabling environment for the adoption of technology in agroforestry. Policies that incentivize the use of climate-smart technologies, promote research and development, and support capacity-building initiatives contribute to the successful integration of innovation into agricultural practices. By aligning policies with the evolving technological landscape, governments can bolster climate-smart agroforestry, ensuring that it remains adaptive, sustainable, and capable of addressing the challenges presented by climate change.

Evaluation of Policy Frameworks and Institutional Support for Climate-Smart Agroforestry

Effective policy frameworks and institutional support play a crucial role in fostering the adoption and sustainability of climate-smart agroforestry practices. The comprehensive evaluation below provides insights into existing policies, their key features, and the role of institutions in promoting climate-smart agroforestry, citing relevant literature.

1. National Agroforestry Policies

- Integration with Climate Goals: Policies that explicitly integrate agroforestry into national climate change mitigation and adaptation strategies, aligning them with broader environmental objectives.
- Incentive Mechanisms: Financial incentives, subsidies, or tax breaks encourage farmers to adopt agroforestry practices, thereby promoting widespread adoption.
- Research and Extension: Policies supporting research, development, and extension services to disseminate climate-smart agroforestry practices. This includes funding for scientific studies and the provision of training programs for extension workers.

Evaluation: National agroforestry policies vary in their comprehensiveness and implementation. Countries like India and Kenya have developed robust policies, acknowledging the potential of agroforestry in achieving sustainable development and climate resilience (Garrity et al., 2010; Place & Garrity, 2016).

2. Institutional Support for Agroforestry

- Extension Services: Supportive institutional structures providing extension services and training programs for farmers on climate-smart agroforestry. This involves disseminating knowledge on best practices, tree-crop interactions, and sustainable management.
- Land Tenure Security: Policies and institutions ensuring land tenure security, promoting long-term investment in agroforestry systems. Clear land tenure encourages farmers to make sustained investments in tree planting and management.
- Farmer Cooperatives: Establishing or supporting farmer cooperatives for collective action and resource-sharing in agroforestry. This fosters a sense of community and enables smallholder farmers to access resources and markets more effectively.

Evaluation: While institutional support is crucial, gaps may exist in some regions. Countries with strong extension services and farmer-focused institutions tend to witness higher rates of adoption and sustainability of climate-smart agroforestry (Place & Weber, 2020; Torquebiau, 2000).

3. Climate Finance Mechanisms

- Access to Finance: Policies facilitating smallholder farmers' access to climate finance for implementing agroforestry practices. This includes creating financial mechanisms tailored to the needs of small-scale farmers.
- Carbon Markets: Mechanisms supporting the inclusion of agroforestry projects in carbon markets, providing additional income streams for farmers through carbon sequestration. This enhances the economic viability of agroforestry.
- Insurance Programs: Integration of agroforestry into climate-resilient insurance programs to mitigate risks associated with climate change. This involves developing insurance products that specifically address the unique challenges faced by agroforestry systems.

Evaluation: The availability and accessibility of climate finance for agroforestry projects vary globally. Countries with successful experiences have integrated agroforestry into climate finance mechanisms, demonstrating the economic viability of these practices (Minang et al., 2015; Foli et al., 2016).

Current Challenges and Constraints in Implementation

- Climate Change Variability: Climate change introduces unprecedented variability, making it challenging to predict and adapt to the dynamic shifts in weather patterns and the frequency of extreme events. Agroforestry systems, reliant on a delicate balance between trees and crops, face heightened vulnerability to erratic climatic conditions. Unpredictable rainfall, prolonged droughts, and increased temperatures pose immediate threats, demanding adaptive strategies for sustained productivity.

- Resource Constraints: The successful implementation of climate-smart agroforestry is hindered by resource constraints at multiple levels. Farmers often lack the financial support necessary for initial investments in climate-resilient tree varieties and sustainable practices. Additionally, limited access to advanced technologies, such as precision agriculture tools and monitoring devices, restricts the capacity to implement and monitor climate-smart initiatives. Comprehensive training programs are essential to empower farmers with the knowledge and skills required for effective adoption.

- Awareness and Education: Inadequate awareness and education among farmers pose a substantial barrier to the widespread adoption of climate-smart agroforestry practices. Many farmers may not fully understand the benefits of integrating trees into their farming systems or lack the technical know-how to implement these practices effectively. Bridging this knowledge gap through targeted awareness campaigns, extension services, and farmer education programs is crucial for overcoming this challenge.

Barriers: Socio-Economic, Cultural, and Institutional Factors

- Socio-Economic Factors: The socio-economic landscape plays a critical role in determining the success of climate-smart agroforestry initiatives. Poverty is pervasive in many agricultural communities, limiting the financial capacity of farmers to invest in climate-resilient practices. Lack of access to credit, insurance, and social safety nets further exacerbates vulnerability, hindering the ability to absorb shocks associated with climate change impacts.

- Cultural Resistance: Cultural resistance to change poses a significant hurdle, as traditional farming beliefs and practices may be deeply ingrained in communities. Farmers often resist adopting new methods, fearing disruption to established practices that have sustained their livelihoods for generations. Overcoming cultural resistance requires an understanding of local beliefs and values, integrating traditional knowledge with modern climate-smart techniques in culturally sensitive ways.

- Institutional Barriers: Institutional barriers, ranging from unclear land tenure systems to insufficient policy frameworks, impede the seamless integration of climate-smart agroforestry. Ambiguities in land ownership and tenure can discourage long-term investments in tree planting. Inadequate policy support, including incentives and regulations, may fail to create an enabling environment for climate-smart practices. Strengthening extension services and creating supportive policies are crucial steps toward overcoming institutional barriers.

Insights into Future Directions, Research Needs, and Opportunities

- Research and Innovation: Strategic investments in research and innovation are essential for the future of climate-smart agroforestry. Research should focus on developing climate-resilient tree varieties, understanding the dynamics of tree-crop interactions, and exploring innovative technologies for sustainable land management. Innovations in precision agriculture, remote sensing, and data analytics can enhance monitoring and decision-making, contributing to the adaptability of agroforestry systems.

- Context-Specific Approaches: Future research needs to emphasize context-specific approaches that consider the diversity of agroecological zones and the socio-economic backgrounds of farmers. Recognizing the unique challenges and opportunities in different regions ensures that solutions are tailored to local conditions, promoting the successful adoption of climate-smart practices.

- Collaborative Partnerships: Opportunities for advancing climate-smart agroforestry lie in fostering collaborative partnerships among researchers, policymakers, and local communities. Knowledge exchange and co-development of solutions enhance the relevance and effectiveness of interventions. Public-private partnerships can facilitate the scaling up of successful initiatives, ensuring that innovations are accessible to a broader range of farmers.

- Comprehensive Strategies: Integrating agroforestry into comprehensive climate change adaptation and mitigation strategies presents a promising avenue for long-term sustainability. This involves aligning agroforestry practices with broader national and international climate goals. By integrating agroforestry into climate change policies, governments can incentivize adoption, promote research and development, and create a conducive environment for sustainable practices.

Navigating the complexities of challenges and opportunities in climate-smart agroforestry demands a multifaceted approach. A comprehensive strategy should include initiatives to address resource constraints, promote awareness and education, overcome socio-economic and cultural barriers, and foster collaborative research and innovation. Integrating agroforestry into broader climate change agendas ensures a holistic and sustainable approach to resilient and adaptive agriculture in a changing world.

Conclusion

Climate-smart agroforestry stands out as a dynamic and transformative approach in the pursuit of sustainable and resilient agricultural practices, particularly in the face of climate change. Despite the intricate challenges such as climate variability, resource constraints, and limited awareness, integrating trees with crops and livestock within agroecosystems offers profound benefits.

The current hurdles faced by climate-smart agroforestry, including the unpredictability of climate change impacts and resource limitations, underscore the need for urgent interventions. Adaptive management practices are essential to address climate variability, while targeted investments empower farmers with financial resources, advanced technologies, and comprehensive training programs, enhancing both adaptive capacity and economic empowerment.

The knowledge gap poses a significant challenge in the widespread adoption of climate-smart agroforestry, emphasizing the crucial role of awareness campaigns and educational programs. Bridging this gap fosters a culture of sustainability, innovation, and informed decision-making among farmers, making them key stakeholders in enhancing the resilience and sustainability of agroforestry systems.

Cultural resistance and institutional barriers further complicate the adoption process, emphasizing the importance of culturally sensitive approaches and strategic reforms. Acknowledging local traditions and values while integrating modern techniques is vital, alongside institutional reforms to streamline governance structures and incentivize sustainable practices.

Collaborative efforts between researchers, policymakers, and local communities are pivotal for successful implementation. Partnerships facilitate knowledge exchange, technology transfer, and solution co-creation, fostering scalable and context-specific approaches that address diverse agroecological challenges.

Looking ahead, ongoing research and innovation are critical for the trajectory of climate-smart agroforestry. Future directions demand a focus on developing climate-resilient tree varieties, understanding tree-crop interactions, and leveraging innovative technologies for sustainable land management, ensuring scalability and relevance across agroecological zones.

Integrating agroforestry into broader climate change agendas aligns sustainable agriculture with global goals, propelling agriculture as a proactive force in combating climate change. In essence, climate-smart agroforestry offers a beacon of hope for sustainable food production, environmental stewardship, and resilient communities.

In the Indian context, where agriculture sustains a significant portion of the population, climate-smart agroforestry is particularly imperative. With diverse agroclimatic zones and pressing climate change impacts, embracing climate-smart practices is urgent. Cultural sensitivity, targeted investments, and collaborative partnerships are essential for success, offering India a pathway to emerge as a global leader in climate-smart agriculture, ensuring food security, environmental sustainability, and the prosperity of farming communities amidst a changing climate.

References

Aayog, N. I. T. I. (2018). Contributing to Sustainable Development in the Indian Himalayan Region. *NITI Aayog: New Delhi, India.*

Balemi, T., & Negisho, K. (2012). Management of soil fertility in acid soils of tropical and sub-tropical regions: A review. Journal of Soil Science and Environmental Management, 3(9), 253-270.

Bebber, D. P., Ramotowski, M. A., & Gurr, S. J. (2013). Crop pests and pathogens move polewards in a warming world. Nature Climate Change, 3(11), 985-988.

Chaplin-Kramer, R., O'Rourke, M. E., Blitzer, E. J., & Kremen, C. (2015). A meta-analysis of crop pest and natural enemy response to landscape complexity. Ecology Letters, 18(7), 520-530.

Chaturvedi, R. K., Joshi, J., Jayaraman, M., Bala, G., & Ravindranath, N. H. (2012). Multi-model climate change projections for India under representative concentration pathways. Current Science, 103(7), 791-802.

Choudhury, D.P., & Jansen, J.H. (1999). Agroforestry potential for shifting cultivation systems in Asia. Agroforestry Systems, 44(1), 31-52.

Dai, J., Wang, B., Xi, H., Wang, X., & Zhang, Z. (2018). Impacts of temperature variation on tree-crop interactions: a case study of apple orchards in China. Agricultural and Forest Meteorology, 253-254, 1-11.

Dawoe, E., Acheampong, E., Martensson, L. M., & Wam, H. K. (2016). Farmers' perception of climate change: a case study from the Savannah zone of Ghana. Climate and Development, 8(5), 420-434.

FAO. (2010). Climate-Smart Agriculture: Policies, Practices, and Financing for Food Security, Adaptation, and Mitigation.

FAO. (2018). State of the World's Forests 2018 - Forest Pathways to Sustainable Development.

Fisher, M., Acuña, T. B., Carrasco, L. R., Giri, K., Goh, T. B., & Milner-Gulland, E. J. (2018). Tree cover in a changing landscape: the combined effects of land use and climate on savanna trees in northern Borneo. Global Change Biology, 24(8), 3557-3570.

Franzel, S., Coe, R., Cooper, P., Place, F., Scherr, S., Schulte-Herbrüggen, B., ... & Jama, B. (2004). Integrated soil fertility management in Africa: principles, practices and development process. IFDC Technical Bulletin, (19), 1-40.

Foli, S., et al. (2016). Climate-smart agriculture: perspectives and contradictions. Climatic Change, 135(2), 233-248.

Garrity, D. P., Akinnifesi, F. K., Ajayi, O. C., Weldesemayat, S. G., Mowo, J. G., Kalinganire, A., ... & Bationo, A. (2010). Evergreen Agriculture: a robust approach to sustainable food security in Africa. Food Security, 2(3), 197-214.

Gomide, J. A., de Godoy, L. J. G., Sousa-Silva, J. C., & Oliveira, T. S. (2007). Agroforestry in the Cerrado Region of Brazil. In Agroforestry - The Future of Global Land Use (pp. 109-123).

Himachal Pradesh Forest Department. (2021). Social Forestry.

Huang, J., Pray, C., & Rozelle, S. (2002). Enhancing the crops to feed the poor. Nature, 418(6898), 678-684.

Isaacs, R., Tuell, J., Fiedler, A., Gardiner, M., Landis, D., & Artz, D. (2017). Maximizing arthropod-mediated ecosystem services in agricultural landscapes: the role of native plants. Frontiers in Ecology and the Environment, 15(4), 195-202.

Jat, M. L., Datta, A., Choudhary, M., & Sharma, P. C. (2019). Climate-smart agriculture practices for climate-resilient farming systems. In Innovations in Rural Extension (pp. 39-53). Springer.

Joshi, P. K., Garg, A., & Chaturvedi, R. K. (2017). Agroforestry policy and its impact on land-use changes: A case study in Gujarat, India. Land Use Policy, 60, 362-375.

Jose, S. (2009). Agroforestry for ecosystem services and environmental benefits: an overview. Agroforestry Systems, 76(1), 1-10.

Kala, P., & Arora, S. (2017). A comprehensive study of classical and hybrid multilevel inverter topologies for renewable energy applications. *Renewable and Sustainable Energy Reviews*, *76*, 905-931.

Kumar, B. M., et al. (2005). Impact of integrated tree crop interactions on productivity and nutrient cycling: case of South India. Agroforestry Systems, 64(3), 241-253.

Kumar, P., Chander, S., & Pandey, S. (2019). Land-use and land-cover change and its drivers in a rapidly urbanizing region in India. Geocarto International, 34(8), 907-926.

Lin, B. B., Huang, J., & Hsu, C. F. (2017). Local and landscape drivers of biodiversity of four groups of ants in Neotropical cocoa farms. Agriculture, Ecosystems & Environment, 248, 51-59.

Luedeling, E., et al. (2014). Climate change impacts on winter chill for temperate fruit and nut production: A review. Scientia Horticulturae, 144, 218-229.

Mckenzie, F. E., & Williams, E. R. (2015). Land, agriculture, and food: A tale of three eras. Agricultural History, 89(3), 414-430.

Mimura, N., et al. (2015). Adaptation planning and implementation. In Climate Change 2014: Impacts, Adaptation, and Vulnerability (pp. 869-898). Cambridge, UK: Cambridge University Press.

Minang, P. A., van Noordwijk, M., Freeman, O. E., Mbow, C., de Leeuw, J., Catacutan, D., & Öborn, I. (2014). Climate-smart landscapes: Multifunctionality in practice. Nairobi, Kenya: World Agroforestry Centre (ICRAF).

Minang, P. A., et al. (2015). Climate-smart landscapes: multifunctionality in practice. Climate, 3(1), 216-233.

Ministry of Jal Shakti, Government of India. (2019). Jal Shakti Abhiyan.

Ministry of Tribal Affairs, Government of India. (2020). Van Dhan Yojana Guidelines.

Mishra, A. K., Mohapatra, P. K., & Hota, M. (2018). Financial viability and constraints in agroforestry practices: A case study of Odisha, India. Forest Policy and Economics, 91, 40-46.

Montagnini, F. (2000). Accumulation in above-ground biomass and soil storage of mineral nutrients in pure and mixed plantations in a humid tropical lowland. Agroforestry Systems, 49(3), 275-291.

Montagnini, F., & Nair, P. K. R. (2004). Carbon sequestration: An underexploited environmental benefit of agroforestry systems. Agroforestry Systems, 61(1-3), 281-295.

Place, F., & Garrity, D. (Eds.). (2016). Agroforestry: Trees and Agriculture Sustainability. Routledge.

Place, F., & Weber, J. (2020). Trees on farms: An update and reanalysis of agroforestry's global extent and socio-ecological characteristics. Global Environmental Change, 61, 102054.

Perfecto, I., Rice, R. A., Greenberg, R., & Van der Voort, M. E. (1996). Shade coffee: A disappearing refuge for biodiversity. BioScience, 46(8), 598-608.

Philpott, S. M., & Dietsch, T. V. (2003). Coffee and conservation: a global context and the value of farmer involvement. Conservation Biology, 17(6), 1844-1846.

Pretty, J., Noble, A. D., Bossio, D., Dixon, J., Hine, R. E., Penning de Vries, F. W. T., & Morison, J. I. L. (2006). Resource-conserving agriculture increases yields in developing countries. Environmental Science & Technology, 40(4), 1114-1119.

Pretty, J., Toulmin, C., & Williams, S. (2011). Sustainable intensification in African agriculture. International Journal of Agricultural Sustainability, 9(1), 5-24.

Remans, R., Flynn, D. F., DeClerck, F., Diru, W., Fanzo, J., Gaynor, K., ... & Willett, W. (2014). Assessing nutritional diversity of cropping systems in African villages. PLoS ONE, 9(5), e96082.

Rice, R. A., & Greenberg, R. (2000). Cacao cultivation and the conservation of biological diversity. Ambio, 29(3), 167-173.

Rosegrant, M. W., et al. (2001). Global food projections to 2020: Emerging trends and alternative futures. Food policy, 25(5), 518-550.

Sanchez, P. A., & Jama, B. (2002). Soil fertility replenishment takes off in East and Southern Africa. In Integrated Soil Fertility Management in Africa: Principles, Practices, and Developmental Process, pp. 51-83.

Saxena, A., Dhyani, D., & Singh, J. (2018). Biodiversity in agroforestry: issues and opportunities. In Agroforestry Systems for Sustainable Land Use (pp. 87-111). Springer.

Sayer, J., Sunderland, T., Ghazoul, J., Pfund, J. L., Sheil, D., Meijaard, E., ... & van Oosten, C. (2013). Ten principles for a landscape approach to reconciling agriculture, conservation, and other competing land uses. Proceedings of the National Academy of Sciences, 110(21), 8349-8356.

Smith, J., Smith, P., Wattenbach, M., & Zaehle, S. (2005). Projected changes in mineral soil carbon of European croplands and grasslands, 1990–2080. Global Change Biology, 11(12), 2141-2152.

Smith, P., Martino, D., Cai, Z., Gwary, D., Janzen, H., Kumar, P., ... & Smith, J. (2008). Greenhouse gas mitigation in agriculture. Philosophical Transactions of the Royal Society B: Biological Sciences, 363(1492), 789-813.

Smith, P., Martino, D., Cai, Z., Gwary, D., Janzen, H., Kumar, P., ... & Smith, J. (2010). Greenhouse gas mitigation in agriculture. Philosophical Transactions of the Royal Society B: Biological Sciences, 365(1554), 2697-2717.

Smith, P., Bustamante, M., Ahammad, H., Clark, H., Dong, H., Elsiddig, E. A., ... & Tubiello, F. (2014). Agriculture, forestry, and other land use (AFOLU). In Climate Change 2014: Mitigation of Climate Change. Contribution of Working Group III to the Fifth Assessment Report of the Intergovernmental Panel on Climate Change.

Torquebiau, E. F. (2000). Are tropical agroforestry home gardens sustainable? Agriculture, ecosystems & environment, 38(3), 163-174.

Van Noordwijk, M., Minang, P. A., Freeman, O. E., Mbow, C., de Leeuw, J., Catacutan, D., & Öborn, I. (2014). Climate-smart landscapes: multifunctionality in practice. World Agroforestry Centre (ICRAF).

Verchot, L. V., et al. (2007). Climate change: linking adaptation and mitigation through agroforestry. Mitigation and adaptation strategies for global change, 12(5), 901-918.

Verma, A., & Negi, A. (2018). Influence of changing climate on tree-crop interactions in agroforestry systems: A review. Agroforestry Systems, 92(3), 725-739.

Watershed Organization Trust. (n.d.). Watershed Development.

Zhang, W., et al. (2007). The sloping land conversion program in China: a 10-year review. Journal of Environmental Management, 85(3), 661-676.

CHAPTER FIVE

SUSTAINABLE LIVELIHOODS THROUGH FRUIT-BASED AGROFORESTRY PRACTICES

NAVJOT SINGH KALER*[1], SHIWANGEE[2], KRITIKA[1], KAMAL SHARMA[1] AND AMIT KUMAR[2]

[1]COLLEGE OF HORTICULTURE AND FORESTRY, NERI-HAMIRPUR, HIMACHAL PRADESH, INDIA-177001

[2]CSIR-INSTITUTE OF HIMALAYAN BIORESOURCE TECHNOLOGY, PALAMPUR-176061

Abstract

As agroforestry benefits farmers socioeconomically, it is one of the most sustainable land management practices in use worldwide. With the increasing prominence of climate change as a global concern, agroforestry has become an important land management strategy for the many issues and limitations we are currently facing as a result of our growing population pressure, rising food and fodder needs, depletion of natural resources, and climate change. It offers a win-win situation by serving as a carbon sink and supporting the attainment of food security, livelihood income, and environmental improvement. Through the adoption of agroforestry, farmers' livelihoods significantly improved since they had higher access to fuel wood, food, and fodder, which is reflected in their increased access to livelihood capital. Practices in agroforestry promote species diversity, guarantee financial gain, and sustain farmers's livelihoods.

Introduction

Agriculture, which has nourished nations, driven economies, and supplied nourishment for endless generations, is today confronting a slew of difficulties that threaten its long-term viability and ability to feed the world's rising population. The main reason behind these manifold challenges and constraints is our growing demographic pressure, increasing food and fodder needs, natural resource degradation, and climate change (Dhyani and Handa 2013). Climate change and the alteration of temperature are going to become our biggest restraints in achieving our goal of food security in the future. Farmers face escalating vulnerability due to intertwined environmental issues like resource depletion and climate shifts. The quality of agricultural food has decreased over the past 60 years as a result of growing environmental concerns around the world and the detrimental impacts of using excessive amounts of pesticides, insecticides, weedicides, and artificial fertilizers (Carvalho, 2017). About 8000 ha of forest land is lost each year due to the ongoing shift of forest land to aquaculture, agriculture, homesteads, and other uses (FAO, 1981; Rahman GMM, 2014). To improve livelihoods and sustain food production while preserving soil health and combating climate change, a robust management system must be devised.

Agroforestry, an integrated approach combining trees, crops, and sometimes livestock within a single land area, presents a promising solution. Acknowledged as a sustainable method, it harmonizes agricultural productivity with environmental preservation while making a significant contribution towards reducing poverty and addressing food insecurity. This ancient practice has garnered renewed scientific interest to meet rising population demands, ensure agroecosystem sustainability, counter deforestation rates and soil degradation, and preserve biodiversity. The implementation of a fruit-based agroforestry system that is centered around fruit trees can significantly enhance the standard of living for smallholder farmers by offering them additional food and different sources of income. Additionally, it reduces poverty by boosting earnings and involving women in productive endeavors (Rahman et al. 2017; Leaky and Simons 1998). This agricultural strategy, which combines conventional crops with fruit-bearing trees, has great potential to improve livelihoods and address the many issues that rural communities confront. It is the link between sustainable agriculture, environmental resilience, and community well-being. Due to the modest pre-production period of fruit trees, the high market value of their products, and the contribution of fruits to household dietary needs, they are very popular among producers with limited resources (Shamebo, 1998). These systems contributed significantly to the farming community's means of subsistence by offering a range of advantages, including the ability to generate revenue through the sale of various agricultural products, including food, fuel, building materials, fodder, and shade for crops that prefer shade.

The many facets of this agricultural paradigm become apparent as we dive more into the investigation of how fruit-based agroforestry might improve livelihoods. Fruit trees have a wide range of benefits for agricultural landscapes, from reducing climate change to improving nutritional quality and supporting biodiversity. This chapter seeks to unravel the complexities of these interactions, providing a nuanced understanding of how fruit-based agroforestry systems emerge as not just agricultural practices but as transformative agents for sustainable livelihoods and their impact on carbon sequestration, climate change mitigation, nutritional importance, and biodiversity enhancement.

Key components of a fruit-based agroforestry system

In a fruit-based agroforestry system, farmers strategically plant fruit trees alongside other crops or livestock, creating a harmonious and mutually beneficial relationship between them. Each and every plant cannot be planted in the combination. We have to choose different components of the system very judiciously for better financial as well as sustainable production. Various components like trees, crops, and animals have to complement and supplement each other instead of creating competition for various resources like light, water, nutrients, etc. in the system. Trees, which have a deep root system, fulfill their nutritional and water requirements from the deeper layer of soil, whereas agricultural crops, which have a shallow root system, obtain nutrients from the upper layers of soil. Livestock, which is an important component of the system, not only provides organic matter to the field but also serves as an asset in case of any natural calamities or any kind of crop failure, sustaining the livelihood of the farmer. Table 1. shows different combinations of fruit-based agroforestry. Prior to the table, we need to understand the key components of fruit-based agroforestry systems which are as follows:

- Fruit trees (selected on the basis of local climate, soil conditions, and market demand)
- Crops (including grains, vegetables, or other cash crops)
- Livestock (sheep, goats, and poultry).

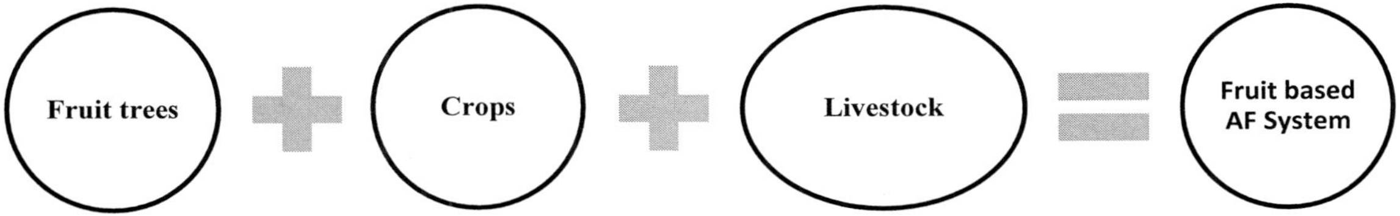

Table 1. Different combinations of fruit-based agroforestry

Sr. no	Fruit-based agroforestry System	Agricultural crops	Fruit trees	Forest trees	References
1.	Agri-horticulture systems	Maize, Rice, Blackgram, Tomato, Okra, Soybean, Wheat, Barley, Mustard, Gram, Pea, Cabbage, Cauliflower, Potato, Garlic, Onion	Mango, Citrus, Plum, Litchi, Guava, Papaya	-	Toky et al., 1989; Gupta et al., 2017; Tiwari et al., 2018; Kumar et al.,2018
2.	Agri-hortisilvopastoral systems	Maize, BlackGram, Tomato, Soybean, Colocasia, Rajmah, Ginger, Shimla Mirch, Wheat, Barley, Mustard, Gram Pea, Cabbage, Cauliflower, Garlic, Onion, Turmeric	Peach, Pear, Plum, Apricot, Apple, Akhrot, Daru (Wild Pomegranate)	*Toona ciliata, Grewia optiva, Bauhinia variegata, Bombax ceiba, Albizia chinensis, Leucaenaa leucocephala, Ficus palmata*	Toky et al., 1989; Kumar et al., 2018; Gupta et al.,2017; Rajput et al., 2017; Tiwari et al., 2018; Goswami et al.,2014

3.	Horti-silviculture systems		Apple, Peach, Pear, Plum, Apricot, Persimon, Daru (Wild Pomegranate), Akhrot, Almond, Pistachio nut	*Populus ciliata, Celtis australis, Cedrus deodara, Salix tetrasperma*	Gupta et al., 2017; Kumari et al., 2008; Rajput et al.,2017; Tiwari et al, 2018; Chisanga et al., 2013; Salve and Bhardwaj, 2020; Kumar et al., 2018; Goswami et al., 2014
4.	Hortisilvopastoral systems		Apple, Apricot, Grapes, Akhrot, Almond, Pistachio nut	*Quercus dilatata, Celtis australis, Populus ciliata, Morus alba, Populus ciliata*	Kaler et al.,2017; Gupta et al, 2017; Kumar et al., 2008; Chisanga et al., 2013; Salve and Bhardwaj, 2020; Kumar et al., 2018

Major factors responsible for the successful raising of fruit-based agroforestry systems include:

- **Microclimate Considerations:** Evaluate the site's microclimate, considering factors such as temperature, rainfall, and sunlight, to ensure compatibility with the chosen fruit tree species. Assess the land's topography for water drainage and soil erosion control.

- **Species Selection and Diversity:** Choose fruit tree species adapted to the local climate and soil conditions for optimal growth and productivity. Introduce a diverse range of fruit species to enhance resilience, market options, and overall system stability.

- **Agroecological Practices:** Incorporate companion plants that support fruit tree growth by providing additional benefits such as pest control or nitrogen fixation. Embrace organic and sustainable farming practices to reduce reliance on synthetic inputs and enhance ecosystem health.

- **Integrated Pest Management (IPM) and Water Management:** Implement IPM strategies to manage pests and diseases in a holistic and environmentally friendly manner, reducing reliance on chemical pesticides. Along with this, develop efficient water management systems, including rainwater harvesting and drip irrigation, to meet the specific water needs of fruit trees. Use water conservation techniques to optimize water use efficiency.

- **Soil Health and Fertility**: Conduct regular soil tests to monitor nutrient levels and adjust fertilization practices accordingly. Implement soil conservation measures to prevent erosion and maintain soil structure.

- **Pruning and Training:** Apply appropriate pruning and training techniques to shape fruit trees, improve light penetration, and enhance overall productivity. Regularly monitor and manage the canopy structure for optimal fruit production.

- **Harvesting and Post-Harvest Handling:** Time harvests based on fruit maturity to ensure quality and marketability. Implement proper post-harvest handling practices to minimize losses and maximize the economic value of the produce.

- **Community Engagement and Knowledge Sharing:** Foster community involvement and knowledge sharing to build a collective understanding of fruit-based agroforestry practices. Establish local networks for experience exchange, training, and mutual support.

- **Financial Planning and Support:** Develop a financial plan that includes an initial investment, ongoing operational costs, and potential returns. Explore available financial support mechanisms, grants, or subsidies for agroforestry initiatives.

Fruit-based agroforestry systems in livelihood upliftment:

Natural resources are under severe threat due to the quickening pace of climate change, resulting in an environmental catastrophe that is certain to worsen in the near future. Agroforestry has garnered significant attention from the scientific community recently due to its potential to achieve sustainable livelihood, food security, and a sustainable means of subsistence by producing a variety of goods, including fruit, fuel wood, fodder, food, and other non-timber forest products, as well as a range of ecosystem and environmental services. It is also able to mitigate problems like deforestation, climate change, and the degradation of fertile land. By quickly increasing productivity per unit area on a sustainable basis, it is the most promising land management system, helping to achieve sustainable livelihoods. By improving the quantity and quality of food products, raw materials, and other types of trees and shrubs, agroforestry techniques help farmers by giving them a source of income (Naiem, 2007).

This land management system is viewed as an alternative paradigm for rural development globally since it is centered on fruit trees and has the potential to meet farmers' nutritional demands in low-input, limited-land settings while also gaining high market value. Additionally, by generating income and work possibilities in rural regions, agroforestry enhances socioeconomic conditions by lowering food production scarcity and enhancing financial standing.

- **Nutritional Importance**: The nutritional significance of fruit-based agroforestry systems extends beyond mere sustenance. The integration of diverse fruit crops into farming practices enriches diets by providing a spectrum of essential vitamins, minerals, and antioxidants. This nutritional diversity not only fosters healthier communities but also addresses broader issues of food security and dietary adequacy. Through a careful examination of the nutritional implications, we unravel the intricate connections between fruit-based agroforestry and human well-being.

- **Employment:** Promotes local employment opportunities through the cultivation, harvesting, and processing of diverse fruit crops. Support the development of small-scale enterprises related to fruit production, such as value-added processing and marketing. Orchards with tourism potential can employ guides to educate visitors about fruit cultivation and offer recreational activities. Establishing nurseries, orchards, or processing units can lead to entrepreneurship opportunities, create jobs, and stimulate local economies.

- **Full utilization of land** enhances overall farm productivity by utilizing vertical space for fruit trees alongside traditional crops, maximizing land use efficiency.

- **Soil Improvement:** The root systems of fruit trees enhance soil structure and fertility, preventing soil erosion and promoting water retention. Organic matter from fallen leaves and fruit residues contributes to soil health, fostering long-term sustainability and resilience.

- **Climate change mitigation:** Fruit trees contribute to carbon sequestration, mitigating the impacts of climate change by absorbing and storing atmospheric carbon dioxide. The shading effect of fruit trees can also contribute to microclimatic regulation, reducing temperature extremes and improving growing conditions.

- **Carbon Sequestration:** Fruit-based agroforestry acts as a carbon sink, sequestering carbon in both aboveground biomass and soil organic matter. It offers numerous ecological advantages, and along with this, it has the ability to help low-income households (Shukla et al. 2018).

- **Biodiversity enhancement:** Diverse fruit tree species increase on-farm biodiversity, attracting beneficial insects and pollinators. These systems maintain the general health of the ecosystem and aid in the preservation of regional plant genetic resources.

Economic aspects of fruit-based agroforestry systems that contribute to livelihood enhancement:

Agroforestry offers numerous ecological advantages, and along with this, it has the ability to help low-income households (Shukla et al. 2018). Due to the country's growing urbanization and economic expansion, farming communities now have an unpredicted chance to supply farming products beyond subsistence levels (ICAR, 2020). In addition to the financial gains, agroforestry has a huge potential for creating jobs and ensuring and improving the standard of living of people who depend on agriculture (Mirjha and Rana, 2016). Within this sustainable agroecosystem, integrating livestock can act as a safety net, providing regular income to households in the event of crop failures due to climate change, insect attacks, or other pests (Sharma et al., 2008). Table 2 shows the Benefit-cost ratio of a fruit-based agroforestry system.

- **Carbon trading (turning carbon into a financial asset):** Agroforestry projects that sequester carbon can participate in carbon trading markets. Carbon credits are an asset available to the farmer and can be banked or sold on the open carbon market at any time. Thus, providing a financial incentive to farmers

- **Diversified Income Streams:** Fruit-based agroforestry introduces a diverse range of fruit crops, enabling farmers to tap into various markets and diversify their income sources. The staggered harvest periods of different fruit varieties contribute to a more stable and reliable income annually.

- **Market Access and Value Addition:** Integration of fruit crops enhances market access for farmers, as fruits are often in demand both locally and globally. Value addition through processing (e.g., jams, juices) further increases economic returns, providing opportunities for small-scale enterprises and agribusiness development.

- **Employment Generation:** The cultivation, harvesting, and processing of fruit crops create employment opportunities within the community. Local labor engagement in various stages of fruit production contributes to a vibrant rural economy.

- **Fruit Tree Products and By-products:** Beyond the fruits themselves, various by-products such as timber, medicinal extracts, and essential oils from fruit trees contribute to additional income streams. Utilization of by-products reduces waste and maximizes the economic potential of the entire agroforestry system.

- **Enhanced Land Productivity:** Integration of fruit trees maximizes land use efficiency, leading to increased overall farm productivity. The effective utilization of land resources contributes to higher yields per unit area, translating into improved economic returns for farmers.

- **Cost Reduction through Agroecological Practices**: Agroforestry systems often involve **natural pest control**, nutrient cycling, and improved water retention, reducing the need for external inputs. Lower input costs contribute to increased profitability and economic sustainability for farmers practicing fruit-based agroforestry.

- **Long-Term Economic Resilience**: The perennial nature of fruit trees contributes to the long-term economic resilience of farmers, as fruit-bearing trees continue to produce for many years. This longevity minimizes the risks associated with annual crop fluctuations, providing a more secure economic foundation for farming communities.

- **Access to Climate Finance and Incentives:** Fruit-based agroforestry, with its positive environmental impacts, may open avenues for farmers to access climate finance and incentives. Participation in carbon offset programs and other environmental initiatives can further enhance the economic viability of agroforestry systems.

In summary, the economic benefits of fruit-based agroforestry systems are substantial, encompassing diversified income streams, employment generation, value addition, and long-term resilience. As we analyze the economic dimensions, it becomes evident that integrating fruit trees into agroecosystems goes beyond mere cultivation and transforms farming into a sustainable and economically viable enterprise.

Table 2. The benefit-cost ratio of a fruit-based agroforestry system:

Location	Agroforestry System	Gross return	Total cost of production	Net return	Benefit Cost Ratio	References
Himachal Pradesh	Peach + *Ocimum sanctum*	385800	206027.28		1.87	Sharma et al., 2021
West Bengal	*Eucalyptus* + Mango+ Cowpea-toria	355610	66889	288721	5.31	Das et al., 2022
Meghalaya	*Psidium guajava* + rice (RCPL-24)	51,248	25,440	25,808	2.96	Bhatt and Misra, 2003
Bangladesh	Coconut + Guava +Misridana	235444	46525	188919	5.06	Bari and Rahim, 2012
Uttar Pradesh	*Psidium guajava* + *Curcuma longa*	-	-		4.74	Chandra, 2014
Gujarat	*Manilkara acharas* + *Jatropha curcas*+Basil	137985	80341.95	57643.05	1.71	Solanki et al., 2014
Karnataka	Sapota+ Grass + *Dalbergia sissoo* + Maize + sunhemp	1170	962	208	1.22	Patil et al., 2010

Carbon sequestration through a fruit-based agroforestry system:

Climate change is a crucial problem affecting the entire globe. The potential impacts of climate change have been more damaging and devastating, including sea level rises, increased frequency, and intensity of floods, changing patterns of precipitation, a new problem of climate change refugees, changes in vegetational patterns, gender inequity, decreasing working efficiency, effects on livestock rearing, etc. A significant sector of the economy, agriculture is vulnerable to climate change because of unsustainable management practices that degrade land and resources, cause crop failure and insect pest attacks, and result in financial losses that pose a serious danger to food security. (SIWI, 2018) and our various sustainable development goals like no poverty, zero hunger, good health and well-being, clean water, etc. It is imperative to identify and decide on a strategy for climate change adaptation and mitigation, food security, biodiversity conservation, and sustainable land resource management that will reduce the risk of climate change because a significant portion of the world's population depends on agriculture, which is predicted to be severely impacted by climate change (Turral, 2011).

Agroforestry is a sustainable and efficient land management system that contributes to achieving nine out of the 17 SDGs, according to a report titled "Achieving the Global Goals through Agroforestry." According to this report, agroforestry has the strongest impact potential on poverty reduction, hunger alleviation, climate action, biodiversity conservation, sustainable land management, gender equality, health, increasing access to clean water, sustainable energy solutions, and responsible agriculture production. It differs from other land uses by incorporating woody perennials into farms, which results in significant economic and ecological interactions between woody and non-wood components (Nair, 1993).

Agroforestry is the combination of crops with trees that are important for carbon sequestration (Takimoto et al., 2009). The UNFCCC defines carbon sequestration as "the process of extracting carbon (C) from the atmosphere and depositing it in reservoirs or transferring atmospheric CO_2 to secure storage in long-lived pools" (UNFCC, 2007). According to projections from the Intergovernmental Panel on Climate Change (IPCC, 2000), global temperatures might rise by 1.4 to 5.8 degrees Celsius by 2100. Activities like vegetation fires and land clearing for agriculture and forestry, although releasing substantial CO_2 into the atmosphere, often see recapture through subsequent vegetation regrowth (Lorenz and Lal, 2010; Nair et al., 2010). Agroforestry holds significant promise for carbon sequestration both above and below ground (Goswami et al., 2017). Integrating fruit trees alongside timber trees in agricultural production systems has the potential to enhance carbon storage on agricultural lands. This approach enables the cultivation of food crops alongside the growth of fruit and timber trees (Kursten, 2000). The only terrestrial carbon reservoir, the soil organic carbon (SOC) pool, is deliberately expanded by using various fruit-based agroforestry practices (Lorenz and Lal, 2014). In terms of SOC content, land use systems are typically ranked as follows: Forests > Agroforests > Tree Plantations > Arable Crops (agroforests are multi-strata systems similar in complexity to home gardens but larger in area; Nair et al., 2009a, b).

Carbon sequestration through agroforestry involves several scientific processes (Kaler et al., 2024):

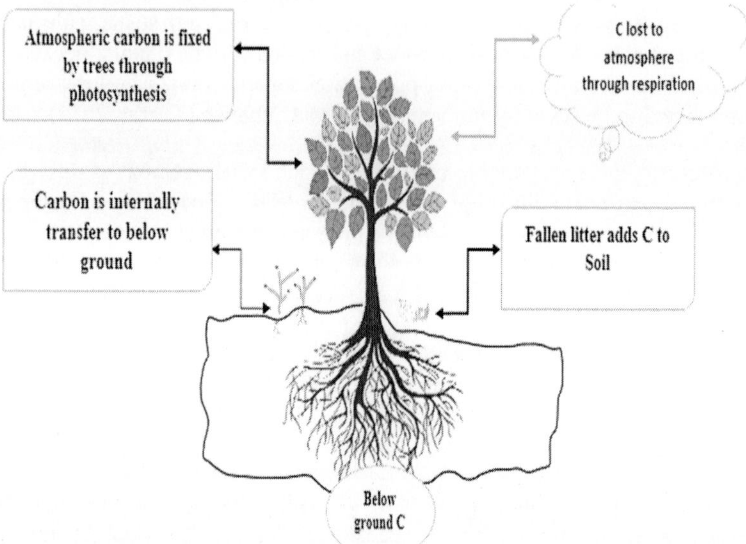

- **Photosynthesis** is the process by which agricultural crops and trees in an AFS turn CO_2 from the atmosphere into organic molecules that are needed for growth.

- **Biomass accumulation:** As roots, stems, leaves, and branches grow, trees in the AFS store carbon from the atmosphere.

- **Decomposition of litter:** Organic matter, such as branches and leaves that have fallen, adds to the soil's carbon pool. This organic matter (OM) is broken down by microorganisms, which release carbon into the soil.

- **Carbon Sequestration in Soil:** By increasing the amount of organic matter in the soil, the AFS system helps to sequester carbon. Additionally, the presence of trees lessens the loss of topsoil, which is rich in carbon, by limiting soil erosion. Exudates from roots increase soil organic carbon (SOC) and promote microbial activity, which improves carbon sequestration.

Constraints faced in adopting a fruit-based agroforestry system:

- **Inadequate extension activities:** The dissemination of agroforestry research findings to farmers is hindered by an inefficient extension system, primarily resulting from the absence of an institutional framework (NAFP, 2014; Tiwari et al., 2018). Additionally, there is no universally applicable, one-size-fits-all model for conducting on-farm research, which largely depends on the nature, objectives, and conditions of conducting research.

- **Climate change:** seasonal variations, temperature alteration, changing rainfall, and snow patterns can impact fruit tree growth and production.

- **Financial constraints:** The initial investment cost for establishing agroforestry systems can be high, which might deter adoption. Slow economic returns from fruit trees.

- **Inefficient transport and a lack of cold storage facilities**

- **Tree crop interaction studies:** There are studies that depict the negative influence, such as the shading effect of tree canopies on the performance of crops grown under them and competition for light, water, and nutrients between agriculture and crop farmers, sometimes in adopting these systems.

- **Market constraints:** With the exception of a few states that have specially created marketing infrastructure for agroforestry produce or joined the regulated agriculture commodity marketing system, the country largely lacks marketing infrastructure, including price discovery methods for agroforestry produce. Remote locations can present challenges in accessing markets for the sale of fruits, affecting economic viability.

Solution/way forward for constraints currently facing:

- **Efficient Extension Services: An efficient** communication channel, collaboration, and effective extension services are needed in order to channel the research results from lab to land, which have been helpful to farmers in achieving their personal as well as national Sustainable Development Goals.

- **Farmer training and capacity building:** conduct training programs to educate farmers about best agroforestry practices, including tree planting techniques, pruning, pest management, and various combinations of intercrops grown along with fruit crops for different agroclimatic zones.

- **Financial support:** Provide financial incentives, subsidies, or low-interest loans to small-scale farmers in order to encourage them to adopt fruit-based agroforestry systems. This can help overcome initial investment costs and encourage more farmers to participate.

- **Climate-smart practices:** Integrate climate-smart practices into fruit-based agroforestry systems such as water conservation, mulching, fertilization, and integrated pest management. These practices can enhance climate resilience and reduce vulnerability to extreme weather events. Although we cannot halt climate change, we can minimize it through climate-smart practices.

- **Policy support:** Develop and implement supportive policies and regulations that promote and incentivize fruit-based agroforestry systems. This may include tax incentives, land tenure security, and recognition of agroforestry in national agricultural strategies.

- **Developing marketing infrastructure and linkages:** A market that is effective, competitive, and readily available, along with suitable post-harvest and marketing infrastructure, has the ability to meet its main goal of delivering a net positive return to the producer (GOI, 2018).

- **E-commerce platforms** help eliminate middlemen and bring farmers and consumers together directly; online marketplaces can boost earnings.

- **Fairtrade and certification:** Farmers can benefit from better prices and ethical sourcing by supporting fair trade practices and sustainability certifications.

- **Value chain development: by** creating processing facilities, cold storage facilities, and cooperatives, fruit value will increase, and farmers will receive greater prices.

Case Studies:

Sustained productivity:

Bellow, Hudson, and Nair (2008) investigated fruit-tree-based agroforestry systems in low-intensity home gardens in northwest Guatemala. These systems, which include apple, peach, and pear trees, have the potential to benefit resource-constrained farmers. Adoption is more likely in reasonably rich households with bigger land holdings, but increased productivity is critical among smaller landholders. Food shortages, fruit quality, and market infrastructure are all constraints. Complementarity with maize crops, household fruit consumption, and additional revenue creation on restricted land are all factors that favor adoption.

Ecosystem service:

Contribution of trees to the conservation of biodiversity and ecosystem services in agricultural landscapes (Barrios et al. 2018)

The transition from monocropping to tea agroforestry in China

Tea production in China increased by close to 50% when the tea was grown with *Alnus nepalensis.* Ectomycorrhizal biomass, as well as that of bacteria, was found to be greater in the soils surrounding *A. nepalensis*.

Aspect	**Monoculture Tea**	**Agroforestry (with *A. nepalensis*)**
Tea Production Increase	No change	Close to 50% increase
Microbial Biomass	No change	21–23% increase
Effect on Tea Plantation Productivity	Standard productivity	Improved productivity

Coffee agroforestry transition in Indonesia (Sumatra)

Effect on Earthworms: Significant loss of four native large-bodied earthworm species, replaced by six exotic smaller-bodied earthworm species.

Soil Physical Properties: 71–76% lower soil macro porosity in coffee monoculture compared to natural forest.

Aspect	**Forest Conversation to Coffee Monoculture**	**Coffee Agroforestry**
Effect on Annual Litterfall	71% reduction in annual litterfall in coffee monocrops.	41% reduction
Effect on Standing Litter	43% reduction in standing litter in coffee monocrops.	29% reduction

Highly economic:

Bari and Rahim 2012 evaluated the economic and yield performance of some medicinal plants *viz* Aloevera (*Aloe indica*), Asparagus (*Asparagus racemosus*), and Misridana (*Kaempferia angustifolia*) in Coconut based AFS. Results revealed that among three medicinal plants, Asparagus and Misridana produced the maximum under the Medicinal plant-based Agroforestry System, while Aloe vera yielded the maximum in sole cropping. Highest B:C ratio (3.54, 3.21, 5.06) was recorded from the Coconut + Guava MAF under Aloevera (*Aloe indica*), Asparagus (*Asparagus racemosus*) and Misridana (*Kaempferia angustifolia*).

Carbon sequestration:

Zahoor et al.2021 evaluated apple-based agroforestry systems in the central Kashmir region of India with the aim of acknowledging the potential of fruit-based agroforestry systems in biomass production and carbon stock assessment for climate change mitigation. They recorded maximum fruit biomass under apple + rajma + oats (3.69 t/ha) and lowest in sole apple (3.11 t/ha). The highest ecosystem carbon was stored in the combination of apple + rajma + oats (64.18 t/ha) and lowest in french bean

+ oats (37.06 t/ha) which is around 1.5 to 2 times higher than the agriculture-based system

Soil carbon sequestration:

Zahoor et al.2022 investigated the influence of intercropping on soil properties and nutrient status of fruit-based agroforestry systems in the Kashmir valley of India. Four intercrops of rajma, green gram, French bean, and oats were intercropped with an 11-year orchard of apples. A maximum volume of soil carbon was observed in apple + rajma and oats and apple + green gram and oats (1.74 %) followed by apple + French bean and oats (1.65%) at the time of harvesting of oats and overall highest values of nitrogen (387.7 kg/ha) and soil moisture (12.68%) under apple+ green gram and oats and lowest nitrogen (334.88 kg/ha) and soil moisture (9.82%) in sole apple. It was concluded from the study that the tree-based land use frameworks are more sustainable and environment benevolent than agriculture-based land use systems for the temperate zone of the Himalayan region.

Conclusion

Fruit tree-based agroforestry systems are gaining popularity and acceptance globally, which is evident from the literature cited above. Fruits, along with various products harvested from the agroforestry system, have the potential to sustain livelihoods, gain high market value, and also meet the nutritional and dietary needs of farmers in low-input and limited-land situations. Overall, the future of fruit-based agroforestry systems is bright, offering a pathway toward sustainable food security, income generation, and improved livelihoods for communities worldwide. Carbon sequestration through fruit trees in an agroforestry system not only contributes to climate change resilience but also provides a major economic opportunity through carbon trading in the international market, along with addressing social demands through multiple products. Thus, it is the key path to wealth and prosperity for farmers, generating employment and revenue, meeting other basic needs on a sustainable basis, and cushioning farmers from the harness of climate change. Fruit-based agroforestry appears as a feasible answer in view of the rising demand for sustainable agricultural practices and the necessity for climate-resilient farming systems. As a result, the need of the hour is to concentrate on agroforestry practices and models that would boost the productivity and profitability of small landholders in the state. This chapter provides valuable information regarding the potential and capability of fruit-based agroforestry land use systems with respect to carbon sequestration, food security, and livelihood improvement along with the desirable data for farmers, researchers, scientists, policymakers, etc.

References

1. Bari MS and Rahim MA. 2012. Economic evaluation and yield performance of some medicinal plants in coconut-based Multistoried Agroforestry Systems. *A scientific Journal of Krishi Foundation.* 10(1):71-80
2. Barrios E, Valencia V, Jonsson M, Brauman A, Hairiah K, Mortimer PE and Okubo S.2018. Contribution of trees to the conservation of biodiversity and ecosystem services in agricultural landscapes. *International Journal of Biodiversity Science, Ecosystem Services & Management.* 14(1):1-16.
3. Bellow, J. G., Reggie L. Hudson, and P. K. R. Nair. 2008. Adoption Potential of Fruit-Tree-Based Agroforestry on Small Farms in the Subtropical Highlands. *Agroforestry Systems*. 73 (1): 23–36.
4. Bhatt BP and Misra LK. 2003. Production Potential and cost-benefit analysis of Agrihorticulture agroforestry system in northeast India. *Journal of Sustainable Agriculture* 22(2):99-108.
5. Carvalho FP. 2017. Pesticides, environment, and food safety. Food and energy security 6:48-60.
6. Chandra KK. 2014. Evaluation of growth and economic parameters of *Curcuma longa* and *Amorphophallus paeonifolius* intercrops in medium-aged *Psidium guajava* orchard. *Caribbean Journal of Science.* 2(1): 392-398.
7. Chisanga K, Bhardwaj DR and Sharma S. 2013. Bio-economic appraisal of agroforestry systems in Dry temperate Western Himalayas. *Journal of Tree Scien*ces. 32(1&2): 1-7.
8. Das Pinakesh, Panda Subhabrata and Dhara PK. 2022. Fruit-based agroforestry systems are potential means for sustaining carbon sequestration and improving soil health and diet of communities in red and lateritic zones of West Bengal, India. *International Journal of scientific and engineering research.* 13(12):832-842.
9. Dhyani SK and Handa AK. 2013. The area under agroforestry in India: An assessment for present status and future perspective. *Indian Journal of Agroforestry.* 15(1):1-11.

10. FAO. 1981. FAO Production Yearbook. Food and Agriculture Organization of United Nations. Rome. Vol. 34. 115pp.
11. GoI.2018. Operational guidelines Agricultural Marketing Infrastructure sub-scheme of ISAM. Ministry of Agriculture and Farmer Welfare, GoI.63pp.
12. Goswami, S. B., Verma, K., & Kaushal, R. 2013. Biomass and carbon sequestration in different agroforestry systems of a Western Himalayan watershed. *Biological Agriculture & Horticulture*. 30(2): 88–96
13. Goswami, S. B., Verma, K., & Kaushal, R. 2013. Biomass and carbon sequestration in different agroforestry systems of a Western Himalayan watershed. *Biological Agriculture & Horticulture,* 30(2): 88–96.
14. Gupta B, Sarvade S and Singh M. 2017. Species composition, biomass production and carbon storage potential of agroforestry systems in Himachal Pradesh. In; Agroforestry for increased Production and livelihood security (Gupta SK, Panwar P and Kaushal R, eds.), New India Publishing Agency.242-270pp.
15. ICAR .2020. Agroforestry for income enhancement, climate resilience, and ecosystem services. Indian Council of Agricultural Research, New Delhi. 30pp.
16. IPCC. 2000. IPCC Special Report: Emission Scenarios.
17. Kaler NS, Gupta B and Negi V. 2017. Traditional agroforestry systems in the cold desert region of Himachal Pradesh, India. *Popular Kheti* 5(2): 35-36
18. Kaler NS, Shiwangee, Kritika, Sharma SD and Saralch H. 2024. Agroforestry's Role in mitigating climate change and nurturing sustainable farming systems. *Farm Chronicle*. 03(01):16-19.
19. Kumar BM, Handa AK, Dhyani SK, and Arunchalam A. 2018. Agroforestry in the Indian Himalayan region: Overview. In: Temperate agroforestry systems 2nd Edition (AM Gordon, SM Newman and BRW Coleman, Eds.), CAB International. 153-172pp.
20. Kumar P, Thakur CL, Rai Pand Attri K. 2018. Identification of existing agroforestry systems and socio-economic assessment in Kandhaghat block of Solan District, HP, India. *International Journal of Current Microbiology and Applied Sciences*. 7(4): 3815-3826
21. Kumari A, Seghal RN and Kumar S. 2008. Traditional agroforestry systems practices in Lahaul (Lahul & Spiti) and Kinnaur Districts of Himachal Pradesh. *Indian Forester*.8:1003-1010.
22. Leakey RR, Simons AJ. (1998) The domestic and commercialization of indigenous trees in agroforestry for the alleviation of poverty. *Agroforestry Systems*. 38: 165-176.

23. Lorenz K, Lal R (2014) Soil Organic carbon sequestration in agroforestry system. A review. *Agronomy for sustainable development.*34(2); 443-454
24. Mirjha PR, Rana DS (2016). Yield and yield attributes, system productivity, and economics of mango *Mangifera indica*-based intercropping systems as influenced by mango cultivars and nutrient levels. *Indian Journal of Agronomy.* 61 (3):307-31
25. Na'iem M. 2007. Preface. In: Sabarnurdin M S (Ed.) The role of agroforestry education in the revitalization of agriculture, fishery, and forestry program; proceeding international seminar. Gadjah Mada, SEANAFE, Department of Forestry, and Perum Perhutani. Yogyakarta, 7-8 february 2006. 167p.
26. NAFP. 2014. National Agroforestry Policy. Ministry of Agriculture, GOI. 14p.
27. Nair PKR, Nair VD, Kumar BM, Showalter JM 2010 Carbon sequestration in agroforestry systems. *Advances in Agronomy.*108:237-307
28. Patil SJ, Mutanal SM, Patil HY, Shahapurmath G, Maheswarappa.2010. Performance of Sapota- Teak-based agroforestry system in hill zone of Karnataka.*Indian Journal of Agroforestry*.12(1): 27-34.
29. Rahman GMM, Wadud MA, Shahjahan M and Jewel KNA .2014. Fruit tree-based agroforestry practices in char land farming system. *Journal of Agroforestry and Environment..* 8 (1): 1-6
30. Rahman MA, Tani M, Asahiro K, Ullah SA. (2017) Species composition, diversity and productivity of homesteads in southeastern Bangladesh. *Small Scale Forestry.*16(3):295-309.
31. Rajput BS, Bhardwaj DR and Pala NA. 2017. Factors influencing biomass and carbon storage potential of different land use systems along an altitudinal gradient in temperate northwestern Himalayas. *Agroforestry* Systems. 91: 479-486.
32. Rajput BS, Bhardwaj DR and Pala NA. 2017. Factors influencing biomass and carbon storage potential of different land use systems along an altitudinal gradient in temperate northwestern Himalayas. *Agroforestry Systems*. 91: 479-486.
33. Salve A and Bhardwaj DR. 2020. Soil Carbon stock and nutrient study in different agroforestry systems at Kinnaur district, Himachal Pradesh. *Plant Archives*. 20(2):4251-4260
34. Shamebo D. 1998. Banana in the southern region of Ethiopia (SRE). Bananas and Food Security. 119pp.
35. Sharma K, Thakur S, Sharma R, Kashyap SD. 2008. Production and economics of Kinnow cultivation with wheat and gobhi sarson in

Himachal Pradesh. *Indian Journal of Soil Conservation..* 36(2) : 114-117

36. Sharma S, Pant KS, Dev Kapil. 2021. Economic analysis of *Ocimum sanctum* under Peach and apricot-based agroforestry system. *Journal of Pharmacognosy and Phytochemistry.* 10(1S):93-96
37. Shukla S, Pandey VV and Kumar V. 2018. Agroforestry systems as a tool in sustainable rural development, food scarcity, and income generation. Indian Forester 144(5): 435-441.
38. SIWI. 2018. Water for productive and multifunctional landscapes. Stockholm International Water Institute, Report no. 38.
39. Solanki VK, Jadeja DB and Tandel MB.2014. performance of herbal medicinal crops under sapota-jatropha-based three-tier agroforestry system. *Indian Journal of Agricultural Sciences*. 10(1):267-271.
40. Tiwari P, Pant KS and Singh R. 2018. System units under Prevalent agroforestry systems in North-Western Himalayas and their constraints. *Journal of Pharmacognosy and Phytochemistry*. 7(6):2758-2764.
41. Toky OP, Kumar P and Khosla PK. 1989. Structure and function of traditional Agroforestry system in Garhwal Himalayas Biomass and productivity. *Agroforestry Systems.* 9: 47-70
42. Turral H, Burke and Faures JM. 2011. Climate change, Water and Food Security; Water Report; Food and Agriculture Organization of the United Nations (FAO): Rome, Italy, 200pp.
43. UNFCCC 2007 Report of the conference of parties on its thirteenth session, Bali, Indonesia. In: United Nations Framework Convention on Climate Change, Geneva, Switzerland, UN.
44. Zahoor S, Dutt V, Mughal AH, Rashid M and Khan PA. 2022. Influence of intercropping on soil properties and nutrient status of fruit-based agroforestry systems in the northwestern region of Indian Himalaya. *Indian Journal of Agroforestry.24(1)* 85-92
45. Zahoor S, Dutt V , Mughal AH , Pala NA , Qaisar KN and Khan PA. 2021. Apple-based Agroforestry systems for biomass production and carbon sequestration: implication for food security and climate change contemplates in the temperate region of Northern Himalaya, India. *Agroforest Systems*. 95: 367-382.

CHAPTER SIX

POTENTIALS OF BAMBOO-BASED AGROFORESTRY SYSTEMS FOR CLIMATE CHANGE AND INCOME GENERATION

NARESH SINGH[1], R.S. BALI[1], PRIYANKA THAKUR[1], PANKAJ LAL[1] AND RAM GOPAL[2]

[1]COLLEGE OF FORESTRY, RANICHAURI, TEHRI GARHWAL, UTTARAKHAND

[2]NORTH EASTERN HILL UNIVERSITY, TURA CAMPUS, MEGHALAYA

Abstract

Bamboo-based agroforestry systems are a sustainable and multifunctional approach to land management that combines the cultivation of bamboo with other agricultural or forestry practices. These systems offer a range of benefits, including the production of bamboo biomass for various applications, the provision of ecosystem services, and the diversification of income sources for farmers. This abstract explores the key aspects of bamboo-based agroforestry systems, including their potential for carbon sequestration, soil conservation, and the enhancement of biodiversity. The chapter also discusses the challenges and opportunities associated with the implementation of these systems, such as the selection of appropriate bamboo species, the integration of compatible crops or livestock, and the development of effective management strategies. By highlighting the advantages and considerations of bamboo-based agroforestry systems, these abstract aims to promote their adoption and contribute to the development of sustainable land use practices worldwide.

I. Introduction

A. Definition of Bamboo Based Agroforestry Systems

Bamboo-Based Agroforestry Systems involve integrating bamboo with crops or pasture to provide economic, environmental, and social benefits. Bamboo's fast growth, early maturity, and ability to regenerate vegetatively make it a suitable crop for agroforestry systems (Montagnini & Finney, 2018). Bamboo can be used as live stakes to support crops like yams, potatoes, and beans, reaching maximum productivity in 7 to 8 years (Simons, 2010). It can improve soil health, reduce run-off and erosion, and provide alternative livelihood support for farmers [3]. Despite potential competition among components, bamboo-based agroforestry systems offer a promising approach to sustainable land use, offering benefits such as increased farm productivity, reduced soil erosion, wildlife habitat, and carbon sequestration (Nair, 2012).

B. Importance of Agroforestry in Sustainable Agriculture

Agroforestry plays a crucial role in sustainable agriculture by offering a range of ecological, economic, and social benefits. It integrates trees and shrubs with crops and/or livestock to create a multi-functional land-use system that enhances biodiversity, improves soil health, and increases productivity and quality of crop yields (Nair, 2012). Agroforestry contributes to landscape restoration, ecological stability, food security, economic development of local farmers, and sustainable livelihood enhancement of rural communities (Garrity *et al.*, 2010). By combining trees with agricultural activities, agroforestry systems help prevent soil erosion, increase soil fertility, and add nutrients like nitrogen to the soil, thereby promoting sustainable land management practices (Kumar & Nair, 2004).

C. Significance of Bamboo in Agroforestry

Bamboo holds significant importance in agroforestry due to its diverse ecological, economic, and social benefits. In the context of sustainable agriculture, bamboo plays a crucial role in enhancing the functionality and sustainability of agroforestry systems. Here are the key points highlighting the significance of bamboo in agroforestry

1. Ecological Benefits:

- Bamboo serves as a renewable biomass source for energy production, aiding in reducing reliance on fossil fuels and mitigating deforestation pressures (Scurlock *et al.*, 2000).

- Its ability to thrive in challenging environments makes it valuable for land restoration efforts by preventing erosion and restoring degraded lands (Shackleton *et al.*, 2018).

- Bamboo's strong root systems reduce soil erosion, enhance soil health, and contribute to carbon sequestration, making it a valuable tool for climate change mitigation (Shackleton *et al.*, 2018).

2. Economic Opportunities:

- Bamboo offers economic opportunities for communities through diverse applications, including food, construction materials, handicrafts, pulp, and paper (Sharma *et al.*, 2020).

- Its growth conditions and sustainability credentials make it a green building material that reduces the need for harmful pesticides and fertilizers (Sharma *et al.*, 2020).

- Bamboo-based industries contribute to economic growth, poverty reduction, and livelihood security, with various sectors utilizing bamboo resources extensively (Sharma *et al.*, 2020).

3. Social Empowerment:

- Bamboo cultivation, processing, and trade can lead to economic development, environmental sustainability, and social empowerment when supported by government policies and initiatives (Chakravarty *et al.*, 2019).

- Bamboo's versatility and multiple uses provide communities with alternative livelihood options, particularly in regions where resources are strained due to climate change (Chakravarty *et al.*, 2019).

- By fostering the use of bamboo in agroforestry, food production, and rural economic development, communities can benefit from enhanced livelihoods and sustainable economic growth (Shackleton *et al.*, 2018).

D. Objectives of the Chapter

The objectives of the Bamboo-based Agroforestry System could be to:

1. Highlight the potential of bamboo as a component of agroforestry systems: Bamboo has several qualities that make it a suitable agroforestry component, including its fast growth and early maturity, ability to reestablish degraded lands, maintain soil health, and sequester carbon (Kumar & Joshi, 2020). It can be used for various purposes, such as charcoal production, erosion control, slope stabilization, and as live stakes for supporting crops like yams, potatoes, and beans (Joshi *et al.*, 2018).

2. Discuss the ecological benefits of bamboo-based agroforestry systems: Bamboo-based agroforestry systems can provide several ecological benefits, including landscape restoration, improved soil health, reduced erosion, and enhanced biodiversity (Oelbermann *et al.*, 2018). Bamboo's strong root systems reduce soil erosion, enhance soil health, and contribute to carbon sequestration, making it a valuable tool for climate change mitigation (Nghiem *et al.*, 2021).

3. Explore the economic opportunities of bamboo-based agroforestry systems: Bamboo-based agroforestry systems can provide economic opportunities for communities through diverse applications, including food, construction materials, handicrafts, pulp, and paper (Lahkar *et al.*, 2019). The growth conditions and sustainability credentials of bamboo make it a green building material that reduces the need for harmful pesticides and fertilizers (Khan *et al.*, 2020).

4. Examine the social empowerment potential of bamboo-based agroforestry systems: Bamboo cultivation, processing, and trade can lead to economic development, environmental sustainability, and social empowerment when supported by government policies and initiatives (Sharma *et al.*, 2017). Bamboo's versatility and multiple uses provide communities with alternative livelihood options, particularly in regions where resources are strained due to climate change (Wasswa *et al.*, 2019).

5. Discuss the challenges and potential solutions for bamboo-based agroforestry systems: Bamboo-based agroforestry systems face several challenges, including interspecific competition, tradeoffs between bamboo and intercrop yields, and limited available data to verify the suitability of the technology in SSA and Africa in general (Gardner *et al.*, 2016). Addressing these challenges requires further research and the development of cost-effective sources of protein, vitamins, and minerals for many households (Groom & Palmer, 2020).

6. Explore the potential of bamboo-based agroforestry systems for sustainable development in Sub-Saharan Africa: Bamboo-based agroforestry systems have the potential to contribute to sustainable development in Sub-Saharan Africa by providing a renewable energy source, restoring degraded forestlands, and serving as a sustainable carbon sink (Nair *et al.*, 2018). Incorporating bamboo into mixed-use agroforestry complexes can maximize its functionality while integrating it with other production crops (Devi *et al.*, 2021).

II. Historical Perspectives

The historical perspectives of bamboo-based agroforestry systems reveal a rich tradition of utilizing bamboo for sustainable land management and agricultural practices (Joshi, 2018). Bamboo, known for its fast growth and versatility, has been integrated into agroforestry systems for centuries, offering a range of ecological, economic, and social benefits (Raman, 2014). Historically, bamboo has been valued for its role in erosion control, slope stabilization, and as a source of income through charcoal production (Sinha *et al.*, 2020). Its ability to thrive in challenging environments and restore degraded lands has made it a valuable component of sustainable agriculture (FAO, 2017).

A. Origins of Bamboo Cultivation

Bamboo cultivation has a long history, with evidence of bamboo used in construction and agriculture dating back to the Neolithic period in China (Li *et al.*, 2020). The origins of bamboo cultivation can be traced to tropical and subtropical regions, mainly in the Far East, although wild species can also be found in Africa, Oceania, and America (Ohrnberger, 1999). China is where bamboo has best prospered, with the Moso species being the most cultivated and covering about 3 million hectares of the territory (Zhang *et al.*, 2018).

Bamboo cultivation has been documented in Western gardens since the early 19th Century or possibly the late 18th Century, with tropical species such as Bambusa vulgaris and B. bambos from India being the first to be introduced to Europe (Fryer & Henderson, 2010). Temperate bamboos are from more distant shores and inaccessible mountains, with some arriving from China and Japan in the mid-19th Century (Crouzet, 2019).

In Malaysia and the Philippines, there are similar stories of a man dreaming of a beautiful woman while sleeping under a bamboo plant and discovering her inside the bamboo stem when he wakes up and breaks it (Gomez *et al.*, 2015; Sangha *et al.*, 2017). In Japan, bamboo is a symbol of prosperity and is used to make New Year's decorations called kadomatsu (Fukushima *et al.*, 2019). Bamboo forests surround Shinto shrines and Buddhist temples as part of a sacred barrier against evil (Takizawa *et al.*, 2018).

Bamboo cultivation is also associated with the birthplace of paddy fields in China, with evidence of a bamboo mud cabin found in the Baodun culture, which is considered the birthplace of rice cultivation in China (Li *et al.*, 2020). The discovery of the Baodun bamboo house evidence lies in the heart of the upper reaches of the Yangtze River on the Chengdu Plain, which is also considered to be "the birthplace of rice cultivation" in China (Li *et al.*, 2020).

B. Evolution of Agroforestry Practices

The evolution of agroforestry practices can be traced back to ancient times in various regions, where the integration of trees with agriculture or livestock has been a common practice (Nair, 2018). Agroforestry has a rich historical background, with evidence of its existence in European countries since ancient times, as well as in regions like Greece and the Philippines (Dupraz *et al.*, 2018). Traditional agroforestry systems have been prevalent in Greece, contributing significantly to land use practices in the country (Plieninger *et al.*, 2012). In the Philippines, community-based forest management projects have played a role in shaping agroforestry practices, with farmers implementing diverse agroforestry systems to enhance sustainability and biodiversity (Mercado Jr *et al.*, 2014).

The evolution of agroforestry practices reflects a dynamic process influenced by historical, cultural, and environmental factors (Tsonkova *et al.*, 2018). Over time, agroforestry systems have adapted to changing landscapes and societal needs, incorporating traditional knowledge and modern innovations to optimize land use efficiency and sustainability (Lasco *et al.*, 2014). The integration of trees with agricultural activities has evolved to address contemporary challenges such as climate change, biodiversity loss, and land degradation, emphasizing the importance of agroforestry in promoting ecological resilience and sustainable land management practices (Jose, 2009).

C. Traditional Bamboo Agroforestry Systems in Different Cultures

Traditional bamboo agroforestry systems have been practiced in various cultures across the world, including India, Ghana, and Ethiopia. In India, bamboo has been used as boundary plants around agricultural fields to protect food crops from high winds and for water and soil conservation during water-stressed periods (Kumar *et al.*, 2018). In Ghana, bamboo-based agroforestry is relatively new, but there is potential for its use in improving the productivity of agricultural systems. Bamboo can be integrated into farming systems as live stakes for growing crops such as yam, potatoes, and beans, and can also be used for water and soil conservation (Asase *et al.*, 2018). In Ethiopia, bamboo plants have many traditional and cultural uses, including food crop production, live stakes, and fodder for livestock (Yitaferu *et al.*, 2008).

Bamboo-based agroforestry systems can provide various benefits, including reducing soil erosion, contributing to environmental sustainability, and providing alternative livelihood support for farmers during off-seasons (Asase *et al.*, 2018). Bamboo is a fast-growing species and can be harvested repeatedly, making it a valuable resource for farmers. In India, bamboo-based agroforestry has been found to raise the overall production and productivity of farming systems in wasteland conditions, with higher monetary returns compared to sole cropping systems (Kumar *et al.*, 2018).

In Ghana, bamboo-based agroforestry is relatively new, but there is potential for its use in improving the productivity of agricultural systems. Bamboo can be integrated into farming systems as live stakes for growing crops such as yam, potatoes, and beans, and can also be used for water and soil conservation (Asase *et al.*, 2018). Studies have shown that integrating bamboo into smallholder agricultural intercropping systems can contribute to food security, diversification of income sources, and sustainable bio-energy production (Kwaw *et al.*, 2020).

In Ethiopia, bamboo plants have many traditional and cultural uses, including food crop production, live stakes, and fodder for livestock (Yitaferu *et al.*, 2008). Bamboo-based agroforestry systems have the potential to improve the productivity of agricultural systems in Ethiopia, as they can provide various benefits, including reducing soil erosion, contributing to environmental sustainability, and providing alternative livelihood support for farmers during off-seasons.

C. Impact of Historical Practices on Modern Approaches

The impact of historical practices on modern approaches in bamboo-based agroforestry can be summarized in the following table:

Table 1: Impact of historical practices on modern approaches

Historical Practices	Modern Approaches	Impact	References
Integration of trees	Agroforestry systems with diverse species mix	Increased biodiversity and ecosystem resilience	Nair, P. K. R. (2018); Dupraz *et al.* (2018)
Traditional knowledge	Incorporation of Indigenous knowledge	Enhanced sustainability and local adaptation	Lasco *et al.* (2014); Mercado Jr *et al.* (2014)
Land management	Sustainable land use practices	Soil conservation and improved soil fertility	Kumar *et al.* (2018); Tsonkova *et al.* (2018)
Resource utilization	Optimized use of renewable resources	Economic benefits and reduced environmental impact	Joshi, G. (2018); Fukushima *et al.* (2019)
Environmental ethics	Promotion of eco-friendly practices	Conservation of natural resources	Plieninger *et al.* (2012); Takizawa *et al.* (2018)

Advantages of Traditional Bamboo-Based Agroforestry Practices:

The advantages and disadvantages of traditional bamboo-based agroforestry practices compared to modern approaches can be summarized as follows:

Advantages of Traditional Bamboo-Based Agroforestry Practices
1. **Integration of Traditional Ecological Knowledge**: Traditional practices often incorporate local wisdom and knowledge, enhancing the sustainability and resilience of agroforestry systems (Nair, 2018).

2. **Long-Term Sustainability**: Historical practices like intercropping bamboo with food crops have shown increased productivity and economic returns, contributing to the long-term sustainability of farming systems (Kumar *et al.*, 2018).
3. **Diversification of Livelihoods**: Traditional bamboo-based agroforestry systems have provided communities with diverse products and services, improving income sources and food security (Asase *et al.*, 2018).
4. **Environmental Benefits**: Bamboo in agroforestry systems can help reduce soil erosion, enhance soil fertility, and contribute to environmental sustainability (Yitaferu *et al.*, 2008).

Disadvantages of Traditional Bamboo-Based Agroforestry Practices
1. **Time-Intensive**: Traditional practices may take longer to yield results as trees, including bamboo, require time to mature before fulfilling their purpose in the system (Nair, 2018).
2. **Complex Maintenance**: Growing healthy trees for profit in traditional systems requires regular maintenance work that is highly specialized and time-consuming (Mercado Jr *et al.*, 2014).
3. **Limited Mechanization**: The presence of trees or shrubs in traditional agroforestry systems may hinder complete mechanization of farm production, posing challenges for some farmers (Mercado Jr *et al.*, 2014).
4. **Competition Among Components**: In traditional systems, there may be competition among components, with bamboo potentially out-competing agricultural or food crop components due to its higher root densities (Kumar *et al.*, 2018).

III. Ecological Benefits of Bamboo Based Agroforestry

The ecological benefits of bamboo-based agroforestry include enhancing productivity, sustainability, and resource conservation in agricultural systems. Bamboo plays a crucial role in reducing soil erosion, contributing to environmental sustainability, and providing economic benefits to farmers (Nair, 2018). Additionally, bamboo-based agroforestry systems have been reported to raise overall production and productivity in farming systems, especially in wasteland conditions, demonstrating the potential for ecological restoration and improved land use efficiency (Kumar *et al.*, 2018). Furthermore, bamboo can serve as a boundary plant around agricultural fields, protecting crops from wind and soil erosion, while also providing additional income sources through the sale of bamboo products and fodder for livestock (Asase *et al.*, 2018). Overall, bamboo-based agroforestry systems offer a range of ecological benefits that contribute to sustainable land management practices and environmental conservation.

A. Soil Conservation and Erosion Control

The use of bamboo in agroforestry systems plays a crucial role in soil conservation and erosion control. Bamboo's extensive root system acts as a natural erosion control agent, effectively stabilizing slopes, protecting riverbanks, and preventing soil erosion (Yitaferu *et al.*, 2008). The net-like root system of bamboo creates an effective mechanism for watershed protection by binding the soil together, especially in areas prone to earthquakes, mudslides, and deforestation (Kumar *et al.*, 2018). Additionally, bamboo-based agroforestry systems contribute to sustainable land management practices by enhancing soil fertility, reducing erosion, and conserving resources, making them valuable tools for environmental conservation and sustainable agriculture.

B. Carbon Sequestration and Climate Regulation

1. Bamboo is a fast-growing plant that can absorb and store large amounts of carbon dioxide (CO2) from the atmosphere, making it an effective carbon sink (Bhattacharya *et al.*, 2018; Nair, 2018; Tilak *et al.*, 2019). Bamboo forests have been found to have higher carbon storage per hectare compared to fast-growing tropical and subtropical trees under similar conditions (Bhattacharya *et al.*, 2018).

2. The rapid growth and short harvesting cycle of bamboo (3-5 years) allow for efficient carbon sequestration, as the captured carbon is quickly stored in the biomass and can be harvested and utilized, rather than remaining in the ecosystem for decades like many tree species (Bhattacharya *et al.*, 2018; Nair, 2018; Tilak *et al.*, 2019).

3. Bamboo-based agroforestry systems can contribute to climate change mitigation by enhancing carbon storage in the soil, aboveground biomass, and bamboo products. This helps to offset greenhouse gas emissions and regulate the climate (Bhattacharya *et al.*, 2018; Nair, 2018; Tilak *et al.*, 2019).

4. Bamboo can be grown in degraded tropical forests, agroforestry systems, and other land-use types, making it a versatile and valuable resource for environmental sustainability and climate change adaptation (Bhattacharya *et al.*, 2018).

5. Ethiopia, with its extensive highland bamboo forests, has significant potential to leverage bamboo-based agroforestry for carbon sequestration and climate regulation, contributing to the country's efforts to mitigate the impacts of climate change (Bhattacharya *et al.*, 2018).

C. Biodiversity Conservation

The sources provided emphasize the importance of bamboo-based agroforestry systems in promoting biodiversity conservation:

1. Bamboo forests play a crucial role in supporting biodiversity by providing habitats for a variety of plant and animal species, contributing to ecosystem diversity and resilience (Bhattacharya *et al.*, 2018).

2. Agroforestry systems that incorporate bamboo help create diverse and complex ecosystems that support a wide range of flora and fauna, enhancing overall biodiversity levels within agricultural landscapes (Nair, 2018).

3. Bamboo-based agroforestry practices can serve as a protective system that ensures resource conservation, providing social benefits and contributing to biodiversity conservation efforts (Nair, 2018).

4. The integration of bamboo into farming systems, such as using bamboo as boundary plants around agricultural fields, not only protects crops but also supports biodiversity conservation by creating additional habitats for wildlife (Kumar *et al.*, 2018).

D. Water Resource Management

The sources provided emphasize the role of bamboo-based agroforestry systems in water resource management:

1. Bamboo plays a crucial role in water conservation efforts, especially in areas prone to water stress. Bamboo plants can reduce runoff, contribute to water infiltration, and help restore water resources by conserving water and improving groundwater storage (Bhattacharya *et al.*, 2018).

2. Bamboo-based agroforestry systems have been found to enhance water resource sustainability by promoting the conservation of water sources, such as springs, and improving soil quality through bamboo planting projects (Nair, 2018).

3. The integration of bamboo into farming systems for water conservation purposes can benefit communities by ensuring water availability, supporting water sustainability, and providing additional socio-economic opportunities through bamboo ecotourism and other related activities (Nair, 2018).

IV. Economic Viability

The economic viability of bamboo-based agroforestry systems is highlighted in the provided sources:

1. Bamboo-based agroforestry systems have the potential to enhance productivity, sustainability, and resource conservation in agricultural systems, leading to higher monetary returns compared to sole cropping systems (Kumar *et al.*, 2018).

2. Studies have shown that bamboo-based agroforestry systems can raise overall production and productivity in farming systems, contributing to economic benefits for farmers (Kumar *et al.*, 2018).

3. Bamboo shoots could serve as a cost-effective source of protein and vitamins, offering economic opportunities for farmers in Sub-Saharan Africa (Eilu *et al.*, 2019).

4. Bamboo culms from agroforestry systems have been identified as good alternatives to depleting and expensive timber resources, providing a more affordable option for various uses (Kumar *et al.*, 2018).

5. The economic feasibility of bamboo-based agroforestry systems in contributing to rural poverty reduction and improved livelihoods in Sub-Saharan Africa is recognized, highlighting the potential for economic benefits and sustainable development (Eilu *et al.*, 2019).

A. Bamboo Market Trends

The bamboo market trends in India and globally indicate a positive economic viability for bamboo-based products and agroforestry systems. In India, the bamboo market is estimated to be USD 314.67 million in 2023, with significant growth potential driven by the rising demand for eco-friendly products, sustainable construction materials, and versatile bamboo-based goods (Research and Markets, 2021). The market is expected to witness impressive growth with a CAGR of 5.82% through

2029, reflecting the economic opportunities presented by the bamboo sector in India. The diverse applications of bamboo in construction, infrastructure, furniture, handicrafts, and other industries contribute to its economic viability and market growth.

Globally, the bamboo products market is experiencing a surge in revenue growth, with a focus on sustainability and the availability of affordable bamboo-based products (Grand View Research, 2021). The global market for bamboo products is expanding rapidly, driven by the shift towards renewable and eco-friendly materials. Bamboo's versatility, rapid growth, adaptability, and low maintenance characteristics make it a suitable alternative to conventional resources, leading to its increased adoption in various sectors. The economic viability of bamboo products is further supported by the growing demand for eco-friendly and sustainable goods, positioning the bamboo market as a key player in the global economy.

B. Income Generation Opportunities for Farmers

The sources provided highlight various income generation opportunities for farmers through bamboo-based agroforestry systems:

1. Higher Monetary Returns: Bamboo-based agroforestry systems have been reported to result in higher monetary returns compared to sole cropping systems, indicating the economic viability and potential profitability of integrating bamboo into farming practices (Kumar *et al.*, 2018).

2. Bamboo Products: Bamboo offers a wide range of products that can generate income for farmers, including bamboo shoots for food, fodder, furniture, building materials, paper, particle board, energy, and medicine. The versatility of bamboo as a resource provides various opportunities for income generation and livelihood improvement (Eilu *et al.*, 2019).

3. Bamboo Handicraft Industry: The bamboo handicraft industry creates a considerable casual labor market, benefiting villagers economically. This industry provides employment opportunities and income for rural communities, contributing to economic development and poverty alleviation (Eilu *et al.*, 2019).

4. Bamboo Farming: Bamboo farming can contribute to doubling farmers' income by utilizing indigenous knowledge and skills to establish suitable location-specific bamboo-based agroforestry systems. This approach aims to enhance productivity, sustainability, and resource conservation while providing economic benefits for farmers (Kumar *et al.*, 2018).

5. Bamboo as a Trade Commodity: Bamboo has become an important trade commodity globally, with significant contributions to forest exports in countries like China. The economic importance of bamboo at a national level reflects its potential for income generation and economic growth through trade and export activities (Eilu *et al.*, 2019).

Bamboo-based agroforestry systems that have generated income for farmers:

1. Intercropping bamboo with maize, cowpea, and cassava in Ghana:

- This example revealed a greater advantage of integrating bamboo with food crops over monoculture systems, leading to higher soil moisture, pH, and crop productivity levels.
- The bamboo-based agroforestry system was found to have higher monetary returns compared to sole cropping systems (Asante *et al.*, 2020).

2. The bamboo cum cereal farming system in Ethiopia:

- Farmers plant bamboo as boundary plants or fence rows between croplands and grazing lands, interspersed with crops like barley, wheat, millet, and potato.
- Bamboo is harvested year-round and used to make various products like mats, carpets, brooms, and furniture, providing farmers with a continuous cash income with minimal inputs (Feyisa, 2018).

3. The "talun-kebun" system in Indonesia:

- This modified shifting cultivation system involves a 6-7-year management cycle with a 4-5-year fallow period of clustering bamboo followed by 2 years of crop cultivation.
- The success of the system is based on bamboo's ability to reverse soil nutrient leaching and deposit them near the soil surface, benefiting the subsequent crop cultivation (Maryanto *et al.*, 2017).

4. Bamboo-based agroforestry in India:

- Bamboo is planted around agricultural fields as boundary plants to protect crops from high winds, and the bamboo can also be used for charcoal production to generate additional income.

- Studies have shown that bamboo-based agroforestry systems in India can raise overall production and productivity, leading to higher monetary returns for farmers compared to monoculture systems (Pandey *et al.*, 2017).

The key factors that contribute to the success of bamboo-based agroforestry systems in generating income for farmers include:

1. Versatility of Bamboo: Bamboo's versatility as a resource allows for the production of a wide range of products such as bamboo shoots, furniture, building materials, paper, and charcoal, providing multiple income streams for farmers (Fauziyah, 2023).

2. Shorter Harvesting Time: Bamboo's rapid growth and early maturity enable farmers to generate income relatively quickly compared to traditional tree crops, allowing for more frequent harvesting cycles and continuous cash flow (Arjumand and Rani, 2023).

3. Soil and Water Conservation: Bamboo's role in soil and water conservation enhances agricultural productivity, leading to increased crop yields and improved soil fertility, which in turn contributes to higher income for farmers (Fauziyah, 2023).

4. Integration with Food Crops: The possibility of integrating bamboo into food crop production, such as using bamboo as live stakes for yam production, offers additional income opportunities for farmers through diversified agricultural practices (Arjumand and Rani, 2023).

5. Market Availability and Demand: The economic success of bamboo-based agroforestry systems is also influenced by factors like market availability for produce, demand for bamboo products, and secure tenure, ensuring a sustainable market for farmers' products and income generation (Addo-Danso and Amankwaa-Yeboah, 2021).

D. Cost-effectiveness of Bamboo Agroforestry Systems

The statement you provided highlights the economic and environmental benefits of bamboo agroforestry systems. Here's the revised version with added references:

The cost-effectiveness of bamboo agroforestry systems is evident from their ability to provide sustainable income-generation opportunities for farmers while requiring relatively low inputs and maintenance costs. Bamboo's rapid growth, early maturity, and versatility as a resource contribute to the cost-effectiveness of these systems (Rahman *et al.*, 2019). The multiple income streams generated from bamboo products, such as bamboo shoots, furniture, building materials, and charcoal, offer farmers a diverse range of economic opportunities (Shi *et al.*, 2016).

Additionally, the environmental benefits of bamboo, including soil conservation, water resource management, and carbon sequestration, further enhance the cost-effectiveness of bamboo agroforestry systems by promoting sustainable land management practices and reducing the need for expensive inputs like fertilizers and pesticides (Gao *et al.*, 2020).

D. Economic Benefits in Comparison to Conventional Farming Practices

The economic benefits of bamboo-based agroforestry systems, in comparison to conventional farming practices, stem from several key factors outlined in the provided sources:

1. Higher Monetary Returns: Bamboo-based agroforestry systems have been reported to result in higher monetary returns compared to sole cropping systems, indicating the economic viability and potential profitability of integrating bamboo into farming practices (Addo-Danso, and Amankwaa-Yeboah, 2021).

2. Diversified Income Streams: Bamboo offers a wide range of products that can generate income for farmers, including bamboo shoots, furniture, building materials, paper, and charcoal, providing multiple income streams and economic opportunities beyond traditional crops (Phimmachanh *et al.*, 2015).

3. Quick Turnaround Time: Bamboo's rapid growth and early maturity enable farmers to generate income relatively quickly compared to traditional tree crops, allowing for more frequent harvesting cycles and continuous cash flow, which can be advantageous over longer-term tree crops (Addo-Danso, and Amankwaa-Yeboah, 2021).

4. Environmental Benefits: Bamboo's role in soil and water conservation enhances agricultural productivity, leading to increased crop yields and improved soil fertility, which can translate into higher income for farmers. Additionally, bamboo's ability to sequester carbon contributes to environmental sustainability and may offer additional economic benefits through carbon credits or eco-friendly branding (Jeffery *et al.*, 2023).

5. Cost-Effectiveness: Bamboo is known for its cost-effectiveness as a resource, being sufficiently cheap and plentiful to meet various human needs. This affordability and availability make bamboo a valuable asset for income generation and livelihood improvement, especially in rural areas (Phimmachanh *et al.*, 2015).

The economic benefits of agroforestry systems compared to conventional farming practices include:

1. Diversified Income Streams: Agroforestry systems offer multiple income streams for farmers through the cultivation of trees, shrubs, and crops, providing a more diverse and stable source of revenue compared to monoculture farming (Vinodhini *et al.*, 2023).

2. Long-Term Profitability: While trees in agroforestry systems take time to mature, once established, they can provide long-term economic benefits through timber, fruit, nuts, and other valuable products, contributing to sustained profitability over time (Vinodhini *et al.*, 2023).

3. Enhanced Soil Fertility: Agroforestry practices improve soil health and fertility through the interaction of trees, shrubs, and crops, leading to increased agricultural productivity and reduced reliance on external inputs like fertilizers, ultimately resulting in cost savings for farmers (Obsa *et al.*, 2015).

4. Environmental Sustainability: Agroforestry systems contribute to environmental sustainability by promoting soil and water conservation, reducing erosion, enhancing biodiversity, and sequestering carbon, aligning with sustainable land management practices that can lead to long-term economic gains (Lu *et al.*, 2018).

5. Resilience to Climate Change: The diverse and integrated nature of agroforestry systems provides resilience to climate change impacts, such as extreme weather events, by offering a buffer against crop failures and income losses, thereby ensuring economic stability for farmers in the face of changing environmental conditions (Addo-Danso, and Amankwaa-Yeboah, 2021).

A. Case Studies from Different Continents

Continent	Country	Location	Bamboo Species	Description	Reference
Asia	China	Anji County, Zhejiang Province	Moso Bamboo (Phyllostachys edulis)	Anji County practices an extensive agroforestry system integrating Moso bamboo with tea, fruits, and vegetables. Bamboo serves multiple purposes including soil conservation and additional income.	(Li, 2020).
Africa	Ethiopia	Kafa Biosphere Reserve	Various Bamboo Species	Indigenous communities in the Kafa Biosphere Reserve implement bamboo-based agroforestry systems alongside coffee, maize, and beans. Bamboo aids in soil conservation and provides additional income.	(Feyisa and Ango, 2019).
South America	Ecuador	Manabí Province	Guadua Bamboo (Guadua angustifolia)	Farmers in Manabí Province cultivate Guadua bamboo alongside crops like cacao and plantains. Bamboo provides shade, windbreaks, and erosion control, enhancing farm productivity.	(Vargas *et al.*, 2018).
Europe	Italy	Tuscany	Various Bamboo Species	In Tuscany, farmers integrate bamboo with vineyards and olive groves. Bamboo regulates microclimate, reduces erosion, and serves as a sustainable source of biomass for energy.	(Cantiani *et al.*, 2016).

B. Challenges and Solutions in Various Regions

Challenges and solutions in various regions regarding bamboo-based agroforestry systems are outlined in the provided sources:

1. Southern China: The main challenge in growing trees and shrubs for profit in agroforestry systems is the time factor. Trees, unlike crops, take a long time to mature before they can fulfill their purpose in the system. This time constraint poses a challenge for those seeking quick returns from agroforestry practices (Sollen-Norrlin, 2020).

2. India: In India, one of the challenges faced in bamboo-based agroforestry systems is the need for careful consideration of factors like the overall goal of the system, specific needs to be met, financial and technical capabilities, market availability for produce, and secure tenure to ensure successful implementation and sustainability of effects. Solutions involve integrating bamboo into farming systems with a focus on enhancing productivity, sustainability, and resource conservation (Addo-Danso, and Amankwaa-Yeboah, 2021).

3. Vietnam: Competition among components in bamboo-based agroforestry systems can be a challenge, with bamboo potentially out-competing agricultural or food crop components due to its higher root densities. To address this, individual plant care is essential to ensure the success of each component in the system (Addo-Danso, and Amankwaa-Yeboah, 2021).

4. Ethiopia, Madagascar, and Tanzania: The challenge of competition among components is also noted in these regions, emphasizing the need for careful planning and management to prevent bamboo from out-competing other components. Lessons from successful bamboo-based agroforestry systems outside the tropical African region can be integrated to address this challenge and optimize the benefits of bamboo in agricultural systems (Addo-Danso, and Amankwaa-Yeboah, 2021).

C. Lessons Learned for Global Implementation

Here are some key lessons learned for the global implementation of bamboo-based agroforestry systems:

1. Addressing Time Constraints: One of the main challenges in growing trees and shrubs for profit in agroforestry systems is the time factor, as trees take a long time to mature before they can fulfill their purpose. Bamboo's rapid growth and early maturity (3-4 years to production, 7-8 years to maximum productivity) can help overcome this challenge and provide quicker returns for farmers (Singh *et al.*, 2021).

2. Careful Planning and Management: Successful implementation of bamboo-based agroforestry systems requires careful consideration of factors like the overall goal of the system, specific needs to be met, financial and technical capabilities, market availability for produce, and secure tenur. Proper planning and management are essential to ensure the sustainability of these systems (Murthy *et al.*, 2016).

3. Addressing Competition among Components: Competition between bamboo and other crop components can be a challenge, with bamboo potentially out-competing agricultural or food crops due to its higher root densities. Careful selection of compatible species and individual plant care are necessary to address this issue and optimize the benefits of the system (Murthy *et al.*, 2016).

4. Leveraging Bamboo's Versatility: Bamboo's versatility as a resource, with its ability to be used for a wide range of products (construction, furniture, handicrafts, food, etc.), can provide diverse income streams and economic opportunities for farmers, making it a valuable component of agroforestry systems (Dev *et al.*, 2020).

5. Integrating Lessons from Successful Examples: Lessons from successful bamboo-based agroforestry systems in regions like Southern China, India, Vietnam, and tropical Africa can be integrated and adapted to other locations to promote the successful global implementation of these systems (Addo-Danso and Amankwaa-Yeboah, 2021).

6. Addressing Myths and Lack of Awareness: Addressing myths and misconceptions about bamboo, as well as raising awareness among farmers about the economic potential and utilization of bamboo, can help drive the adoption and success of bamboo-based agroforestry systems (Addo-Danso and Amankwaa-Yeboah, 2021).

VI. Bamboo Agroforestry in India: Current Scenario

The current scenario of bamboo-based agroforestry systems in India can be summarized as follows:

1. Bamboo Abundance and Diversity:

- India has the largest forest area of bamboo, estimated at 13.96 million hectares, and is rich in bamboo diversity with 124 indigenous and exotic species under 23 genera (Dev *et al.*, 2020).

- The most commonly found bamboo species in semi-arid India is *Dendrocalamus strictus*, accounting for 53% of India's total bamboo area (Dev *et al.*, 2020).

2. Potential for Agroforestry Integration:

- Bamboo is considered a promising option for sustainable land management and integration into agroforestry systems in India due to its versatility, multiple uses, and quick returns (Kittur *et al.*, 2016).

- However, the full potential of bamboo in agroforestry systems has not been harnessed, and the success stories of bamboo-based agroforestry are limited in the region (Dev *et al.*, 2020).

3. Challenges and Barriers:

- Lack of awareness among farmers about the utilization and economic potential of bamboo, non-availability of quality planting material, and limited market access are some of the key challenges hindering the widespread adoption of bamboo-based agroforestry systems in India (Dev *et al.*, 2020).

- Competition between bamboo and other crop components, as well as the time factor (long maturation period of trees), are additional challenges that need to be addressed.

4. Ongoing Efforts and Initiatives:

- The National Bamboo Mission in India has initiated various steps to increase the availability of quality planting material and strengthen the marketing of bamboo products to promote bamboo cultivation and utilization (Dev *et al.*, 2020).

- Research and development efforts are underway to optimize the management of bamboo-based mixed-species production systems, including understanding bamboo spacing, root activity, and soil nutrient dynamics (Kittur et.al., 2016).

A. Overview of Bamboo Cultivation in India

In India, bamboo cultivation plays a significant role in the economic and cultural spheres, with the country having a rich bamboo resource base. The area under bamboo cultivation in India is estimated to be more than 13.96 million hectares, encompassing over 136 distinct species. Bamboo is mainly grown through vegetative propagation, although it can also be grown through culms and rhizomes. The cultivation of bamboo is vital for economic growth, rural livelihoods, and environmental sustainability in India. The country's diverse bamboo species, such as Beema Bamboo, Dendrocalamus hamiltonii, Melocanna Bacciferra, Tulda Bamboo, and Gracilis, contribute to its status as a significant bamboo producer globally. Additionally, bamboo cultivation in India benefits from the country's warm, humid climate, with bamboo thriving in wet and tropical conditions. The soil requirements for successful bamboo farming include slightly acidic pH levels and well-drained sandy-loam or clay-loam soils. Bamboo farming in India is supported by various irrigation methods like drip irrigation, flood irrigation, and natural/rainwater irrigation. The cultivation of bamboo in India not only contributes to the economy but also aids in carbon sequestration efforts and environmental management. The government's initiatives, such as the National Bamboo Mission, aim to promote bamboo cultivation, increase farmers' income, and enhance the green cover of the country (Tewari *et al.*, 2019; Bareja;2010; Rathour et. al.,2022).

Bamboo Species	**Local Names**	**Height**	**Diameter**	**Uses**
Bambusa polymorpha	Assam—Betua-, Beng—Jamabetua	20-27 m	15 cm	Construction, thatching, roofing, paper, agarbatti sticks, fiberboards, edible shoots
Bambusa striata	Bengali—Basni Bans; Marathi-Kalaka; Tamil—Ponmungil	6-10 m	5-8 cm	Furniture, toys, cages, construction, ornamental

Bambusa tulda	Bengali—Tulda; Hindi—Santal—Mak	Varies	Varies	Building, scaffolding, roofing, mats, baskets, paper pulp
Dendrocalamus giganteus	-	24-35 m	20-30 cm	Building, water buckets, boxes, masts of boats, flower vases, edible shoots
Dendrocalamus hamiltonii	Assam—Kokua; Bengali—Pecha; Hindi—Kaghzi Bans	-	-	-
Indocalamus wightianus	Tamil—Chevani	2-3 m	1.2 cm	Mats, baskets, fences
Melocanna baccifera	Bengali—Muli; Assam—Tarai	10-20 m	4-7 cm	Construction, baskets, mats, food containers, umbrella sticks, hats, toys, paper pulp
Ochlandra scriptoria	Malayalam—Ammei, Ottal	Up to 5 m	2.5 m	Mats, baskets, paper
Ochlandra travancorica	Malayalam—Eetta; Tamil—Eeral, Odai	Up to 6 m	5 cm	Paper pulp, mats, baskets, huts, soil binder
Pseudoxytenanthera Ritchey	Kannada—Chiwa, Choua-Maraihi—Huda, Manga	-	-	-

[2] Seethalakshmi, K. K., *et al.* *Bamboos of India: A compendium.* Vol. 17. Brill, 1998.

E. Traditional Practices and Indigenous Knowledge

Based on the information provided in the sources, here are the key points regarding traditional practices and Indigenous knowledge related to bamboo-based agroforestry:

1. Indigenous Propagation Techniques:

- Local communities have extensive experience in developing bamboo stands using their indigenous knowledge of propagation techniques, with bamboo offsetting being the most preferred method (Cédric *et al.*, 2021).

- The source of bamboo offsets is primarily obtained from individual farmers' existing bamboo stands (Cédric *et al.*, 2021).

2. Bamboo Stand Management Practices:

- Indigenous communities have traditional practices for managing bamboo stands, such as fencing the bamboo stands to protect them (Cédric *et al.*, 2021).

- Farmers have traditional habits of caring for bamboo roots and rhizomes, which helps sustain bamboo resource utilization (Cédric *et al.*, 2021).

3. Bamboo Identification and Selection:

- Local communities use various indigenous criteria to identify different bamboo landraces, such as bamboo diameter, length of internodes, stem color, and splitting nature (Cédric *et al.*, 2021).

- This traditional knowledge of bamboo identification and selection is an important aspect of managing and utilizing bamboo resources.

4. Diverse Bamboo Utilization:

- Indigenous communities have a wide range of traditional applications for bamboo, including house construction, fencing, fodder, furniture, utensils, fuel wood, and as a cash source (Cédric *et al.*, 2021).

- The traditional use of bamboo is deeply integrated into the lifestyle and culture of these communities (Kanglin *et al.*, 2020).

5. Sustainable Practices:

- Indigenous communities exhibit sustainable practices in bamboo cultivation, harvesting, and resource replenishment, which align with modern sustainable resource management strategies (Kittur *et al.*, 2021).

- Their traditional knowledge and practices have enabled the coexistence of indigenous groups with bamboo-rich ecosystems over generations (Kittur *et al.*, 2021).

C. Government Initiatives and Policies Supporting Bamboo Agroforestry

The Government of India has implemented several initiatives and policies to support bamboo agroforestry in the country:

1. National Bamboo Mission (NBM):

- The National Bamboo Mission, launched by the Ministry of Agriculture & Farmers Welfare, focuses on expanding bamboo plantations in government and private farmlands, community lands, and culturable areas. It aims to enhance research and development, plantation development, extension services, education, skill development, micro-irrigation, post-harvest storage, infrastructure development for the bamboo market, market research, and production development and processing (Baksy, 2013).

2. National Agroforestry Policy:

- The National Agroforestry Policy recognizes the importance of agroforestry as a traditional land-use system in India and aims to promote its benefits in improving farmers' livelihoods. The policy addresses issues related to quality planting material, tree insurance, transit, and harvesting restrictions, marketing of agroforestry produce, research, and extension services to encourage the successful implementation of agroforestry practices in the country (Tambe *et al.*, 2021).

3. State-Level Initiatives:

- States like Nagaland have introduced policies focusing on the sustainable development of bamboo resources. These policies include strategies to develop natural bamboo forests and bamboo plantations, and the removal of transit pass requirements to facilitate the growth of the bamboo sector within the state (Tambe *et al.*, 2021).

D. Role of Bamboo in Rural Livelihoods

Bamboo plays a crucial role in rural livelihoods in India by providing sustainable income opportunities, creating employment, reducing poverty, and promoting economic stability in rural communities. Here are the details of the role of bamboo in rural livelihoods in India based on the provided sources:

1. Significance in Livelihoods:

- Bamboo is a vital forest product with high relevance as a livelihood option for forest-dependent people in India. It contributes significantly to poverty alleviation and generates subsistence income for the rural poor (Jamatia, 2012).

- Employment generation from bamboo cultivation and processing makes a substantial contribution to people's incomes, especially in rural areas where bamboo is grown (Jamatia, 2012).

2. Economic Impact:

- Bamboo is considered the "poor man's timber" and is utilized for various livelihood options, including construction, handicrafts, agricultural implements, and more. It serves as a valuable resource for rural communities, providing income opportunities and economic stability (Jamatia, 2012).

- The wide gap between the demand and supply of bamboo in India highlights the significant opportunity for the development of the bamboo sector, especially in states like Madhya Pradesh, which have a substantial bamboo resource base (Jamatia, 2012).

3. Employment Generation:

- Bamboo cultivation and processing offer employment opportunities for rural families, contributing to poverty reduction and improving the quality of life for forest-dependent communities. The sector has the potential to create jobs and enhance economic opportunities for men and women in rural areas (Wang *et al*, 2021).

- Through small businesses, social enterprises, and community groups, bamboo can generate new sources of income and improve the livelihoods of rural people, ultimately leading to a more socially and economically stable environment (Wang *et al*, 2021).

VII. Potential for Expansion and Scaling Up

The potential for expansion and scaling up bamboo-based agroforestry systems is significant, as highlighted in the provided sources. Bamboo agroforestry offers various benefits and opportunities for sustainable development, rural livelihood improvement, and environmental conservation. Here are the key points regarding the potential for expansion and scaling up bamboo-based agroforestry:

1. Economic Opportunities:

- Bamboo-based agroforestry systems present a promising avenue for enhancing economic opportunities, especially in rural areas. The versatility of bamboo as a fast-growing and renewable resource allows for the production of a wide range of products, creating diverse income streams for farmers and communities (Solomon *et al.*, 2021).

2. Environmental Sustainability:

- Bamboo agroforestry contributes to environmental sustainability by promoting soil health, biodiversity, carbon sequestration, and erosion control. The integration of bamboo into agroforestry systems helps restore degraded lands, enhance ecosystem services, and mitigate climate change impacts (Nfornkah, 2013).

3. Livelihood Improvement:

- Scaling up bamboo-based agroforestry can lead to improved livelihoods for rural communities through job creation, income generation, and sustainable land management practices. Bamboo cultivation offers a sustainable income source and provides alternative livelihood support for farmers (Kittur *et al.*, 2016).

4. Resource Efficiency:

- Bamboo's fast growth and early maturity make it a resource-efficient option for agro-forestry systems. Its ability to regenerate vegetatively, be harvested repeatedly, and grow on marginal lands unsuitable for agriculture or forestry enhances its potential for expansion and scaling up in diverse agroecological settings (Partey *et al.*, 2017).

A. Opportunities for Integrating Bamboo in Agroforestry Systems

Opportunities for integrating bamboo in agroforestry systems in India include:

1. Economic Diversification:

- Bamboo cultivation offers economic diversification opportunities for farmers by providing additional income streams through the production of various bamboo products like construction materials, handicrafts, furniture, and more. Integrating bamboo into agroforestry systems can enhance income generation and economic stability for rural communities (Partey *et al.*, 2017).

2. Environmental Benefits:

- Bamboo's fast growth rate, high productivity, and ability to sequester carbon make it an environmentally sustainable option for agroforestry systems. By integrating bamboo, farmers can contribute to environmental conservation, soil health improvement, and biodiversity enhancement in their agricultural landscapes (Bhardwaj *et al.* 2021).

3. Soil Health and Erosion Control:

- Bamboo's extensive root system helps in soil stabilization, erosion control, and watershed protection. Integrating bamboo into agroforestry systems can help prevent soil erosion, improve soil structure, and enhance overall soil health in agricultural lands (Bhardwaj *et al.* 2021).

4. Livelihood Improvement:

- Bamboo agroforestry systems can provide livelihood improvement opportunities for rural communities by creating jobs, enhancing income, and offering sustainable land management practices. The integration of bamboo can lead to poverty reduction, improved livelihoods, and socio-economic development in rural areas (Kittur *et al.*, 2016).

5. Resource Efficiency:

- Bamboo's ability to regenerate vegetatively, grow rapidly, and be harvested repeatedly makes it a resource-efficient option for agroforestry systems. Integrating bamboo into mixed-use agroforestry complexes can maximize its functionality while promoting sustainable land management practices (Partey *et al.*, 2017).

B. Research and Development Needs

Research and development needs for bamboo-based agroforestry systems include:

1. Identification of Suitable Agroforestry Systems:

- Research is needed to identify and develop suitable agroforestry systems where bamboo can be effectively integrated with other crops. Understanding the ecological requirements and interactions of bamboo with different plant species is crucial for optimizing productivity and sustainability (Partey, 2017).

2. Enhancing Productivity and Sustainability:

- Research should focus on enhancing the productivity and sustainability of bamboo-based agroforestry systems. This includes studying optimal planting densities, spacing, and management practices to maximize yields, improve soil health, and enhance ecosystem services within these systems (Kittur *et al.*, 2016).

3. Economic Viability and Market Access:

- Studies are required to assess the economic viability of bamboo agroforestry, including cost-benefit analyses, market assessments for bamboo products, and value chain development. Research can help identify potential markets, value-added products, and income-generation opportunities for farmers (Kittur *et al.*, 2016).

4. Soil Health and Nutrient Dynamics:

- Research on soil health, nutrient cycling, and the impact of bamboo on soil fertility is essential for sustainable agroforestry practices. Understanding the nutrient requirements of bamboo, its effects on soil structure, and nutrient dynamics in agroforestry systems can guide effective management strategies (Kittur *et al.*, 2016).

5. Climate Resilience and Adaptation:

- Research is needed to explore the role of bamboo-based agroforestry in climate resilience and adaptation. Studying how bamboo agroforestry systems can contribute to climate change mitigation, water conservation, and biodiversity conservation will be crucial for sustainable land management practices in the face of changing climatic conditions (Divakara *et al.*, 2001).

C. Scaling Up Bamboo Agroforestry for Commercial Production

Scaling up bamboo agroforestry for commercial production in India requires a comprehensive approach that addresses various aspects of cultivation, management, and market integration. Here are the key strategies for scaling up bamboo agroforestry for commercial production in India based on the provided sources:

1. Research and Development:

- Conduct research to optimize bamboo cultivation techniques, including spacing, management practices, and intercropping methods to maximize productivity and commercial viability (Kumar *et al.*, 2015).

- Invest in research to understand the interactions between bamboo and other crops, soil health, nutrient dynamics, and the impact of different agroforestry systems on overall production and profitability (Fahad *et al.*, 2022).

2. Policy Support:

- Develop policies that incentivize and support commercial bamboo agroforestry, including providing financial incentives, technical assistance, and regulatory frameworks that promote sustainable and profitable bamboo cultivation practices (Alemayehu and Hido, 2023).

- Ensure that policies address land tenure issues, market access, value chain development, and certification standards to enhance the commercialization of bamboo products (Alemayehu and Hido, 2023).

3. Market Integration:

- Establish linkages between bamboo producers and markets to ensure a steady demand for bamboo products. This includes developing fair-value chains, connecting farmers to consumers, and promoting the marketing of bamboo products domestically and internationally (Oduro et al., 2020).

- Support the development of inclusive market systems, cooperatives, and partnerships that enable smallholder farmers to access fair and trusted markets for their bamboo products (Oduro et al., 2020).

4. Capacity Building:

- Provide training, capacity building, and extension services to farmers on best practices for bamboo cultivation, management, and value addition. This includes promoting sustainable land management practices, efficient harvesting techniques, and post-harvest processing methods to enhance product quality and market competitiveness (Jacob *et al.*, 2019).

D. Collaboration and Partnership for Sustainable Growth

Collaboration and partnership are essential for sustainable growth in bamboo-based agroforestry in India. By working together, various stakeholders can leverage their expertise, resources, and networks to promote the development and expansion of bamboo agroforestry systems. Here are some key points based on the provided sources:

1. Research Collaboration:

- Collaborating with research institutions, universities, and agricultural experts can help in conducting studies, trials, and experiments to optimize bamboo cultivation techniques, improve productivity, and enhance the sustainability of agroforestry systems.

- Research partnerships can also focus on identifying suitable agroforestry systems, and understanding soil health, nutrient dynamics, and root interactions to maximize the benefits of integrating bamboo into agricultural landscapes.

2. Government-Industry Partnerships:

- Partnerships between government agencies, industry stakeholders, and bamboo farmers can facilitate the development of supportive policies, financial incentives, and market linkages to promote commercial bamboo cultivation and value-added product development (Partey *et al.*, 2017).

- Collaboration with government bodies can help create a conducive regulatory environment, providing extension services, and ensuring access to markets for bamboo products, thereby fostering sustainable growth in the sector (Binfield *et al.*, 2022).

3. Community Engagement:

- Engaging local communities, farmers, and cooperatives in collaborative initiatives can enhance the adoption of bamboo agroforestry practices, promote knowledge sharing, and empower rural stakeholders to actively participate in sustainable land management and economic development (Nandini and Rahayu, 2023).

- Community partnerships can facilitate capacity building, training programs, and the establishment of cooperative structures that enable smallholder farmers to benefit from bamboo cultivation and agroforestry practices (Ekawati *et al.*, 2023).

VIII. Challenges and Future Directions

The key challenges and future directions for agroforestry-based agroforestry systems include:

Challenges:

1. Time factor: The main disadvantage of agroforestry is the long time it takes for trees and shrubs to mature and fulfill their purpose in the system, unlike annual crops. This time constraint poses a challenge for those seeking quick returns (Budowski, 1981).

2. Complexity of social-ecological systems: Agroforestry systems are complex social-ecological systems that interact with institutional, national, and international scales. Understanding and managing these complex interactions is a significant challenge (Hastings Silao *et al.*, 2023).

3. Barriers to adoption and scaling up: Barriers such as insufficient access to secure land tenure, high start-up costs, lack of financial incentives, supply chain limitations, and limited market opportunities can hinder the widespread adoption and scaling up of agroforestry (Hastings Silao *et al.*, 2023).

Future Directions:

1. Optimizing system design: Research is needed to optimize the design of agroforestry systems for improved production and environmental benefits, including understanding the interactions between different components (Hastings Silao *et al.*, 2023).

2. Addressing social dimensions: Examining the social dimensions of agroforestry transitions, such as barriers and opportunities, is crucial to inform policy and programs that can effectively expand agroforestry adoption (Hastings Silao *et al.*, 2023).

3. Enhancing equity and sustainability: Developing strategies to equitably increase and sustain agroforestry on a global scale to meet climate commitments is a key future direction, considering the complex social-ecological nature of these systems (Hastings Silao *et al.*, 2023).

4. Leveraging indigenous knowledge: Collaborating with local communities and integrating their traditional knowledge and sustainable practices can contribute to the development of more effective and culturally sensitive agroforestry approaches (Paudel *et al.*, 2022).

A. Constraints in Bamboo Agroforestry Adoption

Based on the information provided in the search results, the key constraints in the adoption of bamboo-based agroforestry systems in India include:

1. Time Factor:

- The long maturation period of trees and shrubs in agroforestry systems, unlike annual crops, poses a challenge for farmers seeking quick returns. This time constraint can hinder the adoption of agroforestry practices (Jerneck and Olsson, 2013)

2. Poverty and Lack of Financial Resources:

- Farmers who are struggling to make a living and rely on irregular cash income are often not in a position to invest in agroforestry, which requires patience and upfront costs before yielding benefits (Sharma *et al.*, 2017).

3. Insecure Land Tenure:

- Ambiguities around land tenure and use rights can act as a barrier to the adoption of agroforestry, as farmers may be hesitant to invest in long-term tree-based systems without secure land tenure (Sharma *et al.*, 2017).

4. Lack of Awareness and Knowledge:

- Limited awareness among farmers about the potential benefits and management practices of bamboo-based agroforestry systems can constrain their adoption (Partey *et al.*, 2017).

.5. Access to Inputs and Markets:

- Lack of access to quality planting material, financial capital, and well-developed market linkages for bamboo products can hinder the widespread adoption of bamboo agroforestry (Partey *et al.*, 2017).

6. Competition with Annual Crops:

- The competition between bamboo and annual crops for resources like light, water, and nutrients can be a challenge in integrating bamboo into agroforestry systems (Kittur *et al.*, 2016).

7. Regulatory Barriers:

- Restrictive policies and regulations related to the harvesting, transportation, and marketing of bamboo can act as barriers to the adoption of bamboo agroforestry (Sharma *et al.*, 2016).

B. Technical, Social, and Environmental Challenges

Bamboo-based agroforestry systems face several technical, social, and environmental challenges that need to be addressed for successful implementation and sustainability:

Technical Challenges

- Lack of genetic diversity among bamboo species, with many being clonal (Jeffery *et al.*, 2023). This limits the ability to select for desirable traits like fast growth, high biomass yield, and adaptability to different environments.

- Lack of robust scientific findings on the suitability and ecological principles of integrating bamboo into indigenous cropping systems in Sub-Saharan Africa (Partey *et al.*, 2017). More research is needed to define the best planting densities and management practices.

- Experimental designs with different bamboo planting spacing and densities along with crops are crucial in defining the best potential bamboo-based agroforestry systems (Hailu, 2023).

Social Challenges

- Achieving both socioeconomic and environmental needs of land users at all levels is dependent on the woody component, which is the most integral part of agroforestry systems (Partey *et al.*, 2017). Ensuring equitable distribution of benefits is important.

- Secure land tenure and market accessibility for bamboo products are key factors needed for successful implementation and sustainability of bamboo agroforestry (Hailu, 2023).

- General objectives of the systems, specific needs to be met, and the economic and technical capacities of farmers should be considered (Hailu, 2023).

Environmental Challenges

- Sole bamboo plantations may negatively impact food security unless established on degraded lands (Partey *et al.*, 2017). Careful integration with food crops is needed.

- Competitive and complementary interactions within a bamboo agroforestry system depend on planting density and level of management (Hailu, 2023).

- Erratic rainfall, high frequency of drought, undulating topography, poor groundwater resources and shallow, infertile soils in some regions like

Bundelkhand, India pose challenges to bamboo growth and productivity (Dev *et al.*, 2020).

IX. Conclusion

Bamboo-based agroforestry systems present a promising avenue for sustainable land management and economic development in both India and globally. The integration of bamboo into agroforestry practices offers numerous advantages, including its fast growth rate, versatility of use, and ability to thrive in diverse climatic conditions. In India, bamboo agroforestry has shown potential in enhancing productivity, sustainability, and resource conservation, benefiting farmers and the environment alike. The utilization of bamboo in agroforestry systems can lead to improved soil health, erosion control, and watershed protection, contributing to environmental sustainability. Moreover, the diverse applications of bamboo, from construction materials to erosion control, highlight its multifaceted role in agroforestry systems. As the world increasingly focuses on sustainable land management practices and climate-resilient agriculture, bamboo-based agroforestry emerges as a valuable and innovative approach that can enhance livelihoods, promote biodiversity, and mitigate environmental degradation. By leveraging the unique qualities of bamboo and integrating it into agroforestry systems, India and the world can harness the full potential of this versatile plant for sustainable agricultural practices and economic growth.

References

Addo-Danso, A., & Amankwaa-Yeboah, P. (2021). The Potentials of Bamboo-Based Agroforestry Systems in Improving the Productivity of Tropical African Agricultural Systems. *Agricultural and Food Science Journal of Ghana*, *14*(1), 1468-1482.

Akoto, D. S., Denich, M., Partey, S. T., Frith, O., Kwaku, M., Mensah, A. A., & Borgemeister, C. (2018). Socioeconomic indicators of bamboo use for agroforestry development in the dry semi-deciduous forest zone of Ghana. *Sustainability*, *10*(7), 2324.

Alemayehu, A., & Hido, A. (2023). Determinates and governance of bamboo production and marketing in Ethiopia: A critical review. *Advances in Bamboo Science*, 100047.

Arjumand, T., & Rani, B. R. (2023). The Case study on Bamboo farming, Centre for Climate Change and Adaptation (CCA), National Institute of Agricultural Extension Management (MANAGE), Hyderabad, India.

Asante, P.K., Opoku-Ameyaw, K., Asare, R. et al. Intercropping Bamboo with Food Crops Enhances Soil Fertility, Soil Moisture, and Crop Productivity. Agric Res 9, 445–454 (2020). https://doi.org/10.1007/s40003-020-00489-8

Asase, A., Amponsah, A., & Asare, E. (2018). Bamboo-based agroforestry for sustainable development in Ghana: Opportunities and challenges. Sustainable Development, 26(5), 489-497.

Baksy, A. (2013). The bamboo industry in India: supply chain structure, challenges and recommendations. *Challenges and Recommendations (July 2013)*.

Bareja, G. B. (2010). Bamboo production and propagation methods. *Researched March*, *5*, 2013.

Bhattacharya, S. S., Das, S., Choudhury, P., Dey, A., & Bhattacharya, S. (2018). Bamboo as a Potential Carbon Sink: A Review. Journal of Forestry Research, 29(6), 1529–1539.

Binfield, L., Britton, T. L., Dai, C., & Innes, J. (2022). Evidence on the social, economic, and environmental impact of interventions that facilitate bamboo industry development for sustainable livelihoods: a systematic map protocol. *Environmental Evidence*, *11*(1), 33.

Budowski, G. (1981). Advantages and disadvantages of agroforestry systems (during or after the cultivation of trees with crops and/or forage plants) in comparison with monoculture without trees.

Cantiani, P., Mattioli, W., & Chiesi, M. (2016). Bamboo Agroforestry Systems for Bioenergy: A Review and Case Study in Central Italy. BioEnergy Research, 9(1), 189-204.

Cédric, C. D., Nfornkah, B. N., Forje, G. W., Princely, A. N., Réné, K., Jovis, N. A., & Louis, Z. (2021). Indigenous knowledge of Bamboo products and uses in the Western Highlands of Cameroon. *Asian Journal of Research in Agriculture and Forestry*, *7*, 22-30.

Chakravarty, S., Pillay, S., & Moges, A. (2019). Bamboo for Sustainable Development: An Economic and Ecological Security Provider. Springer Nature.

Chavan, S. B., Keerthika, A., Dhyani, S. K., Handa, A. K., Newaj, R., & Rajarajan, K. (2015). National Agroforestry Policy in India: a low hanging fruit. *Current Science*, 1826-1834.

Crouzet, Y. (2019). Gardening with bamboo. Crowood Press.

Dev, I., Ram, A., Ahlawat, S. P., Palsaniya, D. R., Singh, R., Dhyani, S. K., ... & Prasad, J. (2020). Bamboo-based agroforestry system (Dendrocalamus strictus+ sesame–chickpea) for enhancing productivity in semi-arid tropics of central India. *Agroforestry Systems*, *94*, 1725-1739.

Divakara, B. N., Mohan Kumar, B., Balachandran, P. V., & Kamalam, N. V. (2001). Bamboo hedgerow systems in Kerala, India: Root distribution and competition with trees for phosphorus. *Agroforestry Systems*, *51*, 189-200.

Dupraz, C., Burgess, P., Gavaland, A., & Graves, A. (2018). Agroforestry and the design of climate-resilient farming systems. In Climate-Smart Agriculture: Building Resilience to

Eilu, G., Okia, C. A., Obua, J., Akampurira, P., & Telford, A. (2019). Bamboo as a Rural Livelihood in Sub-Saharan Africa: An Overview. Forests, 10(10), 861.

Ekawati, D., Karlinasari, L., Soekmadi, R., & Machfud. (2023). A Model of Integrated Community-Based Bamboo Management for the Bamboo Industry in Ngada Regency, East Nusa Tenggara, Indonesia. *Sustainability*, *15*(2), 977.

Fahad, S., Chavan, S. B., Chichaghare, A. R., Uthappa, A. R., Kumar, M., Kakade, V., ... & Poczai, P. (2022). Agroforestry systems for soil health improvement and maintenance. *Sustainability*, *14*(22), 14877.

Fauziyah, E., Sanudin, S., & Hani, A. (2023). Impact of Bamboo-Based Agroforestry Demonstration Plots Based on Farmers' Perceptions. In *BIO Web of Conferences* (Vol. 80, p. 03012). EDP Sciences.

Feyisa, G.L., Asfaw, Z., Lemma, T., & Mekonen, K. (2018). The Roles of Bamboo in Livelihood Diversification and Income Generation in Bonga District, Southwestern Ethiopia. Small-scale Forestry, 17, 177–188. https://doi.org/10.1007/s11842-018-9395-9

Food and Agriculture Organization of the United Nations (FAO). (2017). Bamboo for restoration and livelihoods. Rome, Italy.

Fryer, J. C., & Henderson, A. (2010). Bambusa bambos and B. vulgaris. In Field Guide to the Palms of New Guinea (pp. 8-11). Royal Botanic Gardens, Kew.

Fukushima, T., Noguchi, T., & Sakai, S. (2019). Cultural services of bamboo forests in Japan: Historical perspectives, current status, and future possibilities. Sustainability, 11(20), 5790.

Gao, Z., Zhang, C., Zhao, H., Wang, M., Sun, J., & Li, X. (2020). Research progress on bamboo agroforestry systems: A review. Journal of Forestry Research, 31(4), 1177-1191.

Gardner, T. A., et al. (2016). *A framework for integrating biodiversity concerns into national REDD+ programs.*

Garima, Bhardwaj, D. R., Thakur, C. L., Kaushal, R., Sharma, P., Kumar, D., & Kumari, Y. (2021). Bamboo-based agroforestry system effects on soil fertility: Ginger performance in the bamboo subcanopy in the Himalayas (India). *Agronomy Journal*, *113*(3), 2832-2845.

Garrity, D.P., Akinnifesi, F.K., Ajayi, O.C., Weldesemayat, S.G., Mowo, J.G., Kalinganire, A., & Larwanou, M. (2010). Evergreen Agriculture: A robust approach to sustainable food security in Africa. Food Security, 2(3), 197-214.

Gomez, M. F. A., Lucero, M. R. B., & Padilla, N. E. (2015). Folklore and mythology in the Babaylan tradition. In Philippine Folk Literature (Vol. 3, pp. 1-13). Anvil Publishing.

Grand View Research. (2021). Bamboo Products Market Size, Share & Trends Analysis Report By Product (Structural, Non-structural), By Application (Industrial, Commercial), By Region, And Segment Forecasts, 2021 - 2028.

Groom, R. J., & Palmer, R. E. (2020). *The contribution of wildlife management areas to household income and dietary diversity in the Rukwa Rift Valley, Tanzania.*

Hailu, B. K. (2023). Potentials of bamboo-based agroforestry for sustainable development in Ethiopia. *Journal of the Selva Andina Biosphere*, *11*(2), 159-171.

Hastings Silao, Z., Ocloo, X. S., Chapman, M., Hunt, L., & Stenger, K. (2023). Trends in agroforestry research over 4 decades. *Elementa: Science of the Anthropocene*, *11*(1).

Jacob, Vinu, Arvind Bijalwan, Pooja Verma, and M. J. R. Dobriyal. "Need for managerial interventions in scaling up traditional agroforestry practices: An insight." *Indian Journal of Agroforestry* 21, no. 1 (2019): 13-20.

Jamatia, S. (2012, August). Livelihood of the Bamboo base: Challenges and Opportunities. In *Proceedings of 54th Society of Wood Science and Technology conference on sustainable development of wood and biomass in our new global economy, Beijing China: International Bamboo and Rattan* (Vol. 20).

Jeffery, I. E., AkinyemI, O. O., Adedoyin, A. A., & Matthew, U. F. (2023). Potentials of bamboo and its ecological benefits in Nigeria. *Advances in Bamboo Science*, 100032.

Jerneck, A., & Olsson, L. (2013). More than trees! Understanding the agroforestry adoption gap in subsistence agriculture: Insights from narrative walks in Kenya. *Journal of Rural Studies*, *32*, 114-125.

Jose, S. (2009). Agroforestry for ecosystem services and environmental benefits: An overview. Agroforestry Systems, 76(1), 1-10.

Joshi, G. (2018). Bamboo agroforestry in India: An overview. International Journal of Agriculture, Environment, and Biotechnology, 11(6), 861-865.

Kanglin, W., Jianchu, X., Shengji, P., & Sanyang, C. H. E. N. (2000). Folk classification and conservation of bamboo in Xichuangbanna, Yunnan, Southwest China. *Journal of Ethnobiology*, *20*(1), 113-127.

Kittur, B. H., Sudhakara, K., Mohan Kumar, B., Kunhamu, T. K., & Sureshkumar, P. (2016). Bamboo-based agroforestry systems in Kerala, India: performance of turmeric (*Curcuma longa* L.) in the subcanopy of differentially spaced seven-year-old bamboo stand. *Agroforestry Systems*, *90*, 237-250.

Kumar, A., Kumar, A., & Verma, S. (2018). Bamboo-based agroforestry for wasteland reclamation: A case study from Bihar, India. International Journal of Current Microbiology and Applied Sciences, 7(5), 2149-2154.

Kumar, A., Kumar, A., & Verma, S. (2018). Bamboo-based agroforestry for wasteland reclamation: A case study from Bihar, India. International Journal of Current Microbiology and Applied Sciences, 7(5), 2149-2154.

Kumar, B.M. & Nair, P.K.R. (2004). The enigma of tropical home gardens. Agroforestry Systems, 61(1-3), 135-152.

Kumar, H., Mehera, B., Lal, S. B., & Umrao, R. (2015). The potential of bamboo cultivation as a way forward in improving livelihood: a case study.

Kwaw, N. A., Akromah, R., & Appiah, M. (2020). Bamboo agroforestry: A pathway to sustainable agriculture and livelihood improvement in Ghana. African Journal of Agricultural Research, 15(10), 1488-1497.

Lasco, R. D., Delfino, R. J. P., Pulhin, F. B., Visco, R. G., Rabelo, C. L., & Sales, R. F. (2014). An assessment of the potential of agroforestry for carbon sequestration in the Philippines. Mitigation and Adaptation Strategies for Global Change, 19(5), 591-609.

Li, H., Feng, C., & Ren, H. (2020). Bamboo: An important non-timber forest product for the Neolithic Period in East China. Vegetation History and Archaeobotany, 29(2), 227-238.

Lu, H. F., Cai, C. J., Zeng, X. S., Campbell, D. E., Fan, S. H., & Liu, G. L. (2018). Bamboo vs. crops: An integrated energy and economic evaluation of using bamboo to replace crops in south Sichuan Province, China. *Journal of cleaner production*, *177*, 464-473.

Maryanto, I., Widyani, N., Nurhayati, T., & Suhartoyo. (2017). Development of bamboo-based agroforestry system in Indonesia: A case study of talun-kebun system. IOP Conference Series: Earth and

Environmental Science, 89(1), 012033. https://doi.org/10.1088/1755-1315/89/1/012033

Mercado Jr, A. R., Pulhin, J. M., Enters, T., & Ballesteros, M. A. L. (2014). Social capital and the adaptive capacity of community-based forest management in the Philippines. Society & Natural Resources, 27(6), 615-631.

Montagnini, F., & Finney, C. (2018). Restoring diverse forests: The key to sustainable land use in agricultural landscapes. In World Scientific Handbook of forest economics (Vol. 2, pp. 201-233). World Scientific

Murthy, I. K., Dutta, S., Varghese, V., Joshi, P. P., & Kumar, P. (2016). Impact of agroforestry systems on ecological and socio-economic systems: a review. *Glob J Sci Front Res: H Environ Earth Sci*, *16*(5), 15-27.

Nair, P. K. R. (2012). Agroforestry for ecosystem services and environmental benefits: an overview. Agroforestry systems, 86(1), 1-10.

Nandini, R., & Rahayu, A. A. D. (2023, May). Potential development of partnership agroforestry as carbon farming in KHDTK Rarung, Central Lombok. In *IOP Conference Series: Earth and Environmental Science* (Vol. 1180, No. 1, p. 012026). IOP Publishing.

Nfornkah, B. N., Nath, A. J., Kaam, R., Chimi, C. D., & Mezafack, K. L. (2023). Bamboo-Based Forest Landscape Restoration: Practical Lessons and Initiatives to Upscale in Africa. In *Bamboo Science and Technology* (pp. 329-356). Singapore: Springer Nature Singapore.

Obsa, O., Kassa, M., & Tajebu, L. (2015). Income contribution of bamboo (Arundinaria alpina) based agroforestry practice in Dawuro Zone, South West Ethiopia. *Journal of Economics and Sustainable Development*, *6*(9), 155-161.

Oduro, K. A., Obeng, E. A., Abukari, H., & Pentsil, S. (2020). Bamboo Policy Integration Analysis.

Pandey, D., Misra, A.K., Kumar, A., & Rai, S.K. (2018). The potential of bamboo-based agroforestry systems for higher economic returns and carbon storage in India. Journal of Environmental Management, 211, 298–307. https://doi.org/10.1016/j.jenvman.2018.01.

Partey, S. T., Sarfo, D. A., Frith, O., Kwaku, M., & Thevathasan, N. V. (2017). Potentials of bamboo-based agroforestry for sustainable development in Sub-Saharan Africa: a review. *Agricultural research*, *6*(1), 22-32.

Paudel, S., Baral, H., Rojario, A., Bhatta, K. P., & Artati, Y. (2022). Agroforestry: Opportunities and challenges in Timor-Leste. Forests 13: 41.

Phimmachanh, S., Ying, Z., & Beckline, M. (2015). Bamboo resource utilization: A potential source of income to support rural livelihoods. *Applied Ecology and Environmental Sciences*, *3*(6), 176-183.

Plieninger, T., Bieling, C., Ohnesorge, B., Schaich, H., Schleyer, C., & Wolff, F. (2012). Exploring futures of ecosystem services in cultural landscapes through participatory scenario development in the Swabian Alb, Germany. Ecology and Society, 17(3), 5.

Rahman, M., Islam, K., Rahman, M., & Hossain, M. (2019). Bamboo Agroforestry: A Sustainable Agroforestry Practice for the People of Bangladesh. Journal of Forestry Research, 30(6), 2205-2211.

Raman, T. R. S. (2014). Bamboo and Agroforestry. In: Agroforestry for the Management of Waterlogged Saline Soils and Poor-Quality Waters (pp. 259-272). Springer, Dordrecht.

Rathour, R., Kumar, H., Prasad, K., Anerao, P., Kumar, M., Kapley, A., & Singh, L. (2022). Multifunctional applications of bamboo crop beyond environmental management: an Indian perspective. *Bioengineered*, *13*(4), 8893-8914.

Research and Markets. (2021). India Bamboo Market: Prospects, Trends Analysis, Market Size and Forecasts up to 2029.

Sangha, K. K., Ibrahim, A. F., & Sabariah, B. (2017). Environmental knowledge, attitudes and practices of student teachers at a Malaysian teacher training institution. International Journal of Environmental & Science Education, 12(7), 1475-1493.

Scurlock, J. M. O., Dayton, D. C., & Hames, B. (2000). Bamboo: An overlooked biomass resource? Biomass and Bioenergy, 19(4), 229-244.

Shackleton, S., Shackleton, C., Shanley, P., Ndoye, O., & Sajo, M. (2018). Bamboo in Climate Change and Livelihoods. Routledge.

Sharma, P., Singh, M. K., Tiwari, P., & Verma, K. (2017). Agroforestry systems: Opportunities and challenges in India. *Journal of Pharmacognosy and Phytochemistry*, *6*(6S), 953-957.

Sharma, S., Roy, M. M., & Bisht, S. (2020). Bamboo-Based Industries: An Appraisal of Potential and Prospects. Springer Nature.

Shi, G., Xia, C., & Wu, J. (2016). Bamboo as a source of raw materials and its sustainable utilization. Environmental Science and Pollution Research, 23(22), 22533-22544.

Simons, A. J. (2010). Agroforestry: science, policy, and practice: selected papers from the agroforestry sessions of the IUFRO 21st world congress, Kuala Lumpur, Malaysia, August 2000. Springer Science & Business Media.

Singh, Dr-Vikram & Johar, Vishal & Kumar, Rupender & Chaudhary, Mukesh. (2022). Agroforestry Systems: A Review. International Journal of Agriculture Environment and Biotechnology. 14. 521-533.

Sinha, R. K., Bhardwaj, R. L., & Mohan, V. (2020). Bamboo Agroforestry for Climate Change Mitigation and Adaptation. In Bamboo: Conservation, Diversity, Ecogeography, Germplasm, Resource Utilization and Cultivation (pp. 281-290). Academic Press.

Sollen-Norrlin, M., Ghaley, B. B., & Rintoul, N. L. J. (2020). Agroforestry benefits and challenges for adoption in Europe and beyond. *Sustainability*, *12*(17), 7001.

Solomon, T., Moon, H., Abebe, S., Minale, A. S., & Teketay, D. (2020). Promoting bamboo-based agroforestry for enhancing ecosystem services from degraded lands. *Agroforestry for Degraded Landscapes: Recent Advances and Emerging Challenges-Vol. 2*, 423-444.

Takizawa, S., Oku, H., & Tanaka, T. (2018). Changes in traditional landscape management and its impacts on forest composition in the mid-mountain area of the Tanzawa Mountains, central Japan. Landscape and Ecological Engineering, 14(1), 33-46.

Tambe, S., Patnaik, S., Upadhyay, A. P., Edgaonkar, A., Singhal, R., Bisaria, J., & Surkar, P. P. (2020). Research Trends: Evidence-based policy for bamboo development in India: From "supply push" to "demand pull". *Forest policy and economics*, *116*, 102187.

Tewari, S., Negi, H., & Kaushal, R. (2019). Status of bamboo in India. *International Journal of Economic Plants*, *6*(1), 30-39.

Tilak, M. S., Ganesan, M., Anitha, K., Purnachandra Rao, P., & Anitha, P. (2019). Contribution of Bamboo towards Carbon Sequestration and Climate Change Mitigation. Carbon Management, 10(5), 621–632.

Tsonkova, P., Quinkenstein, A., & Böhm, C. (2018). Agroforestry in Europe: A land management policy tool to combat climate change. In Agroforestry for the Management of Waterlogged Saline Soils and Poor-Quality Waters (pp. 299-316). Springer, Cham.

van Noordwijk, M., Speelman, E., Hofstede, G. J., Farida, A., Abdurrahim, A. Y., Miccolis, A., ... & Teuling, A. J. (2020). Sustainable agroforestry landscape management: Changing the game. *Land*, *9*(8), 243.

Vargas, G., Ríos, V., & García, R. (2018). Bamboo agroforestry systems for ecological restoration and rural development in the coastal region of Ecuador. Sustainability, 10(2), 421.

Vinodhini, S. M., Manibharathi, S., Pavithra, G., & Sakthivel, S. (2023). Agroforestry: Integrating Trees into Agricultural Systems. *Recent Approaches in Agriculture Volume*, *2*, 246-272.

Wang, R., Guo, Z., Cai, C., Zhang, J., Bian, F., Sun, S., & Wang, Q. (2021). Practices and roles of bamboo industry development for alleviating poverty in China. *Clean Technologies and Environmental Policy*, *23*, 1687-1699.

Wasswa, P., et al. (2019). *Impacts of Bamboo Utilization on Livelihoods of Rural Communities in Uganda.*

Wilson, M. H., & Lovell, S. T. (2016). Agroforestry—The next step in sustainable and resilient agriculture. *Sustainability*, *8*(6), 574.

Yitaferu, B., Abewa, A., & Amsalu, A. (2008). The role of bamboo in soil and water conservation and sustainable land use: The case of Ethiopia. Journal of Bamboo and Rattan, 7(3-4), 253-265.

Zhang, X., Li, M., & Gao, J. (2018). Bamboo forest dynamics and natural regeneration in southwestern China. Forests, 9(7), 421.

CHAPTER SEVEN

UNDERUTILIZED EDIBLE AGROFORESTRY TREES: A HIDDEN CULINARY TREASURE

DINESHA S[1*], AND SHRIDEVI B. TELI[2]

[1]DEPARTMENT OF FORESTRY, NORTH-EASTERN HILL UNIVERSITY, TURA CAMPUS, TURA-794002, MEGHALAYA, INDIA

[2]SRI KRISHNADEVARAYA COLLEGE OF HORTICULTURAL SCIENCES, ANANTAPUR-515002, ANDHRA PRADESH, INDIA

Abstract

In the changing climate scenario, sustainable and climate-resilient farming practices are crucial for ensuring food and nutritional security. One such practice is the agroforestry system, where the tree component under the agroforestry system modifies the microclimate, improves soil health, reduces soil erosion, increases water holding capacity, and provides food, fuel, and fodder. Trees contribute significantly to sustainable development goals by promoting land degradation neutrality (LDN), carbon sequestration, and increasing agrobiodiversity. Additionally, many trees are used in folk medicine and are worshiped by local communities. Beyond economic, ecological and cultural importance, trees also offer food and nutritional security to many local communities and are integral to their traditional cuisines. However, the culinary uses of products from many trees, such as fruits, pods, seeds, leaves, etc. are often utilized or underutilized in many regions. For instance, the culinary importance of *Parkia timoriana* is well-known in the Northeastern region of India but not in other parts of the country. Similarly, nutrient-rich Khejri is popular in Rajasthan, which is underutilized in other parts. Therefore, it is imperative to raise awareness and invest in the cultivation and utilization of these valuable resources for a more sustainable and flavorful future. This chapter aims to bring awareness about the nutritional values and culinary importance of underutilized trees in the agroforestry system.

Keywords: Agroforestry, culinary uses, multipurpose, nutritional security, underutilized species

7.1 Introduction

Agroforestry, a sustainable land use practice that integrates trees and shrubs into agricultural landscapes. It offers a range of socio-economic and environmental benefits such as soil conservation, agrobiodiversity promotion, and income diversification for various stakeholders (Verma et al., 2023; Kumar et al., 2020; Kumar et al., 2021e). One fascinating aspect of agroforestry lies in its utilization of tree components, often overlooked in the modern agricultural system. Commonly, fast-growing, multipurpose and compatible tree species are selected for meeting diversified needs like food, fodder, fuelwood and fertilizer (Kumar et al., 2020a&b: Kaushal et al., 2021). However, some of the edible underutilized tree species are having potential nutritional values often neglected in our food basket. However, they possess unique flavors, textures, and nutritional profiles that can enrich culinary experiences. Embracing these lesser-known species not only adds diversity to plate but also contributes to food and nutritional security and resilience in the face of climate change (Kumar et al., 2022; Pandey et al., 2018; Bisht et al., 2022). Exploring the culinary potential of underutilized trees presents a pathway towards sustainable food systems. These lesser known edible tree species increase biodiversity, improve food and nutritional security, reduce malnutrition and enhance human health and wellbeing (Tewari et al., 2024; Kumar et al., 2021a&b; Kumar et al., 2022). By incorporating these trees into local cuisines, we not only celebrate cultural diversity but also promote the conservation of these valuable genetic resources. Furthermore, integrating these trees into agroforestry systems enhances ecosystem services while providing additional income streams for farmers. In essence, agroforestry and the culinary exploration of underutilized trees go hand in hand, offering a holistic approach to sustainable agriculture and food security. Through research, education, and culinary innovation, we can unlock the full potential of these overlooked botanical treasures, enriching both our plates and our planet (Tewari et al., 2022; Kumar et al., 2018; Kumar et al., 2021c&d).

7.2 Why underutilized potential Agroforestry trees?

Plenty of potentially valuable agroforestry trees in India remain untapped despite their rapid growth rate, capacity to fix nitrogen, sparse canopy, high compatibility with agricultural crops, and capacity to produce food, fodder, fuelwood, fertilizer, and medicinal products (Tewari et al., 2021; Kumar et al., 2021 Kumar et al., 2024). Countless examples include *Parkia timoriana, Elaeocarpus floribundus, Prosopis cineraria, Buchanania lanzan* and many more.

This chapter focuses on under-utilized fruit trees, which are an integral part of the established agroforestry system but a precise explanation and categorization of fruits may be challenging due to their complexity and it depends on several aspects like availability, awareness, consumption, distribution, exploitation, knowledge, popularity, production, and use (Suresh et al. 2014). The definition of UEFS excludes species that fall under widespread to moderate commercial importance category and species that have been widely naturalized, cultivated and well known in SI. However, this review includes only a lesser known, lesser distributed, underutilized, undomesticated and wild edible fruit species from species from forests, agroforestry systems, gardens, vacant and other marginal lands.

7.3 Role of agroforestry trees in culinary practices

7.3.1 *Artocarpus lakoocha* (Monkey Jackfruit)

Artocarpus lakoocha, commonly known as monkey jack fruit, is a wild edible lesser known and underutilized tropical fruit tree indigenous to the Indian subcontinent and Southeast Asia. As a member of the Moraceae family, this deciduous tree is notable not only for its distinctive and nutritious fruit but also for its varied uses in traditional medicine and agroforestry. The tree is a natural component of the traditional agroforestry system in the eastern part of India. Monkey jack trees are prized for their hardiness and adaptability to different soil conditions, making them a valuable asset in tropical agriculture and providing significant economic and ecological benefits to local communities. Every part of the tree is utilized as edible fruits, timber, fuel, fodder, dye, shade etc. and also used in medicinal purposes. The fruit, also called 'lakoocha' or 'lokhat,' resembles jackfruit in appearance but is smaller and has a unique flavor. The fruits are an edible and rich source of vitamin C, P- carotene and micronutrients (zinc, manganese, copper and iron. The antioxidants such as flavonoids and phenolic acid are also present in the fruit. The fruits are generally sweet sour pulp eaten fresh but mostly made into curries. Different parts of this fruit and a spike of male flowers are used to prepare pickles and delicious sauce. Different parts of this tree like fruits and male flowers are eaten raw, boiled, steamed or roasted. According to Sarala and Krishna Murthy, some of the recipes prepared from monkey jack fruit are mentioned below.

Preparation of powder

The processed or preserved slices are ground into fine powder along with some cooking salt. The product is known as *Vatte pudi or Esluli pudi.* The powder is used as a substitute for tamarind to prepare curry and other dishes.

Preparation of juice

The juice is prepared by grinding deseeded pulp mixed with equal quantities of water as well as some known amount of sugar. The slurry is filtered and a pinch of cardamom and black pepper are added.

Preparation of papad

The aqueous extract of fresh fruits of monkey jack is prepared and mixed with the paste paste with raw rice, green gram, black gram along with a small amount of jeera, pepper powder, chili powder, salt and oil. The paste is spread on banana leaf or thick plastic sheets and dried under direct sunlight.

Preparation of sambar

The vegetables along with dhal are boiled separately. Simultaneously, sufficient quantities of preserved slices of monkey jack or powder or fresh fruits as well as copra are added to fried coriander, black gram, and fenugreek, chili and curry leaves and then ground into paste. This entire paste is transferred into the boiling vegetables and dhal.

Preparation of fish curry

The clean and dried metallic container is taken with a sufficient amount of oil and mustard seeds are added. When mustard seeds crackle, a pinch of fenugreek, garlic and curry leaves are added. Simultaneously, sufficient quantities of chopped onion are added on a medium flame until the content turns into a golden yellow color. The chopped tomatoes and salt are also added. The content is fried until the oil separates from the mixture. A sufficient amount of chili powder, coriander powder, and turmeric powder are added and fried for a few minutes. Then sufficient quantities of aqueous extract of fresh or preserved monkey jack fruits are added under low flame and continued cooking for 10 minutes. The milk of copra and the fish pieces are slowly added. The cooking is continued for another 10 minutes or the required time or desired time. The container is removed from the flame and kept for five minutes.

7.3.2 Mahua (*Madhuca longifolia*)

Madhuca longifolia commonly known as Mahua belongs to the family sapotaceae and originated from different regions like India, Sri Lanka, Myanmar, and Nepal. Mahua is a frost-resistant tree that can grow in marginal areas of tropical and subtropical regions of India. It is found in Uttar Pradesh, Madhya Pradesh, Orissa, Jharkhand, Chattisgarh, Andhra Pradesh, Maharashtra, Bihar, West Bengal, Karnataka, Gujarat and Rajasthan. It is an important multipurpose agroforestry tree that fulfills the three fundamental needs such as food, fodder, and fuel of tribal communities. Mahua tree acts as a soil binder, it reduces soil erosion due to its broad spreading superficial root system that helps to retain soil together. It also increases soil fertility by fixing atmospheric nitrogen and adding leaf litter into the soil. Apart from its ecological benefits, Mahua has great significance in the food industry. The seeds, fruits, and flowers are edible. Flowers of the Mahua tree have a great significance in traditional shifting cultivation as liquor prepared from flowers is offered to local deities before sowing. Flowers are edible and possess high nutritive value, rich in sugar (sucrose, glucose, fructose, and arabinose), vitamins (vitamin C), carotene, protein, minerals (Ca and P), and fats. The higher amount of sugar in flowers is good natural sugar used in the preparation of traditional dishes like *Halwa, Kheer, Meethi Puri and Burfi*. In addition, tribals use Mahua flowers for curing skin diseases, headache, pitta and bronchitis. Flower juice is supplemented to the lactating mother to increase breast milk.

Utilization of Mahua flowers as food products (Pinakin et al. 2018)

Due to its valuable composition, the flower is utilized in the preparation of fermented and non - fermented traditional food products.

As a sweetener: Mahua flowers are utilized as sweeteners due to their high sugar content like sucrose, fructose, arabinose, maltose and rhamnose in the preparation of *Halwa, Meethi Puri, Kheer, burfi* etc.

Cake: It is made from Mahua flowers and rice. The pre-soaked rice and Mahua flowers mixed and grinded. The paste is spread over Sal leaves and burned on fire to make cake.

Mahuli: It is traditionally prepared by local people of Odisha. It is an alcoholic beverage prepared from Mahua flowers. The alcoholic content varies between 30-40%.

Mahua Daaru: Mahua Daaru is a fermented beverage prepared by locals in North-West India. The flowers are gathered and dried, which contains 20-40 % of alcohol. The dried flowers are allowed for fermentation, for a few days. During the fermentation process, ammonium chloride and jaggery are added. Black pepper can also be used to give a spiciness to the dish. The mixture is placed in a typical distillation container after fermentation. This process produces 300-400 ml of daaru from a kilogram of dried flowers.

Sugar syrup: Before concentrating the dried flower water extract to the appropriate concentration, it is decolorized with various decolorizing agents such as lime and activated charcoal. The syrup derived from the Mahua flower is used for a variety of purposes, including the production of chocolate and as a sweetening agent.

Jam, jelly, marmalade and pickle: Jam, jelly, marmalade, and pickle with the addition of citric acid, mature fruits are transformed into jam. To hide the astringent flavor, jelly can be made from the pulp alone or in conjunction with guava.

Edible oil: The seed of Mahua has 30-45% oil content and each tree yields up to 20-40 kg of seeds each year (Jha et al. 2013). The oil extracted from seeds is utilized for culinary purposes by tribal people from Odisha, Chhattisgarh, Maharashtra, etc. The seed oil is commonly known as "Mahua butter".

7.3.3 *Docynia indica* (Assam Apple)

Docynia indica is commonly known as Assam Apple belonging to the family Rosaceae. The tree is indigenous to South and Southeast Asia. It is an important multipurpose fruit tree in the agroforestry system. It is an evergreen wild fruit tree species that bears edible fruits that resemble small apples. Its fruits are rich in nutrients, vitamin C, and antioxidants. Fruits are also used in herbal medicine and processed into various food products such as juice, wine, vinegar, syrup, jam, etc. The acidic fruits are eaten both raw and cooked. In addition, the tree offers timber for making household tools and fuelwood, and material for mulching. The wood is used for making tool handles and the branches make good walking sticks. Sometimes, the stem is also used as rootstock for grafting apples. Economically, the grated fruits are sold in local markets in packets and provide additional income to women folks.

***Carrissa carandas* (Karonda)**

Carrissa carandas is a flowering shrub commonly known as Karonda, belonging to the Apocynaceae family. It is hardy, drought tolerant, and thrives well in a wide range of soils. The plant grows well in high-temperature regions, commonly found in the Western Ghats of Konkan in the western coastal states of Maharashtra, Karnataka, and Goa. It is also grown in the temperate regions of the Himalayan Siwalik Hills of India and Nepal. Karonda tree is popular for its edible fruits, which are rich in iron (39 mg /100g), vitamin C, vitamin A, calcium (160 mg/100g) and phosphorus (60 mg/100g). The dried fruits contain 2.3 % of protein, 67.1% of carbohydrates, and 9.6 % of fat. The fruits are also a good source of Thiamine (B1), Riboflavin (B2), Pantothenic acid (B5), Pyridoxine (B6), Biotin (B7) and Folic acid (B9) (Tripathi et al. 2014). The fruits are also used in folk medicine and Ayurveda to treat acidity, indigestion, fresh and infected wounds, skin disorders, urinary disease and diabetic ulcers, anemia, stomach pain, etc. The decoction from the leaf is used to treat fever, diarrhea, and earache. The roots are also used as insect repellent. The mature fruits are used for making pickles. The fruits contain pectin and hence are used in the preparation of chutney.

Value-added products from Karonda:

Unripe fruits are used for making pickles (*Achar*), sabji, candy, and chutney. Jam, RTS, nectar, squash, canned fruits, syrup, and fruit powder are prepared from fully ripe fruits. The fruits are also used as a faux cherry in cakes, puddings, and other preparations. In addition, colonial British in India also made jelly, jams, and syrups from it.

7.3.4 *Syzygium fruticosum* (Kat Jamun)

Syzygium fruticosum, commonly known as Kat Jamin or Indian Jambul or wild jamun, is a fruit-bearing evergreen shrub that belongs to the Myrtaceae family and originated from India. Trees are suitable for windbreaks and have straight growth habits. All parts of the species are reported to have medicinal value, the bark is used as an astringent, and fruits are used to treat liver problems (Gaur and Lale, 2015). It is also used to treat stomach aches, diabetes, and bronchitis (Jain et al. 2010). The fruits are rich in vitamin C, fiber, and iron, and poor in Ca and total carotene. The seed extract is traditionally used for controlling diabetes and dysentery. The fruits are eaten fresh also.

7.3.5 *Ziziphus mauritiana* (Indian Jujube / ber)

Ziziphus mauritiana, jujube tree, or ber belonging to the family Rhamnaceae. All parts of the plant are very effective against various diseases. The leaves are used to treat diarrhea, wounds, abscesses, liver, asthma, fever, swelling, and gonorrhea (Michel, 2002). The fruits contain carbohydrates (17g/100g), protein (0.8g/100g), fat (0.07g /100g), thiamine (0.02- 0.024 mg / 100g), riboflavin (0.02-0.038 mg / 100g), iron (0.76-1.8 mg /100g), calcium (25.6 mg /100g) and phosphorus (26.8 mg /100g) (Morton, 1987).

7.3.6 *Aegle marmelos* (Bael)

Aegle marmelos (Bael), mainly found in tropical and subtropical regions, belongs to the family Rutaceae. It is one of the most important holly trees of Hindu mythology and has great medicinal value. It is spread all over India, found in sub- Himalayan tract, central, and south India, and Myanmar. Its trifoliate leaves are presented to Lord Shiva as offerings by the devotees. The unripe fruits are greenish white and turn to yellowish brown when ripen. The fruits contain Marmelosin (furocoumarins), which acts as a laxative and diuretic in lower doses. The fruits are a rich source of carbohydrates (30 %), protein (1.2 %), vitamin A (186 IU), sugars (4.6 %), tannins (9% in the pulp and 20% in the rind), and other minerals such as iron, calcium, phosphorous, potassium (Purohit and Vyas, 2004). Bael is used to cure cholera, stomach ache, dog and snake bites, asthma, scurvy, vomiting, constipation and piles.

Value addition in Bael

A large number of bael processed products such as candy, panjiri, toffee, jam, squash RTS, etc. can be prepared from fruits. Preserve and candy is prepared from mature fruits in which sugar is impregnated till it becomes tender and transparent. An ideal composition of bael squash is found to be a half portion of the extracted pulp, 50 ° Brix, and 1 % acidity. Mature green fruits are dehydrated at 55-60°C after fumigating with sulfur dioxide fumes for an hour in a sulfur box and kept for future use. The beal powder is also prepared by grinding fruit slices in a grinder. Beal panjiri is nutritive, restorative, and recommended for stomach ailments. The beal toffees are prepared by mixing 40 parts of sugar, 4.5 parts of glucose, 10 parts of skimmed milk powder, and 6 parts of hydrogenated fat to 100 parts of extracted pulp (Rakesh et al, 2005).

7.3.7 *Spondias dulcis* (Indian Hog Plum / Amra)

Spondias dulcis commonly known as Indian Hog Plum Amra or Malay Apple or Golden Apple, belongs to the family Mangiferaceae. It is a tropical tree that originated from India. The tree bears edible fruits, which are eaten raw. The fruits are also used to treat dysentery and dyspepsia. The fruit contains vitamin C (128 mg), beta-carotene (2377 micrograms), sodium (23 mg), potassium (675 mg), iron (7.62 mg), phosphorus (116 mg) and calcium (51 mg) (Jahan et al. 2011). Leaves are used as an ingredient in the preparation of herbal hair lotion. Every part of the plant has many traditional and ethnomedicinal uses.

Value addition in Amra

The unripe fruit is used as a vegetable and added to stews. It can also be used to prepare chutney, pickles, panna etc. The powder is prepared by drying unripe fruits, which is used as a flavoring agent in various cuisines. Ripe fruits are acidic and used to prepare pickles, jam, jelly, chutney, and beverages like squash, nectar, and RTS. Flowers can be used as salads. Flower buds can also be used to prepare pickles.

7.3.8 *Morus Alba*

Morus alba is commonly known as white mulberry and silkworm mulberry. It is fast growing, small to medium-sized grows to 10-20 m in height. This species is native to China and India. It is commonly cultivated to feed the silkworms in the production of silk, which makes the mulberry the most vital component of the sericulture industry that offers employment to a large number of people in many countries such as India, China, Bangladesh, Pakistan, etc. Other than sericulture use, it has enormous economic benefits. The leaves are also used as feedstock for livestock in areas where dry regions restrict the availability of ground vegetation. In Korea, leaves are used to prepare tea. Mulberry fruits contain Carbohydrates (either in the form of simple sugars or starch), soluble and insoluble fibers, iron, vitamin C, potassium, and vitamin E. Leaves also possess almost equal nutritional properties to fruits. The leaves are used in the preparation of delicious food products, which show anti-tyrosinase, anti-diabetic, anti-oxidant, anti- anti-inflammatory activity. The fruits are utilized in the preparation of food items such as cake, pastry, chocolate, etc., beverages, and preservatives due to their anti-cancer, anti-oxidant, and anti-neurodegenerative properties. The stem extracts can be used to increase the potentiality of food items, which possess anti-microbial and anti-inflammatory properties (Acharya et al. 2022). In addition, the root extract also shows antihyperlipidemic activity and potential cardiovascular trouble reliever. Some of the potential food products prepared from *Morus alba* are discussed below.

Value addition in *Morus alba*

Fruits of *Morus alba* are used for the preparation of spiced squash and appetizers. The pastry is prepared from buckwheat flour, hulls, insulin, and chokeberry as well as mulberry fruit extraction. It is rich in fiber and low in calories. The shelf life of minced meat is enhanced by the metabolic extraction of mulberry leaves. *Morus alba* silage is used for the isolation and culture of lactic acid-forming bacteria. These bacteria can stabilize gastrointestinal tract microbial flora. The mulberry fruits and rosella fruits in the ratio of 70:30 are used to prepare mixed fruit jam. The fruit extract can be used for the preparation of syrup and stored for 6 months under refrigerated conditions. The fruits of mulberry can also be used to prepare vinegar. The fruits of *Morus alba are used* as raw material to brew fruit wine.

7.3.9 *Prosopis cineraria*

Prosopis cineraria is a holly and indigenous tree of western India and is locally known as Khejri (Mann and Saxena, 1980). It is the most drought-tolerant tree species and thrives in hot, arid regions with annual rainfall of less than 500 mm. The tree is worshiped by the Bishnoi community and is known as the "Golden Tree" as its all parts are used traditionally as food, fodder, fuel and to cure various ailments. Leaves, flowers, pods, barks, and plant sap possess anti-cancer, anti-diabetic, anti-inflammatory, and anti-microbial properties and are used in the form of paste, gum, smoke, and decoction. The Khejri tree holds significant economic value, as every part of it is utilized. From dehydrated sangria, which can fetch Rs. 150-200/kg, to its rich fodder and grain yields for crops like cowpea, pearl millet, and cluster bean when grown alongside it, the tree offers various benefits. Its immature pods are packed with protein (18%), carbohydrates (56%), phosphorus, and calcium (0.8% each), suitable for both fresh consumption and dehydration, while its ripe dried pods are utilized in bakery items. Additionally, its leaves serve as valuable fodder, its fruits are rich in essential nutrients (Ca and P) and vitamin C, and it plays a crucial role in carbon sequestration, soil fertility, and agrobiodiversity. Moreover, the Khejri tree supports the growth of other fruit trees and holds cultural significance, being worshiped during Janmashtami (the birthday of Lord Krishna as per Hindu traditions) and honored and worshiped by communities like the Bishnoi. The tree is also resilient to climate extremes as it can withstand and adapt to frost and drought and survives under both high and low temperatures.

Vegetables: *P. cineraria* pods are used as vegetables in the dried or green form. In addition, green and unripe pods are used in the preparation of pickles and curries.

Flour: The pods of the Khejri tree have three parts: mesocarp (56% of pods), endocarp (35%), and seeds (9%). The mesocarp is ground to produce flour, which is used to prepare bread, cake, and chapatti by mixing it with wheat flour and sweets. The flour contains a high level of protein (62%), dietary fiber (25%), and a low level of carbohydrates and fats. It is gluten-free and contains Ca, K, Mg, Zn, and Fe as well as an amino acid lysine which is absent in cereals. This flour is mixed with different concentrations to make different bakery products such as biscuits (5%), breads (10%), pancakes (15%) and chapati (50%). *Prosopis* flour is good for diabetic patients as it requires a longer time (4-6 hrs) to digest compared with cereal flour (1-2 hrs) and contains fructose, which the body can process without insulin.

Syrup and drinks: In many places, the Prosopis species are utilized to create both fermented and non-fermented beverages, as well as syrup. To produce a nutritious syrup, clean green pods are broken into small pieces and boiled in water. The beans are simmered for 2 hours, with small amounts of water continuously added to prevent burning. Afterward, the pods are mashed to release more of the sweet pulp and simmered for a few more minutes. The juice is then strained and stored in clean containers for direct consumption as a drink. Alternatively, additional sugar can be added to the juice and boiled further to create a syrup with a unique flavor.

Panchkuta: *Panchkuta* is prepared from sangria and also includes seeds of *kumat* (*Acacia sengal*), fruits of *ker* (*Capparis decidua*), *kachri* (*Cucumis supp*) and *goonda* (*Cordia mixa*). These are crushed, mixed, and fried after adding condiments to taste.

7.3.10 *Parkia timoriana*

Parkia timoriana commonly known as 'Tree bean' is a leguminous, multipurpose and underutilized nutritious tree native to the North-eastern states of India and Southeast Asian countries. People of North-Eastern states of India grow it in home gardens, agroforestry systems, and fallow lands of shifting cultivation for their dietary supplements. The whole tree has economic, medicinal, nutritional and ecological significance. Seeds contain protein, minerals (K, Fe, Mg, Zn, P, Mn), essential amino acids like isoleucine, leucine, phenylalanine and tyrosine and fatty acids (Angami et al. 2018). The entire green pods are also rich in nutrients and protein. Tree beans have been used in traditional medicine since time immemorial by different ethnic groups of the North Eastern region; the decoction of bark, fruit skin and leaves are used to treat diseases (Hopkins, 1994). The pods and kernels are also used to treat leprosy and hypertension in Manipur. The green pod, leaves, seeds, powdery-like material from the pod, flowers and other parts of the tree bean are consumed either raw or boiled with other ingredients. Consumption of seeds without pulp in curries and fried items is also common among people in North Eastern India, and matured flowers and young shoots are also used in salads and curries. In addition, trees also have ecological significance as they fix atmospheric nitrogen, reduce CO_2, and improve soil fertility. Though It is well known in Manipuian cuisine, the species is underutilized in other northeastern states, Tree bean is one of the highly economical and culturally bound with people from Manipur. Many studies reported on the botanical characteristics, distribution patterns, nutritional composition, medicinal properties, and ecological significance of *Parkia timoriana* in North-Eastern India but studies related to culinary uses are less researched and less documented. The effort is made to highlight the food products prepared from *Parkia timoriana.*

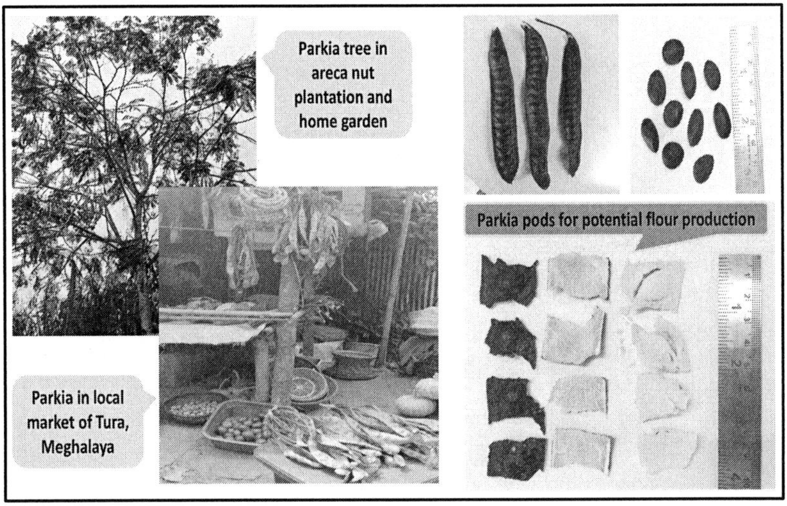

Plate 1. Some glimpses of *Parkia timoriana* prevalent in the Agroforestry, Homegardens and fallow land of Tura, Meghalaya.

Eromba

Eromba is a preparation from the *Meitei* community of Manipur prepared with dried and fermented fish, boiled vegetables, and peeled pods of *Parkia timoriana*. It's typically seasoned with chili and other spices. The main ingredient is fermented dried fish (Ngari). The fish is sun-dried and then fermented in an earthen pot (*Koloh* or *Nagari chapu*) by repeated smearing of oil and no salt. Boiled vegetables including *Parkia timoriana* are added to the mixture of boiled and smashed fish along with chili.

Yongchak Sinju: It is a typical Manipuri salad, which is prepared by mixing scraped and sliced pods with dry fish and dry chili.

Flowers: Yellowish bisexual drumheads like flowers serve as food by preparing a salad-like dish.

Dry mature seeds: Soaked overnight in water, manually de- coated and consumed in different ways. The dry seeds are used in curry, salad, and consumed after roasting and frying with prawns.

Powdery pulp of mature entire pod: Used to thicken the curry

7.3.11 Burmese grapes (*Baccaurea ramiflora* Lour.)

Baccaurea ramiflora Lour. Commonly known as Burmese grapes is an underutilized tropical fruit tree belonging to the euphorbiaceae family. It is native to Southeast Asia. In India, it is distributed along the sub-himalayan region such as Sikkim, Arunachal Pradesh, and Darjeeling hills, Tripura, Assam, Odisha, Meghalaya and Andaman Nicobar. The tree is a slow growing, evergreen and shade-loving plant, mostly found under the areca nut plantation. The tree is multipurpose, and almost all parts have economical, medicinal and religious importance. *Baccaria ramiflora* has been mentioned in the Chinese Dai medicine. It is used to treat various diseases, injuries, rheumatoid arthritis, cellulitis, abscesses, etc. The fruits are a rich source of carbohydrates (51.9%), protein (5%) and fiber (20%). It also contains various minerals such as magnesium (504 mg), potassium (730 mg), phosphorus (132 mg) and Iron (100 mg) per 100 gm of fruit pulp. It has a significant amount of ascorbic acid, which shows antioxidant properties (Raghavan and Ramjan, 2018). The young leaves of the plants are used as vegetable or flavoring agents for curries and minced meat (Raghavan and Ramjan, 2018).

7.3.12 *Buchanania lanzan*

Buchanania lanzan commonly known as Chironji or Charoli has full potential for income and employment generation in Central India through agroforestry interventions. Presently, this multifaceted chironji tree is underutilized in the agroforestry system, which is found in forests in all parts of the country except the eastern Himalayan forests (Singh et al. 2006). Several studies reported that this species is on the verge of extinction and comes under the category of threatened species. In this condition, the adoption of this species at farmers' fields under agroforestry is one of the best ways to protect this species from depletion. A mature tree can give more than 3-4 Kg Chironji kernels per year (Anand, et al. 2014), this way an agroforestry orchard having 100 trees can produce more than 300 kg Chironji kernels per ha per year which can give more than Rs. 2.5 to 3.0 lakh per year gross return. Various parts of the tree are useful in the preparation of food and medicinal products. The flesh of the ripe fruit is eaten raw or roasted; kernels are rich in oil and nutrients and used in the preparation of puddings (Singh, 1982). The fruit is used to prepare juice, and the kernel is rich in protein (19-21 %), lipid or fat (59%), carbohydrate (12.1 %), fiber (3.8 %), minerals like calcium, phosphorus, iron and vitamins such as thiamine, ascorbic acid, riboflavin and niacin (Anupama,

2005). The oil obtained from kernels is sweet in nature and used as an alternative to olive or almond oil. The oil is also used in traditional medicine by tribals in Gujarat as a brain tonic. The paste/ointment from the kernel is used to treat skin diseases, diarrhea, and intercostal pains. In addition, bark is used for tanning purposes in Kerala. Apart from its food and medicinal values, it also gives quality timber for various uses.

Culinary importance

The kernels are highly economic and the most important part of the tree. Hence, kernels have the potential to be processed into several value-added products.

Fruits: The juice recovery from the fruit is less due to large seed size. The juice extracted from the fruits can be used to prepare squash, ready-to-serve (RTS) drinks and nectar. The sun-dried or artificially dried fruits are preserved in the form of powder. In addition, the fruit is also used to prepare fermented beverages (wine).

Chironji kernels and Nuts: Chironji nuts are used in preparation of sweets such as halwa, kheer, laddu, paak, etc.

Chironji oil: The kernels are rich in oil (52 %) (Kumar et al. 2012), and the extracted oil from chironji kernels is used as edible oil by some native people. This is mainly used in cosmetic manufacturing and substitute for oil or almond oil (Siddiqui et al. 2014).

7.3.13 *Syzygium cumini*

It is one of the most widely distributed fruit trees in India, locally known as Jamun belongs to the family Myrtaceae. The tree grows well in moist, damp, or marshy situations, tolerates prolonged flooding, and also drought. It can grow on a wide range of soils even in shallow and rocky areas. These fruit trees show tolerance to frost when it's mature and sensitivity when it's young. The tree is well suited to the agroforestry system, commonly grown in wasteland and dry regions. Jamun is an excellent shade tree that is grown with coffee, banana, and cocoa. The abundant foliage of the Jamun tree produces good shade, hence used to shelter coffee plantations, chicken yards and livestock pastures. Trees can also be planted as windbreakers. Its short rotation multipurpose tree offers multiple uses, provides fodder for animals, excellent firewood and charcoal, timber for making furniture, agriculture implements, plywood, boat building etc., and bark for tanning (yields brown dye) fishnets.

In addition, fruits are rich sources of minerals, vitamins and antioxidants, which contain crude fat 1.02 mg (1.18–4.50%), crude protein 3.84 to 7.17 mg (6.3–8.5%), carbohydrate 22.8 g to 31.6 g (41%), crude fiber 7.01 mg (2.64–16.9%), calcium 0.6 mg (0.41%), and phosphorus 0.072 mg (0.17%) (Rizvi et al. 2022). Jamun fruit is generally acknowledged to be very high quality for its curative function chiefly against diabetes because of its effect on pancreas. Jamun seeds also contain albumen, fat, glycosides, an alkaloid; jambosine, resin, ellagic acid, quercetin, gallic acid as well as elements of zinc, vanadium, chromium, sodium and potassium. Sitoterol is present in unsaponifiable material of seed fat. Fruits are eaten fresh and used to make jams, jelly, juice, pudding and wine. The seeds and bark are well-known in traditional medicines to treat dysentery, and control of hyperglycemia and glycosuria in diabetic patients.

Processed Food Products

Jamun fruit is used when various processed food products can be prepared from Jamun fruits viz. fermented and non-fermented beverages. Besides that, jam, jelly, cheese, toffee, etc. can be prepared. Also, fortified products can be prepared from jamun pulp or its other parts. Procedures for the preparation of different products are given below.

Jamun Juice: Jamun juice is prepared from fruit pulp, pulp is heated at 60°C to extract the juice. Then juice is pasteurized at 85oC and sodium benzoate is used as a preservative to store for a longer period. The pasteurized juice is stored at ambient temperature.

Jamun Squash: Jamun squash is prepared from fruit juice by adding sugar syrup (1.8 kg of sugar mixed with 1L of water and 15 g of citric acid). Sodium benzoate (0.3g) is used as a preservative.

Jamun Wine: The Jamun wine is prepared from fruits dipped in 5% NaCl for 72 hrs after washing. The seeds are separated from the fruit and the pulp is crushed. The extracted juice from the pulp is mixed with sodium metabisulphite (100 micrograms per ml) and TSS is adjusted to 17o Brix with sugar. The acidity of the juice is adjusted to pH 4.5 by adding 1N acetic acid, which is inoculated with yeast culture and kept for fermentation at 32+-2oC for 6 days. Racking and decantation of wine when Brix reaches 2-3o. The fermented wine is stored at ambient temperature in a glass bottle.

Jamun Jam: Jamun jam is prepared by heating fruit pulp and sugar at 110oC, along with some sugar pectin (2 g/kg or 0.2%) is added to heated fruit pulp and stirred constantly to prevent clotting until the mixture starts boiling. At the point of vigorous boiling (approximately 221°C or at 68-70% TSS). To reach the expected acidity acidifier (citric acid) at 4g kg-1 is added; and sodium benzoate is used as a preservative at 1 g kg-1. The hot jam is filled in a bottle, cooled in running water, and stored.

Jamun Jelly: The fruit pulp is homogenized by grinding it with water. The crushed pulp is adjusted with pH 3.4 by adding citric acid, which is mixed with sugar in the ratio of 4:6 and 1% pectin. The mixture is cooked to 67% brix and then stored in a sterilized bottle.

Jamun Cheese: The fruit pulp with water (1 kg of pulp in 100 ml water) is heated up to 70°C. The heated pulp is mixed with 2% pectin, sugar, citric acid and butter. This mixture is cooked to obtain the desired consistency and salt and benzoic acid at 500 ppm are added to this mixture. The prepared product is Spread on butter-smeared trays (0.60 cm thick) and left for cooling and setting for 5-6 hours. Cheese is cut into suitable size pieces and wrapped in butter paper.

Fortified Food Products: The fortification of Chapati with jamun pulp is prepared by preparing dough from whole wheat flour (50g), the dough is left to rest for 30 min at 30°C and 85% relative humidity (RH). The chapati is prepared from the fortified dough.

7.3.14 *Actinidia callosa*

It is also known as the Himalayan kiwi vine, and is a flowering plant in the Chinese gooseberry family Actinidiaceae. It's a deciduous climbing shrub that can grow stems up to 30 meters long, scrambling over the ground and wrapping around other plants for support. It's native to the Himalayas, central and southern China, Taiwan, Myanmar, Vietnam, Peninsular Malaysia, and Sumatra.

7.4 Impact of underutilized trees based agroforestry on local cuisines

Underutilized tree-based agroforestry systems have a significant impact on local cuisines in India. These systems involve integrating lesser-known or traditional tree species into farming practices, providing a range of benefits, including diverse food sources, enhanced nutrition, and preservation of cultural heritage. In western India, particularly in Rajasthan, the Khejri tree holds a significant impact on local cuisines. Khejri pods are used to prepare pickles, curries and syrups. The flour obtained from the mesocarp of the pod is used to prepare bread, cakes and chapati by mixing it with wheat flour. The popular dishes of Khejri pods prepared during festivals are "Ker sangri" and "Panchkuta", the former is dry curry prepared with Khejri pods and Ker berries and the latter is prepared by mixing five deserts vegetables, including sangria. In southern India, monkey jack is used to prepare *Vatte pudi or Esluli pudi* (powder), which is a substitute for tamarind to prepare curry. The fruit is also used for making sambar, fish curry, juice and papad. Karonda is also an important underutilized agroforestry tree rich in iron that is used to prepare pickles, chutney, curries, mixed jams, juice etc. Mahua plays a significant role in the local cuisines and culture of tribes in central India. Its flowers, seeds, and oil are integral to the local cuisines and culture of these regions. The flowers are used as sweeteners in the preparation of *Halwa, Meethi Puri, Kheer, burfi*, etc. Flowers are also fermented and made into beverages, containing 20-40 % of alcohol. The oil extracted from the seeds is used as cooking oil. In the North Eastern region, many underutilized fruit trees such as *Parkia timoriana*, Jalpai, etc. are becoming a part of their traditional cuisines. *Parkia timoriana* pods are used to prepare traditional dishes such as *Eromba, Yongchak Sinju* in Manipur. Flowers are used to prepare salad, dried seeds as daal and powdery pulp from the pod is used to thicken curry. These underutilized trees based agroforestry not only impacted on food and nutritional security of local people but also improved their economic and ecological status.

7.5 Challenges and Solutions of integration of underutilized trees in agroforestry

Farmers' awareness and attitude towards the integration of underutilized trees in the farming system is poor due to the long gestation period of native and natural fruit species which require long-term investment. In many regions, various tangible and intangible benefits and ecosystem services of these trees are not known. Many indigenous fruits such as *Adansonia digitata* and *Citrullus lanatus* in Africa and *Parkia timoriana* and *Pithocelobium dulce* in India have high nutraceuticals but very poorly organized production and marketing systems as well as research and policy support. Size of land, poor irrigation facility and lack of awareness induce negative perceptions and attitudes among farmers towards the adoption of under-utilized native trees. Generally, the productivity of these trees is very low compared with other major tree species grown in India. Hence, there is a critical need to develop improved varieties, quality planting materials, the package of practices, transfer of technologies, post-harvest management, storage, marketing, and transport to overcome these challenges and popularize these native UEAFS species among growers. Many research organizations related to tree improvement are tremendously working under the GOI assistance to standardize cultivation practices and further improve these trees. Developed varieties and standardized several production technologies like propagation methods, plant spacing, canopy management, crop compatibility, nutrient and water management, crop regulation, plant protection, post-harvest management and value addition (Singh et al. 2020). In order to cope with various abiotic stresses, UEAF species like *Emblica officinalis*, *Aegle marmelos* and *Feronia limonia* are modified or improved to assure critical morpho-physiological functions i.e., strong deep root system, high root-to-shoot ratio, selective absorption, uptake and storage of more water and nutrients, reduced transpiration and heat shocks (Meena et al. 2022). Additionally, synchronized flowering and fast fruiting, biosynthesis of antioxidants, proteins and drought-responsive genes, assist in growth and development under adverse conditions (Meena et al. 2022). Generally, breeding and biotechnological interventions are hindered by major obstacles like genetic heterogeneity, long juvenile phase, self and cross incompatibility, sterility and many more (Gill et al. 2022). Successful approaches to overcome these challenges are genetic diversity and phylogenetic studies, hybridization, omics-based interventions, linkage mapping, marker-assisted selection and mutation breeding (Gill et al. 2022).

Table 1. Constraints and Solutions of Underutilized Edible Agroforestry Trees (Modified from Ashwath et al. 2023)

Criteria	Constraints	Solutions
Lesser-known	Poor availability and utilization.	Awareness, research and extension activities are vital.
Lesser-knowledge	Lack of cultivation practices, and economical, nutritional, and medicinal knowledge.	Linking practitioner, researcher and policymaker. Training and capacity building.
Lesser-researched	Lack of mass cultivation, nutrition, pest and disease management.	Research on nutrition, preference and genetic improvement are vital

Less technology	Un availability of quality planting materials, varieties, production, post harvest technologies.	Need of certification, grading, processing, value addition, packaging and adequate storage.
Less popular	Long gestation period, poor yield, less economic, no value addition and awareness.	These issues can be dealt with the application of scientific interventions and policy support.
Less utilized/exploited	Less market, lower commercial exploitation and perishable. Low consumer preference, improper economics, policy and institutional arrangements.	Low preference can be optimized by natural ripening and genetic improvement. Processed into value-added products by using traditional and advanced techniques.
Less production	Poor yield and quality	Application of scientific and technological interventions, and genetic improvement.
Less distribution	Confined, storage and processing constraints	Creating awareness, research and extension activities through training and capacity building.
Less conservation	Genetic erosion, over-exploitation and habitat loss.	Linking society, economics and ecology, and cultivation-assisted conservation are needed.

Conclusion

In conclusion, UEAF trees represent a hidden culinary treasure with immense potential. These trees not only offer unique and diverse flavors that can enrich our diets, but they also contribute significantly to sustainable and resilient agricultural practices. By incorporating these underutilized tree species into AFS, we can increase biodiversity, enhance soil health, and create more resilient farming landscapes. Encouraging these less utilized species can lead to innovative culinary experiences while supporting environmental sustainability and local economies. Therefore, it is imperative to raise awareness and invest in the cultivation and utilization of these valuable resources for a more sustainable and flavorful future. Developing standard cultivation protocols, supply of quality planting materials, transfer of technology, processing and marketing through small-scale industries, self-help groups and farmer producer organizations are considered necessary. The government should also actively participate in this regard by providing necessary inputs, technical support and policy interventions. As these native trees are highly adaptable and show resistance to various insect pests and diseases, hence pesticide and other chemical requirements are almost negligible. Thus, they can fit well into organic and natural farming, as they require less attention from the cultivars. In addition, the promotion of culinary aspects of these underutilized trees will contribute to the government's emphasis on the promotion of indigenous species through the slogan 'vocal for local' and 'local for global'.

References

Acharya, R., Bagchi, T. and Gangopadhyay, D., 2022. Mulberry as a Valuable Resource for Food

Anand R.K, Rao AP, Khare N, Singh SK. 2021. Prospects and Potential of Chironji (Buchanania lanzan Spreng.) based Agroforestry for Livelihood and Environmental security in Central India. In: Agroforestry Prospective, Strategies and FutureAspects. Tarun Publications.

and Pharmaceutical Industries: A Review. *Medicinal Plants.*

Angami, T., Bhagawati, R., Touthang, L., Makdoh, B., Nirmal, Lungmuana, Bharati, K.A., Silambarasan, R. and Ayyanar, M., 2018. Traditional uses, phytochemistry and biological activities of Parkia timoriana (DC.) Merr., an underutilized multipurpose tree bean: a review. *Genetic Resources and Crop Evolution*, *65*, pp.679-692.

Anupama, P. (2015). Chironji Health Benefits and Medicinal Uses. Retrieved from https://www.bimbima.com/ayurveda/chironji-health-benefits-and-medicinaluses/234/ as accessed on 02/05/2017.

Bisht, S., Kumar, A., Kumar, N., Singh, H. and Kumar, P. (2022). Biofuel Production by Using Biomass and Its Application. In book: Renewable Energy and Green Technology in CRC Press.

Do VH, La N, Mulia R, Bergkvist G, Dahlin AS, Nguyen VT, Pham HT and Öborn I 2020. Fruit Tree-Based Agroforestry Systems for Smallholder Farmers in Northwest Vietnam—A Quantitative and Qualitative Assessment. Land, 9(11), 451. https://doi.org/10.3390/land9110451

Dwivedi, A., Priyadarshini, A. and Induar, S., 2022. Mahua (*Madhuca longifolia*) flower and its application in the food industry: A review. *International Journal of Chemical Studies*, *10*(1), pp.80-84.

Gaur, S. K. and Lale, S. K. 2015. Medico-ethno-botanical Survey of Satna Forest Division, Madhya Pradesh, India. Journal of Drug Research in Ayurvedic Sciences, 1(1): 95-96.

Hopkins HCF (1994) The Indo-Pacific species of Parkia (Leguminosae-Mimosoideae). Kew Bull 49(2):181–234

Jahan, S.; Gosh, T.; Begum, M. and Saha, B. K. 2011. Nutritional Profile of Some Tropical Fruits in Bangladesh: Especially Antioxidant Vitamins and Minerals. Bangladesh Journal of Medical Science, 10:2.

Jain, A.; Sharma, S.; Goyal, M.; Dubey, S.; Jain, S.; Sahu, J.; Sharma, A. and Kaushik, A. (2010). Anti-inflammatory activity of Syzygium cumini leaves. Int. J. Phytomed., 2:124-126.

Kaushal, R., Tewari S., Thapliyal SD., Kumar, A., Roy, T., Islam, S., Lepcha, STS., Durai, J. 2021. Build-up of Labile, non-labile carbon fractions under fourteen-year-old bamboo plantations in the Himalayan foothills. Heliyon 7(3):e07850

Kumar, A., Dwivedi, G. K., Tewari, S., Sah, V. K., Singh, H., Kumar, P., Kumar, N., Kaushal, R. 2020a. Soil organic carbon pools under *Terminalia chebula* Retz. based agroforestry system in Himalayan foothills, India. Current Science (00113891), 118(7).

Kumar, A., Dwivedi, G.K., Tewari, S., Paul, J., Anand, R., Kumar, N., Kumar, P., Singh, H. and Kaushal, R. 2020b. Carbon Mineralization and Inorganic Nitrogen Pools under *Terminalia chebula* Retz.-Based Agroforestry System in Himalayan Foothills, India. Forest Science. 66(5), 634-643

Kumar, A., Kumar, P., Singh, H. and Kumar, N., 2021a. Impact of plant functional traits on near saturated hydraulic conductivity of soil under different forests of Kempty watershed in Garhwal Himalaya. Indian Journal of Soil Conservation, 49(1), pp.38-44.

Kumar, A., Kumar, P., Singh, H., Kumar, N., 2021b. Adaptation and mitigation potential of roadside trees with bio-extraction of heavy metals under vehicle emissions and their impact on physiological traits during seasonal regimes. Urban Forestry & Urban Greening, 58, p.126900.

Kumar, A., Kumar, P., Singh, H., Kumar, N., 2021c. Modulation of Plant Functional Traits under Essential Plant Nutrients during Seasonal Regime in Natural Forests of Garhwal Himalayas. Plant and Soil 465(1):1-16

Kumar, A., Kumar, P., Singh, H., Kumar, N., 2021d. Relationship of physiological plant functional traits with soil carbon stock in the temperate forest of Garhwal Himalaya. Current Science 120(7): 1064-1068.

Kumar, A., Singh, H., Kumari, G., Bisht, S., Malik, A., Kumar, N., Singh, M., Raturi, A., Barthwal, S., Thakur, A. and Kaushal, R., 2022. Adaptive resilience of roadside trees to vehicular emissions via leaf enzymatic, physiological, and anatomical trait modulations. Environmental Pollution, p.120191.

Kumar, A., Tewari, S., Singh, H., Kumar, P., Bisht, S., 2021e. Biomass accumulation and carbon stock in different agroforestry systems prevalent in the Himalayan foothills. Current Science 120(6), 1083-1088

Kumar, A., Tewari, S., Singh, Indra., Panday, R., Kumar, D. and Anand, R. 2018. Effect of nutrient sources on growth, yield and quality of turmeric under harad (*Terminalia chebula*) based agroforestry system Indian J. of Agroforestry, 20(2). 1-6

Kumar, J.; Vengaiah, P. C.; Srivastava, P. P. and Bhowmik, P. K. (2012). Chironji nut (Buchanania lanzan) processing, present practices and scope. Indian J. Traditional Knowledge, 11(1): 202-204

Kumar, N., Jeena, N., Kumar, A., Khwairakpam, R., Singh, H. 2021. Comparative response of rice cultivars to elevated air temperature in Bhabar region of Indian Himalaya: Status on Yield attributes. Heliyon 7(6):e07474

Kumar, N., Kumar, A., Jeena, N., Singh, R., and Singh, H. 2020. Factors Influencing Soil Ecosystem and Agricultural Productivity at Higher Altitudes. Microbiological Advancements for Higher Altitude Agro-Ecosystems & Sustainability. Springer. p. 55- 71.

Kumar, N., Singh, H., Giri, K., Kumar, A., Joshi, A., Yadav, S., Singh, R., Bisht, S., Kumari, R., Jeena, N., Khairakpam, R., and Mishra, G. 2024. Physiological and molecular insights into the allelopathic effects on agroecosystems under changing environmental conditions. Physiology and Molecular Biology of Plants. 248. 119255.

Kumar, N., Verma, S., Kumar, A., Dashila, H., Suyal, D.C., Kumari, G., Rawat, S., Jeena, N., Singh, M., Kumar, A. 2022. Molecular insights into stress-responsive genes in the mitigation of environmental stresses. Unravelling Plant-Microbe Synergy, Edition: 1, Publisher: Elsevier

Lyngdoh N, Ravikanth G, Kumar M, Pandey AK. 2016. Parkia timoriana based agro-forestry system in North-East India. Environment and Ecology 34 (4A): 1950-1952

Michel, A. 2002. Tree, Shrub and Liana of West African Zones. Margraf Publishers, GMBH, Paris, p. 440.

Morton, J. 1987. Indian Jujube. In: Fruits of warm climates, Julia F. Morton (Ed.), Miami. Department of Horticulture and Landscape Architecture at Purdue University, pp. 272-275.

Nair, P.K.R., Kumar, B.M., Nair, V.D. 2021. Multipurpose Trees (MPTs) and Other Agroforestry Species. In: An Introduction to Agroforestry. Springer, Cham. https://doi.org/10.1007/978-3-030-75358-0_13

Pandey R., Dubey S., Tewari L, Kumar A., Kumar D., Sharma N. and Tewari S.K. 2018. Performance of soybean under poplar based agroforestry system in Terai region. Indian J. of Agroforestry, 20(1), 31-34.

Pinakin, D.J., Kumar, V., Kumar, A., Gat, Y., Suri, S. and Sharma, K., 2018. Mahua: A boon for the pharmacy and food industry. *Current Research in Nutrition and Food Science Journal*, *6*(2), pp.371-381.

Purohit, S. S. and Vyas, S. P. 2004. In: Aegle marmelos Correa ex Roxb. (Bael), Medicinal Plant Cultivation- A Scientific Approach. Agrobios, pp. 280-285.

Raghavan, M. and Ramjan, M., 2018. Burmese grape (*Baccaurea ramiflora* Lour.): A promising fruit crop for future generations. *Journal of Medicinal Plants Studies*, *6*(3), pp.50-52.

Rakesh, Dhawan, S.S., Arya, S.S. 2005. Product Development: Processed Products of Bael. Processed Food Industry. 8(10): 25-27

Rizvi, M.K., Rabail, R., Munir, S., Inam-Ur-Raheem, M., Qayyum, M.M.N., Kieliszek, M., Hassoun, A. and Aadil, R.M., 2022. Astounding health benefits of jamun (Syzygium cumini) toward metabolic syndrome. *Molecules*, *27*(21), p.7184.

Siddiqui, M. Z.; Chowdhury, A. R.; Prasad, N. and Thomas, M. (2014). Buchanania lanzan: a species of enormous potential. World J. Pharm. Sci., 2(4): 374-379.

Tewari, S., Tewari, L., Dubey, A., Kumar, A., Kumar, N., Kaushal, R. 2022. Use of the RAPD marker to determine the genetic diversity of various *Dalbergia sissoo* Roxb. (Shisham) genotypes and their evolutionary relationship in Indian subcontinents. Vegetos,

Tewari, S., Tewari, L., Kumar, A., Bisht, S, Kaushal, R. and Kumar, A. 2024. Germination Behaviour of Teak Genotypes and Effect of Different Treatments on Seed Germination Characteristics in Himalayan Foothills, India. Indian Forestry, 150 (3): 244-254

Tewari, S., Tewari, L., Kumar, A., Kaushal, R. 2021. Performance of *Glycine max* L. varieties under *Tectona grandis* L. and *Dalbergia sissoo* Roxb. based agroforestry system Indian J. of Agroforestry, 20(2) 1-6

Tripathi, P.C., Karunakaran, G., Sankar, V. and Senthil Kumar, R., 2014. Karonda-A Potential fresh fruit of the future.

Verma, T., Bhardwaj, D.R., Sharma, U., Sharma, P., Kumar, D., Kumar, A. and Kumar, A. 2023. Aroforestry systems in the mid-hills of the north-western Himalaya: A sustainable pathway to improved soil health and climate resilience. Journal of Environmental Management. 348, 119264.

CHAPTER EIGHT

INTEGRATING LIVESTOCK AND TREES FOR ECOSYSTEM RESILIENCE

AKANKSHA BISHT[1] AND
PRAVEEN KUMAR SINGH[2]
[1]SAM HIGGINBOTTAM UNIVERSITY OF AGRICULTURE, TECHNOLOGY AND SCIENCES, PRAYAGRAJ, (U.P.)
[2]FACULTY OF AGRICULTURAL SCIENCES, MAHAYOGI GORAKHNATH UNIVERSITY GORAKHPUR, (U.P.)

Abstract

Integrated crop-livestock production increases total farm productivity by enhancing nutrient cycle and energy efficiency, which benefits the ecosystem and improves diversification. Biological, temporal, or geographical diversification of agroecosystems has been found to be an essential strategy for resilience building through the preservation of ecosystem services and cascade impacts across scales. By utilizing the synergies between various agroecosystem components and the added complexity resulting from trophic levels (i.e., grazing animals), the integration of livestock into a crop production system intensifies crop yield and improves soil quality instead of depending on increased conventional input levels like fertilizers. Improved soil structure, microbial functional diversity, increased nutrient cycling, and increased organic matter in cattle-grazed fields are examples of ecosystem processes that show enhanced biophysical protection against unfavorable ecological conditions. These processes need to be further examined and compared to native grassland systems. The integrated system offered a productivity buffer against long-term climate stress, but the kind and timing of disturbances determined how resilient it was to yearly weather anomalies.

Introduction

It is recognized that livestock and trees produce a variety of ecosystem services, such as provisioning, supporting, regulating, and cultural (Henderson et al., 2016; Ronnback *et al.*, 2007). On farms, crop resources (land, water, nutrients, and labor) may face competition from both trees and cattle. Therefore, further investigation and quantification of their distinct roles are necessary (Mutoko *et al.*, 2015). Small landholders in underdeveloped nations with little resources dominate the livestock industry. The way animals are raised has a negative impact on the environment; this is partly due to the fact that animal husbandry is not connected with other forms of agriculture and forestry. Thus, achieving the Sustainable Development Goals depends on maintaining and growing the livestock industry. The cultivation of seasonal crops and perennial trees has been separated from livestock raising, resulting in a decoupling of the biogeochemical and biogeophysical cycles of carbon (C), water (H_2O), nitrogen (N), phosphorous (P), and sulfur (S).

Degradation of rangelands, reduction in biodiversity, eutrophication and contamination of water supplies, and a rise in N_2O and CH_4 emissions are all caused by this decoupling. Therefore, minimizing the environmental impact of animal-based food items requires designing and implementing systems that combine cattle with crops and trees. Manure and other animal waste should be handled carefully, animals should be fed with precision their dietary protein needs matched, and CH_4 and N_2O emissions should be processed for other purposes. Agroforestry is a collection of technology that allows trees to be combined with crops, livestock, and/or other elements of a wide range of integrated systems, either simultaneously or sequentially. A food crop is produced in the alleyways between the rows of trees in an alley cropping system, which involves planting trees on the contour at a wide spacing of 4–10 meters. Several rows of shrubs and trees that double as feed can be planted as a way to integrate crops and trees with animals. It is also possible to harvest trees for fuel wood. An agro-silvopastoral system is complicated (Okali and Sumberg, 1985; Kang *et al.*, 1990).

Tree species used in temperate alley cropping systems may include those used for hardwood veneer or lumber, softwood species for the production of fiber, or fruits and nuts for food. In strip cropping, trees planted on the contour can also be employed as a contour farming filter strip. When the trees are still young, grain crops such as corn, soybeans, and cowpeas are planted. Forages can be gathered, cattle can graze, and the trimmings can be utilized as green manure for cereals (corn, for example) when the ground is shaded. Leguminous trees improve soil fertility by providing a source of nitrogen. Restoring rangeland for the purpose of storing atmospheric CO_2 as soil organic matter, cutting down on the use of antibiotics, minimizing gas swapping, and adopting diverse GHG viewpoints are all crucial factors to take into account. While crops are primarily planted for domestic use or sales, trees, and cattle can serve as a kind of insurance; in the event of a crisis, a tree or some livestock can be quickly sold for cash (Van der Ploeg et al., 2009). However as insurance, farmers frequently favor cattle over trees, and others argue that the income from livestock is a prerequisite for investing in agroforestry or trees (Jerneck and Olsson, 2013). Livestock was essential to traditional farms because it kept the land fertile and allowed them to use plant material that would have otherwise been squandered. India serves as an illustration of how integrated systems might decrease the amount of land used for pasture. India, which accounts for 2.3% of the world's land area, is home to 18% of the world's people population and 11% of its livestock population, which as of 2019 consisted of 536 million animals and 740 million fowl maintained on just 12.3 million hectares of land used for grazing and permanent pastures. Livestock can improve the organic matter and fertility of the soil, diversify the product base, offer new revenue streams increase farm resilience, and aid in weed and insect management when added to a crop farm (vegetable, grain, or orchard). The cost of fuel, chemicals, and fertilizers can be reduced while giving the farm a fresh aspect. On the other hand, conventional farming frequently faces issues with low productivity, soil erosion, and wasteful resource use. By utilizing ecosystem services (such as soil structure, nutrient cycling, and rangeland restoration) and ecosystem resilience mechanisms such as soil carbon accrual, sustainable intensification of trees seeks to decrease the rate at which tree land area expands while also improving resource efficiency (Sanderson *et al.*, 2013).

The practice of integrating tree species and animals into small agricultural systems is a prevalent method of diversifying productivity in tropical regions, particularly for smaller farms. Additionally typical are intensive silvopastoral systems. Trees, enhanced pastures, and high densities of fodder bushes are all combined in intensive silvopastoral, a sustainable agroforestry method for cattle production. For instance, trees range in density from fewer than three to more than fifty trees per hectare on 26 to 69 percent of the pastures of each farm located in the Caribbean area of Colombia. Larger trees, including *Albizia caribae* and *Tabebuia rosea*, give farmers shade and lumber, while medium-sized trees, like *Albizia saman* and *Guazuma ulmifolia*, give cattle fodder (Devandra and Ibrahim, 2004).

By recognizing and capitalizing on the interconnectedness of crops and animals, farmers may boost productivity, safeguard ecosystems, and improve the quality of life for people living in rural areas (UN, 2019). To meet the challenge of expanding food production while minimizing negative environmental effects and safeguarding finite resources, creative solutions are required. Among them, combining trees and cattle is one approach that has the potential to assist developing countries in overcoming challenges and advancing sustainable land management. Although farms and agricultural landscapes are coupled social-ecological systems that require holistic assessment of resilience at multiple scales and across domains, the development of simplified, biophysical proxies of resilience and adaptive capacity—that is, the ability to maintain function by changing in response to disturbance—represent important attempts to make the concept of resilience operational for intensive production systems. Traditionally, biodiverse integrated crop and livestock systems have followed a trend of increasing specialization over the last few decades (Garrett *et al.*, 2020). In addition, by using these proxies, planning, and management can be guided by a deeper comprehension of the underlying mechanisms of resilience in these systems. It has been determined that agroecosystem diversification—whether biological, temporal, or spatial—is an essential mechanism for enhancing resilience by preserving ecosystem services and producing cascade effects across scales (e.g., Lin, 2011; Renard and Tilman, 2019). Changes in the biological, physical, and chemical properties of soil, the cycling of water and nutrients (Assmann *et al.*, 2015; Martins *et al.*, 2016), and plant-soil interactions have all indicated that recoupling of specialized crop and livestock operations as integrated crop-livestock systems has shown promise as a sustainable intensification strategy (Peterson *et al.*, 2020). Integrated farming employs a more innovative and holistic approach than traditional farming methods, which often divide crop and animal husbandry (Ziegler *et al.*, 2009). The implementation of sustainable grazing practices within integrated systems may be able to preserve grassland biodiversity, stop soil erosion, and enhance biological health. Increased farming system diversity, such as combining crops with animals and trees, is thought to improve the long-term and steady delivery of ecosystem services. As much research has been done on ES, not much of it has compared or recognized the roles that livestock and trees play in ES when taking smallholder systems' farm priorities into account.

1. Integrated Tree-livestock systems (ITLS) contribution to food security

In the tropics, livestock play a significant role in agroecosystems. By using creative livestock and agricultural techniques, one can increase productivity while lowering environmental impact. In addition to increasing food security and nutrition parameters like food consumption score and household dietary diversity, the varied cropping systems in ITLS can raise the production of the main crop (Romeo et al., 2016). Crop leftovers make up the majority of animal feed in the majority of ITLS. By producing more crop residues under ITLS, livestock can be fed without having to compete with humans for food (Blummel, 2016). Human health can be improved by increasing the protein content of vegetation and improving animal diets through the use of cover crops and nitrogen-fixing crops in ITLS diversity (Wilkins, 2007). Numerous economic, ecological, and other advantages result from the successful integration of crops and animals, particularly in emerging tropical nations. Notable among these are: giving farmers an additional source of income and reducing rural poverty; building a safety net for the underprivileged, particularly women farmers, increasing farmer assets, and reducing malnourishment. Livestock, however, require more space, water, nutrients, and pasture supplies. Thus, it's imperative to carefully control this sector's expansion, particularly in order to minimize environmental footprints. According to Andersen (2007), the goal of the ITLS is to promote sustainable development and ensure that these systems are socially, economically, and environmentally sound. Sheep and goats are raised as capital stock in small-ruminant agriculture, which has been seen as an income-diversification approach that can also generate income in the event of crop failure. Livestock is frequently regarded as a valuable asset in impoverished rural homes. Having more cattle tends to improve food and nutritional security by enabling impoverished households to invest in small businesses, diversify their sources of income, and become less impoverished. The production of livestock and food crops can both steadily rise in marginal areas when there is interaction between the two. Tree crowns generate microclimatic niches that can enhance crop diversification with edible and/or non-edible goods that can be sold to improve farmers' income and enhance their diet (DeSouza *et al.*, 2012).

The profitability of ITLS can be significantly better than that of specialized crop or livestock production. However, the stability of markets, technologies, social cultures, infrastructure, labor availability, regulations, and biophysical and climate issues all affect how beneficial ITLSs are. The cooperative relationship between cattle and crops presents a number of chances to improve production and the effective use of resources, which raises household incomes and ensures food supply and accessibility.

2. Climate resilience through integrated tree-livestock systems

In order to adapt to and mitigate the effects of climate change, grasslands need to be restored and managed sustainably. According to Yanez-Ruiz *et al.* (2018), atmospheric CH_4 and NH3 emissions can be decreased by employing pertinent animal feeding techniques, such as using protein-rich forages and flax seeds. Extreme weather events and human activity are more likely to occur in arid and semiarid locations. Shorter growing seasons, fewer fertile lands for farming, and decreases in agricultural productivity could result from this. A greater reliance on food imports from other nations, increased poverty, food and nutrition insecurity, and unstable food production are all effects of climate change. In order to adapt to and mitigate the effects of climate change in the livestock industry, governmental interventions that eliminate obstacles to the adoption of proven technology are necessary (Smith *et al.*, 2007). Because of their biologically delicate and harsh climatic conditions, emerging nations must prioritize appropriate policy interventions in order to achieve sustainable rangeland management. For instance, enteric fermentation and animal feces in India are projected to produce 9–10 Tg (teragram = 1 million tons) of total yearly CH_4 emissions. These emissions can be decreased by suitable policy interventions, such as paying for the provision of ecosystem services (Sirohi and Michaelowa, 2007). Integrated systems can improve resistance to the effects of climate change. Improved soil health and a diversity of output and revenue streams can help farms become more resilient to climatic changes like altered rainfall patterns and extreme weather. Reduced rangeland productivity, a rise in disease cases, land demand, and stress on water resources are probably the main issues facing cattle. ITLS can shield any industry from outside pressures resulting from shifting climatic conditions and from optimal social, economic, and environmental circumstances in areas with limited resources. Compared to a specific crop or animal production, crop combinations incorporated with a variety of livestock species can produce a larger yield resilience. ITLS enables flexibility in the selection of crops, feed, and livestock husbandry. It can also be enhanced by introducing new technologies in response to evolving circumstances. A thorough grasp of the long-term difficulties of climate change is crucial for successful management, since it may potentially have an impact on the sustainability of livestock production. Through buffering mechanisms in field- and farm-level processes, such as increased crop yield, nutrient cycling, economic risk mitigation, and livelihood diversification, ITLS promotes agricultural resilience to climate change. The close connection between livestock and human-caused climate change cannot be disregarded or dismissed. It is proven that animals have an impact on climate change, and animals are also being impacted by the fast-changing climate. In this particular setting, the integration of cattle with crops and trees has the potential to significantly contribute to the planet's re-greening efforts (Janzen, 2011). Reducing the environmental impact of farming and managing crops and trees sustainably are two outcomes of utilizing the benefits of livestock-based farming systems, including nutrient-rich food and the eradication of malnutrition and hunger (Herrero *et al.*, 2009). Furthermore, the provision of ecosystem services from these delicate, environmentally sensitive, yet commercially significant ecoregions can be strengthened through the adoption of ecologically based animal husbandry methods in rangeland management (Havstad *et al.,* 2007).

As organic carbon sinks, trees in silvopasture systems sequester five to ten times more carbon than pastures devoid of trees. Reducing carbon dioxide emissions by 26–42 Giga tonnes per year might be achieved by extending silvopasture practices to 750 million hectares. Agroforestry systems can be efficient carbon sinks that considerably reduce the amount of carbon in the atmosphere when they are properly designed and managed. Agroforestry is also a useful tactic for improving local livelihoods and lessening the strain on natural forests because it provides goods and services (such as fuelwood and timber) that would otherwise be derived from forests. All species in an agroforestry system are impacted by temperature, precipitation, and ambient carbon dioxide concentrations, albeit potentially in very diverse ways. All of these variables are expected to vary in response to climate change, thus the sequence and geographical arrangements of the various system components will need to be gradually changed. At the collective level, fewer unproductive or underperforming animals can lead to increased efficiency and decreased emissions. For instance, if the average age at first calving was lowered from 4 to 3 years, farmers would need to retain fewer non-productive heifers in order to keep the stock together. Enhancing animal diets by substituting more digestible feed (such as pasture, agricultural leftovers, fodder trees, and bushes) for roughages, or by processing the roughages (such as breaking them up or treating them with urea).

3. Crop–livestock integration enhanced Soil Quality, Soil Aggregate

Trees grown in silvopasture have large root systems that help improve soil quality, stability, and nutrient cycling while also preventing erosion. Compared to open pastures, silvopasture systems have higher soil infiltration rates, which increases their capacity to store water and aids in water conservation initiatives. Integrated tree-livestock systems can improve overall farm profitability, reduce weeds and pests, improve nutrient cycling due to livestock rotation, prevent the buildup of harmful nutrient levels from animal manure, and facilitate diversification, all of which have positive effects on the environment and soil health. Crop-livestock integrated systems have an impact on the chemical, physical, and biological characteristics of soil.7, 8. Manure from livestock can be used as a natural fertilizer for crops because it is an organic matter produced by the animals. This promotes sustainable soil health and fertility, closes nutrient loops, and lessens reliance on artificial chemical fertilizers. In return, livestock can use crop wastes and byproducts as feed and forage, reducing waste and optimizing resource use. Removing animals from crops and trees can have negative effects on the ecosystem and contribute to a reduction in farm-scale variety in certain climates and geographies. The decoupling of the cycles of carbon (C) from those of water (H_2O), nitrogen (N), phosphorus (P), and sulfur (S) is another consequence of this simplification and biodiversity loss (Lal, 2010).

The decoupling of land use and land use change cycles associated with livestock may occur from this simplification of the farming system. Concentrated ruminants release more GHGs (such as CH_4) and tend to decouple the C and N cycle by releasing digestible C in the form of CO_2 and CH_4 and digestible N in the form of N_2O in waste (Soussana and Lemaire, 2014). Integrating cattle with crops and trees might reduce the dangers of uncoupling, which has serious implications for climate change due to the high global warming potential of CH_4 and N_2O. Livestock reared on the same land unit can have a smaller environmental impact if practices like agroforestry or alley cropping, which create vegetation buffers on agricultural fields to improve biodiversity and conserve soil and water, are used (Goldstein *et al.*, 2012).

In order to better moderate coupled biogeochemical cycles and reduce pollution fluxes into the atmosphere and hydrosphere, as well as to create a more diversified and structured landscape mosaic that supports a variety of habitats, site-specific technologies for integrating livestock with crops and trees are required. Additionally, these technologies will improve the system's ability to adapt to extreme events linked to climate change and changes in the socio-economic and human dimensions. By recycling nutrients from the soil, trees help annual crops develop by reducing the demand for fertilizers and enhancing soil fertility and crop water productivity. Crop yields in agroforestry systems are similar to (and more steady than) those in specialized production systems using synthetic fertilizers (Hall *et al.,* 2006). There are likewise fewer greenhouse gas emissions. Many agroforestry techniques use the biological nitrogen-fixation process of leguminous trees to provide organic matter and nitrogen to both annual and perennial crops (Sileshi *et al*., 2014). As trees provide shelter from strong sunshine, crop yields in agroforestry systems are more stable. In the long term, this promotes a more stable microclimate and aids in soil moisture retention. Crops can also be shielded from wind and hail damage by tree crowns. The benefits of animal excrement and urine on soil fertility, the advantages of crop and tree residues on animal diets, and the advantages of tree shelter and shade on animal productivity are all examples of the relationships between livestock, trees, and crops. Allowing cattle to use the vegetative ground cover beneath the tree canopy boosts overall productivity, lowers the need for and expense of weed control inputs, and generates additional revenue from the production of meat and/or milk. Trees act as a buffer against climatic unpredictability, control soil temperature, and moisture, enhance water infiltration following heavy precipitation, and create a variety of ecological niches that promote the growth of various crops. Additionally, some trees can lessen the impacts of drought. Because of the shade the tree canopy provides, transpiration from the trees is typically higher than soil evaporation. Certain trees have the capacity to extract water from deeper soil layers, allowing the extra water to seep into the soil profile's upper layers and become available to plants with shallower root systems. The most effective use of natural resources, such as land and sunlight, is made possible by the intentional use of trees, shrubs, and herbaceous crops in certain spatial arrangements that take advantage of their architecture and sunlight requirements (i.e., shade-tolerant and light-demanding). Commodity diversification lowers yield loss risks, particularly those brought on by the effects of climate change, and enables adjustments in reaction to labor availability and market demands.

The farming system is more resilient to unfavorable weather occurrences when trees or bushes are grown outside of fields to block wind and prevent soil erosion. According to Tamang et al. (2010), windbreaks planted in citrus orchards, for example, can lower wind speeds by 80 to 95 percent and wind damage to crops by up to twice the distance of the windbreak height. Trees stop landslides and soil erosion during the rainy season, especially in hilly locations.

The total amount of carbon that can be stored in soil and standing woody biomass. The quantity of greenhouse gases emitted from agricultural activities, originating from both direct sources (like fuel burning in mechanized systems) and indirect sources (such as soil disturbances that lead to the mineralization of soil organic carbon and the manufacturing and transportation of agrochemicals). For instance, planting leguminous trees (such as *Gliricidia septum*) improves the amount of nitrogen that is available in the soil, reducing the need for artificial fertilizers. In addition to indirect emissions from ammonia volatilization, nitrate leaching, drainage, and runoff losses, as well as carbon dioxide emissions from manufacture and transportation, nitrogen fertilizers are linked to direct emissions of nitrous oxide.

4. Supporting the Intensification and Diversification of Sustainable Production

Depending on how crop residues, feed, and manure are managed, the interchange of energy, manure, and feed between the productive components of integrated crop-livestock systems can increase the efficiency of the carbon and nitrogen cycles. Positive environmental, social, and economic synergies may result from this. Livelihood diversification increases the economic resilience of integrated farming systems in climate change-affected environments by mitigating the risk of losses from particular biotic or abiotic stresses that could affect exotic livestock breeds or genetically uniform crop monocultures. It is possible to prevent parasitism and emerging diseases that are common in animals housed in large units by keeping livestock in smaller units and possibly combining different species. Whenever possible, use pastures in crop rotation to boost seed dispersal and the number of plant species—especially pollinator species—that provide habitats for biodiversity in the wild. Shortening grasslands can also lessen avalanches, bushfires, and erosion. Optimizing labor and other production inputs helps farmers save money, recycle organic matter and nutrients, and boost productivity (Seo, 2010). When it comes to pest control, employing animals to weed has a beneficial impact on the environment because it uses fewer herbicides and their excrement makes organic fertilizer. Goats grazing on sugar cane fields, for instance, eliminate weeds and dry the leaves of the crop, requiring less labor for harvesting and weed control. Animal numbers and the amount of time they spend in fields must always be managed to prevent soil compaction from trampling and overgrazing. It should be mentioned that organic livestock dung may not always be accessible in the necessary proportions for the control of nutrients. For this reason, when maximizing input-use efficiency, integrated soil fertility management—which combines organic inputs with synthetic nutrients given in the right amount and at the right time—as well as sound agronomic techniques in general, should never be disregarded.

5. Enhanced Forage Production, Conservation of Biodiversity, and Sustainable Land Use

As trees contribute more biomass and leaf litter, which enriches the soil with nutrients and organic matter, silvopasture improves the yield of feed. The microclimate that trees produce encourages the establishment of forage species that can withstand shadow, thereby raising total forage yields. Because silvopasture systems naturally cycle nutrients, they require fewer synthetic fertilizers, which makes them a sustainable method of managing soil fertility. Landscapes with silvopasture produce a variety of ecosystems that are home to a large number of plant and animal species. This strengthens ecological resilience and encourages the protection of biodiversity. The conservation of native plant species in silvopasture settings supports the preservation of indigenous knowledge about customary agroforestry methods as well as the local biodiversity. Pollinators and beneficial insects are drawn to silvopasture systems, which improve the health of the ecosystem overall and help to pollinate crops. Systems of silvopasture produce a variety of habitats that promote the preservation of biodiversity and the resilience of ecosystems. By integrating cattle, forage, and trees, the diverse silvopasture method promotes sustainable land use practices and builds resilient agricultural landscapes for the future.

6. Increased Livestock Productivity

While silvopasture systems offer shade and protection from harsh weather, livestock in these systems enjoy higher health and welfare. A balanced diet for cattle is ensured by the availability of a variety of forage options from both grasses and trees, which improves weight gain and reproduction rates. Extended grazing times made possible by silvopasture improve feed use and lessen the strain on conventional grazing pastures. Silvicopastories offer a potential solution to the two concerns posed by climate change: reduced feed supply and increased heat stress on cattle. Due to its ability to buffer the local environment, silvopasture may serve as a "middle ground" between managed open fields and intact woods, maintaining the benefits of a cooler climate while allowing for the spread of agriculture.

Conclusions

The adoption of an integrated trees and livestock system for a farming operation can guarantee significant income generation and diversified food to sustain farmers' livelihoods if effectively managed. Including animals with trees can lower economic risks, support the earnings of smallholder subsistence farmers, and affect the variety of household diets. Reintegrating livestock and crop systems has several potential advantages, including increased socioeconomic output and better environmental conditions. Adopting Integrating trees with livestock, however, necessitates a substantial commitment as well as expertise of both animals and trees. The coexistence of cattle and trees on farms produces nutrient-dense food that is good for human health and plays a significant role in social and cultural aspects by ensuring food security. Government policies must encourage capital expenditures, infrastructure development, and on-the-ground training, particularly for small and marginal farmers. They also need to pay greater emphasis on integrating small ruminants, high-protein crops, fruits, and vegetables into the ITLS system.

By combining livestock with trees, farming operations might potentially achieve economic sustainability through increased productivity, diversified revenue streams, and the utilization of ecosystem services. Tailored approaches to agroforestry implementation are crucial, as the success of particular tree-crop combinations depends on variables such as local climate, soil conditions, market demand, and farmer preferences. Implementing and maintaining agroforestry systems presents farmers with a number of difficulties, including the labor and initial investment needed, choosing the right tree-crop combinations, and the unpredictability of land tenure and policy. Adoption of agroforestry is also hampered by problems like the lack of suitable land, careless tree cutting, and inadequate government assistance.

Additionally, animals and trees can improve ecosystem services and environmental sustainability, which benefits the entire community. Trees in agroforestry systems increase ecosystem health and environmental quality for all community members by retaining water, preserving biodiversity, and conserving soil. Around the world, a lot of neighborhood-based programs support sustainable agroforestry techniques. In order to promote the use of agroforestry techniques, these programs frequently entail community members' sharing of knowledge, training, and capacity building. They might also grant access to resources like tools, seeds, and technical support to aid with implementation. Farmer cooperatives, community forestry projects, and agroecology networks are a few examples of such programs that enable local communities to sustainably manage their natural resources.

References

Assmann, J. M.; Anghinoni, I.; Martins, A. P.; Costa, S. E. V. G. A.; Kunrath, T. R.; Bayer, C. 2015. Carbon and nitrogen cycling in an integrated soybean-beef cattle production system under different grazing intensities. *Pesqui.Agropecu.Bras.* 50, 967–978. doi: 10.1590/S0100-204X2015001000013.

Blummel, M. 2010. Food feed crops, *Anim. Nutr. Feed Technol.*, 10.

Bowles, T.M.; Mooshammer, M.; Socolar, Y.; Calderón, F.; Cavigelli, M.A.; Culman, S.W.; Deen,W.; Drury, C.F.; Garcia, A.; Gaudin, A.C.; Harkrom, W.S.; Lehman, R.M.; Osborne, S.L.; Robertson, G.P.; Salerno, J.; Schmer, M.R.; Strock, J.; and Grandy, A.S. 2020. Long-term evidence shows that crop rotation diversification increases agricultural resilience to adverse growing conditions in North America. *One Earth* 2:284-293.

DeSouza, H.N.; DeGoede, R.G.M.; Brussaard, L.; Cardoso, I.M.; Duarte, E.M.G.; Fernandes, R.B.A.; Gomes, L.C. and Pulleman, M.M. 2012. Protective shade, tree diversity and soil properties in coffee agroforestry systems in the Atlantic Rainforest biome. *Agriculture, Ecosystems & Environment*, 146: 179-196.

Devendra, C.; and Ibrahim, M. 2004. Silvopastoral systems as a strategy for diversification and productivity enhancement from Livestock in the tropics. In: The Importance of Silvopastoral Systems in Rural Livelihoods to Provide Ecosystem Services. Proc. of the Second International Symposium of Silvopastoral Systems. Yucatán, México. pp. 10-24.

Hall, N.M., Bocary, K.; Janet, D., Ute, S.; Amadou, N. and Tobo, R. 2006. Effect of improved fallow on crop productivity, soil fertility and climate-forcing gas emissions in semi-arid conditions. *Biology and Fertility of Soils*, 42: 224-230.

Havstad, K. M.; Peters, D. P. C.; Skaggs, R.; Brown, J.; Bestelmeyer, B.; Fredrickson, E. 2007. Ecological services to and from rangelands of the United States. *Ecol. Econ.* 64, 261–268.

Henderson, B.; Godde, C.; Medina-Hidalgo, D.; Van Wijk, M.; Silvestri, S.; Douxchamps, S.; Stephenson, E.; Power, B.; Rigolot, C.; Cacho, O. and Errero, M.H. 2016. Closing system-wide yield gaps to increase food production and mitigate GHGs among mixed crop-livestock smallholders in Sub-Saharan Africa. *Agric. Syst.*, 143, pp. 106-113.

Herrero, M.; Thornton, P. K.; Gerber, P.; and Reid, R. S. 2009. Livestock, livelihoods, and the environment: understanding the trade-offs. *Curr. Opin. Environ. Sustain.* 1, 111–120.

Garrett, R. D.; Ryschawy, J.; Bell, L. W.; Cortner, O.; Ferreira, J. 2020. Drivers of decoupling and recoupling of crop and livestock systems at farm and territorial scales. *Ecol. Soc.* 25:24. doi: 10.5751/ES-11412-250124.

Goldstein, J. H.; Caldarone, G.; Duarte, T. K.; Ennaanay, D.; Hannahs, N.; Mendoza, G. 2012. Integrating ecosystem-service tradeoffs into land-use decisions. *Proc. Natl. Acad. Sci.* U.S.A. 109, 7565–7570.

Janzen, H. H. 2011. What place for livestock on a re-greening earth? Anim. Feed Sci. Technol. *J.anifeed.sci.* 166–167, 783–796. doi: 10.1016.

Jerneck, A.; Olsson, L. 2013. More than trees! Understanding the agroforestry adoption gap in subsistence agriculture: insights from narrative walks in Kenya. *J. Rural. Stud.*, 32, pp. 114-125.

Kang, B.T.; Reynolds, L.; and Atta-Krah, A.T. 1990. Alley Farming, *Advances in Agronomy*, vol. 43, 1990, Pages 315-359.

Lal, R. 2010. Managing soils and ecosystems for mitigating anthropogenic carbon emissions and advancing global food security. *Bioscience* 60, 708–721.

Lin, B. B. 2011. Resilience in agriculture through crop diversification: adaptive management for environmental change. *Bioscience* 61, 183–193. doi: 10.1525/bio.2011.61.3.4.

Martins, A. P.; Costa, S. E. V. G. A.; Anghinoni, I.; Kunrath, T.R., Cecagno, D.; Reichert, J. M. 2016. Soil moisture and soybean physiology affected by drought in an integrated crop-livestock system. *Pesqui Agropecu Bras.* 51, 978–989.
doi: 10.1590/S0100-204X2016000800010.

Mutoko, M.C.; Hein, L.; Shisanya, C.A. 2015. Tropical forest conservation versus conversion trade-offs: insights from analysis of ecosystem services provided by Kakamega rainforest in Kenya. *Ecosystem Serv.*, 14, pp. 1-11.

Okali, C. and Sumberg, J. E., 1985. Sheep and goats, men and women: Household relations and small ruminant development in southwest Nigeria, *Agricultural Systems, Elsevier*, vol. 18(1), pages 39-59.

Peterson, C. A.; Deiss, L.; and Gaudin, A. C. M. 2020. Integrated Crop-Livestock Systems Achieve Comparable Crop Yields to Specialized Systems: A Meta-Analysis. *PLoS ONE* 15:e0231840. doi: 10.1371/journal.pone.0231840.

Pinstrup, P. and Andersen. 2007. Agricultural research and policy for better health and nutrition in developing countries: a food systems approach.*Agric. Econ*., 37, pp. 187-198.

Renard, D. and Tilman, D. 2019. National food production is stabilized by crop diversity. *Nature* 571, 257–260. doi: 10.1038/s41586-019-1316-y.

Rönnbäck, P.; Crona, B.; Ingwall, L. 2007. The return of ecosystem goods and services in replanted mangrove forests: perspectives from local communities in Kenya. *Environ. Conserv*., 34 , pp. 313-324.

Romeo, A.; Meerman, J.; Demeke, M.; Scognamillo, A.; Asfaw, A. 2016. Linking farm diversification to household diet diversification: evidence from a sample of Kenyan ultra-poor farmers, *Food Secur*ity., 8, pp. 1069-1085.

Sanderson, M. A.; Archer, D.; Hendrickson, J.; Kronberg, S.; Liebig, M.; Nichols, K. 2013. Diversification and ecosystem services for conservation agriculture: outcomes from pastures and integrated crop-livestock systems. *Renew. Agric. Food Syst.* 28, 129–144.

Seo, S.N. 2010. Is an integrated farm more resilient against climate change? A micro-econometric analysis of portfolio diversification in African agriculture. *Food Policy*, 35(1): 32-40.

Sileshi, G.W., Mafongoya, P. L., Akinnifesi, F.K., Phiri, E., Chirwa, P., Beedy, T., Makumba, W., George, N., Njoloma, J., Wuta, M., Nyamugafata, P. & Jiri, O. 2014. Agroforestry: Fertilizer Trees. *Encyclopedia of agriculture and food systems*, 1: 222-234.

Sirohi, S.; and Michaelowa, A. 2007. Sufferer and cause: Indian livestock and climate change. *Clim. Change* 85, 285–298. doi: 10.1007/s10584-007-9241-8

Smith, P.; Martino, D.; Cai, Z.; Gwary, D.; Janzen, H.; Kumar, P. 2007. Policy and technological constraints to implementation of greenhouse gas mitigation options in agriculture. *Agric. Ecosyst. Environ*. 118, 6–28.

Soussana, J. F.; Allard, V.; Pilegaard, K.; Ambus, P.; Amman, C.; Campbell, C. 2007. A full accounting of the greenhouse gas (CO2, N2O, CH4) budget of nine European grassland sites. *Agric. Ecosyst. Environ*. 121, 121–134.

Tamang, B., Andreu, M.G. & Rockwood, D.L. 2010. Microclimate patterns on the lee side of single row tree windbreaks during different weather conditions in Florida farms: implications for improved crop production. *Agroforestry Systems*, 79: 111-122.

United Nations (UN) .2019. UN World Population Prospects.

Van der Ploeg, J.D.; Laurent, C.; Blondeau, F.; Bonnafous, P. 2009 Farm diversity, classification schemes and multifunctionality *J. Environ. Manag*., 90, pp. S124-S131.

Walker, B. H. 2020. Resilience: what it is and is not. *Ecol. Soc.* 25, 25–27. doi: 10.5751/ES-11647-250211.

Wilkins, R.J. 2007. Eco-efficient approaches to land management: a case for increased integration of crop and animal production systems.*Phil. Trans. Biol. Sci*., 363, pp. 517-525.

Yáñez-Ruiz, D. R.; Morgavi, D.; Misselbrook, T.; Melle, M.; Dreijere, S.; Aes, O. 2018. Mini-paper: *Feeding Strategies to Reduce Methane and Ammonia Emissions. EIP-AGRI Focus Group: Reducing Livestock Emissions From Cattle Farming.*

Ziegler, J.; Golay, C.; Mahon, C. 2009.Peasant Farmers and the Right to Food: A History of Discrimination and Exploitation, Human Rights Council Advisory Committee. A/HRC/AC/3/CRP 5.

CHAPTER NINE

INTEGRATION OF AGRONOMIC INTERVENTIONS FOR SOIL AND WATER CONSERVATION

K. DUJESHWER

DEPARTMENT OF RURAL DEVELOPMENT AND AGRICULTURAL PRODUCTION, NORTH-EASTERN HILL UNIVERSITY, TURA CAMPUS, MEGHALAYA-794002

Abstract

Soil and water erosion has been the biggest challenge for the current agro-ecosystem in India and in the world. The exploitation of natural resources is a key cause of the land degradation. Degradation of the soil can have social and economic repercussions in addition to ecological ones. Farmers, especially those in developing nations where agriculture is their primary source of income, may experience decreased income and food insecurity as a result of lower crop yields and decreasing agricultural productivity. In looking for a sustainable natural-based solution to overcome this issue, agronomic intervention plays a pivotal role. Several agronomic practices like conservation tillage, soil organic mulching, cover crops, crop rotation, alley cropping, agro-forestry interventions, contour cropping, etc. are playing significant roles in combating the soil degradation problem. These interventions are well known and adopted among the farming communities due to their cheap management cost. In the severely affected area, an integrated farming system approach including agroforestry, and permanent vegetation cover are suggested as sustainable land practices. In addition, these practices increase the agricultural output per unit land area and diversify farmers' source of revenue, these practices also protect soils from erosion-causing agents and improve agricultural production while addressing land degradation. In this chapter, the role of agronomic interventions has been explored for better management of land degradation and sustainable soil and water conservation in agroecosystems.

Keywords: Sustainable adoption, Agronomic intervention, land degradation

Introduction

Land degradation, for a developing country, is a great challenge and needs immediate attention, especially for a country like India. Nearly, 30% (120.7 MHA) of the geographical area is already been affected by land degradation, especially the resources that reach the North-Eastern states of India (Sengupta 2021; Jinger et al., 2023). Physical, chemical, and biological forms of degradation are prevailing in India, besides the different causes like land use patterns, deforestation, intensive rainfall, uncontrolled grazing, faulty methods of agriculture, abundant use of chemicals in the agricultural land, urban growth, and fragmentation of land, shifting cultivation and mining activities are some important lists playing a major role in causing soil erosion and land degradation (Bhattacharya et al. 2015). The most vulnerable sector affected by degradation and soil erosion is rainfed farmland responsible for food security, and the forest land offers us defense against climate change (Sengupta 2021). There is a significant effect of soil erosion in agriculture, runoff from eroded soil may carry important plant nutrients and agrochemicals. According to an estimate, the combined effect of salinization, soil erosion, and other degradation processes causes an irreversible loss of 5–6 million hectares of arable land every year, worldwide (Hamdy and Aly, 2014). Another estimate says nearly 74 m tonnes of major nutrients are lost due to erosion annually (Kurrey et al. 2016). Furthermore, soil erosion causes approximately 10 million hectares (mha) of cropland worldwide to become unproductive each year at a rate of 30 t/ha/year. An estimate reports, that 28 Pg of sediment are lost globally each year as a result of water erosion (Avasthe et al., 2021). On the other hand, 68% of the total net sown area in India (140 mha) is reported to be affected by drought and severe drought conditions besides the several localized drought situations almost every year (Murthy and Sesha Sai, 2010). In the upcoming years, it is anticipated that this issue will worsen, increasing the scarcity of water for various purposes. The temperature of the atmosphere is rising as a result of global warming and is predicted to rise more in the years to come. Furthermore, when the atmosphere's temperature rises, the rate of evaporation from both land and oceans will increase. For each degree Fahrenheit increase, the atmosphere can contain 4% more water (Triple Pundit, 2017; Tyagi et al., 2020). The type, rate, and severity of soil degradation are influenced by farming and cropping methods, which modify the soil organic carbon status, structural morphology, and other features through various practices like rotations, soil fertility management, erosion control, grazing/stocking rate, and water management. Crop rotations and soil cover in particular can have a big impact on the soil organic carbon pool and the associated soil attributes (Ryan et al., 2008). Among the conservation techniques available to counter land degradation and for soil and water conservation, agronomic practices are crucial because such practices are experienced by the farming community and it is cheaper than other management options.

Need for agronomic practices

Out of many options used for soil and water conservation in agricultural systems, agronomic/biological practices are gaining more popularity among the farming community because they cover a wide range of methods, practices, and procedures that can be implemented in a wide range of agro-ecology, without involving any significant cost. Here are some key reasons that make agronomic practices important for the farming eco-system:

i. Agronomic practice helps to improve soil moisture, soil organic matter, and fertility through various practices such as cover crop, mulching, crop rotation, diversified cropping, green manuring, etc. It also helps to build strong biological activity in the soil which improves soil health and productivity.

ii. Agronomic practices are effective in preventing erosion and loss of topsoil carrying nutrients and salt through various practices like conservation tillage, contour bunds, vegetative strips, cover crops, alley cropping, etc.

iii. Agronomic practices are important to cope with the effects of climate change and global warming. Conservation agriculture and agro-forestry systems are important in soil health management and carbon sequestration. New agro-techniques like integrated nutrient management (INM), integrated water management (IWM), and precision farming are helping to reduce GHG emissions from cropping activity.

iv. With the adoption of conservation agriculture techniques, soil microbial activities are improved, helping to build up a permanent nutrient cycle in the soil. That may help to reduce dependency on chemical fertilizers.

Key agronomic interventions for soil and water conservation:

A. **Conservation agriculture:** The principles of conservation agriculture include, minimum disturbance of soil, permanent soil covers, and need-based crop diversification. Applying these principles in cropping systems helps to conserve soil and water, and reduces soil erosion. In addition, helping in soil fertility enhancement and carbon sequestration (Kumawat et al. 2021). Key practices of conservation agriculture such as zero, tillage, and conservation tillage are done with a minimum of 30% crop residue retention that helps in the reduction of soil erosion as well as conserving soil moisture while maintaining the organic carbon pool in the soil. It also helps to enhance microbial activity in the soil, hence, increasing nutrient cycling within the soil system (Talukdar et al., 2023).

B. **Diversified crop rotation:** It allows to involvement of several different natural crops such as cereals, pulses, oilseeds, fodder, etc. in the same piece of land that helps to maintain soil nutrient pools as well as to utilize it efficiently by the crop. Inclusion of crops such as erosion permitting followed by erosion resisting reduces soil erosion to a greater extent. On the other hand, they also aid nutrients, helping to minimize the effect of pests, diseases, and weeds. Crop rotation and efficient crop diversification are a significant tool for climate change adoption (Ailincai et al., 2009).

Fig.1. Depiction of diversified crop rotation during Kharif-rabi-zaid seasons

C. **Intercropping/Mixed/multiple cropping:** It is the cultivation of different suitable crops is the same piece of land at the same time. It provides insurance against crop failure due to natural causes. Moreover, it hampers the functioning of harmful insect-pest and diseases. It also allows to maintenance of the nutrient cycle within the cropping period and judicious utilization of soil water from different depths, hence the reduction of soil erosion and water as well as runoff flow at a limited level.

D. **Mulching:** Mulching is a covering of soil surface through organic or inorganic substances it may be organic agriculture waste i.e. crop stubbles, rice, and wheat straw/husk, or synthetic products like polythene, etc. Mulching helps to reduce evaporation, increase infiltration, reduce run-off, and control weeds.

E. **Contour cropping:** Crops cultivation along the slope is known as contour cropping. Contour crops as planted across the slope that prevent soil erosion and reduces runoff intensity. Contour bunds help to store runoff water for crop utilization. Contour farming greatly lowers runoff and soil erosion, it works best when paired with other conservation tillage techniques like minimum or no tillage. This approach is suitable on gentle slopes. Contour farming on permanent raised beds paired with residue left on the soil's surface resulted most effective, according to the results of earlier studies (Farahani et al., 2016).

F. **Terracing:** Terracing is an engineering structure made for controlling soil and water erosion on high steep slopes. Often used in a slope range between 16-33%. Zing terracing, bench terracing, and compartmental bindings are used based on the slope. They are very effective in preventing runoff and improve water infiltration (Rutebuka et al., 2021).

G. **Alley cropping:** The agronomic annual crops are planted in the space between the plantation, trees, and shrubs. It allows us to utilize the space as well as to improve soil fertility by adding legume crops. This alley crop improves soil moisture and reduces erosion losses (Mishra et al., 2019). Several trees and hedges viz. *Leucaena and gliricidia* with grass barriers in alley cropping are effective in controlling runoff flow and soil erosion (Mishra and Tripathi, 2013).

H. **Agroforestry:** Agroforestry is one of the emerging agricultural practices to counter current climate change and also for soil and water conservation in degraded land areas. The inclusion of forest trees with agronomic crops benefits each other in several ways. The agro-forestry system improves soil health by maintaining the nutrient cycle, high soil biological activities, boosting infiltration, reducing evaporation, and ensuring substantial production. Organic matter in the soil is maintained due to the continuous supply of biomass of trees and shrubs, helping to strengthen the soil structure and water-holding capacity of the soil (Mbow et al., 2014; Mishra et al., 2019).

Table 1: Commonly followed Agro-forestry systems in India

Agro-forestry systems	Suitable trees/hedge spp.	Reference
Agri-silviculture (crops + trees)	*Prosopis cineraria, Eucalytus sp, Acacia tortilis, Hardwickia binata Ziziphus jujuba, Azadirachta indiaca* etc.	Reddy, 2019; Jinger et al., 2023
Agri-horti-silviculture (crops + multi-purpose trees + fruit trees)	*Leucaena leucocephala, sesbania sp.*	

Silvi- pastoral (Tree + pasture + animal)	*Leucaena latisiliqua, Bauhinia variegate, Albizzia babbek, Albizzia amara, Moringa olerifera, sesbania sesban, Hardwickia binate* etc.	
Horti- pastoral (Fruit tree + pasture)	*Cenchrus sp., Lasiurus ap., Chtysopogon sp., Stylosanthes sp., Sirato sp.*, etc. as forage and fodder purpose	
Agri-silvi-pasture Silvi- pastoral (Crop+ tree + pasture)	*Acacia senegal, Azadirachta Indica, and Hardwickia binate* along with grass and legume sp. like *Cenchrus sp., Stylosanthes sp., Stylosanthes sp.*, etc.	

I. Natural vegetative barrier: A vegetative barrier of 50-60 cm across the slope is planted in the hilly area. In this strip, perennial fodder, shrubs, hedges, or any native flora is added. These barriers restrict the movement of intensive runoff, improve infiltration, and control soil erosion effectively (Richet et al., 2017). Some important perennial species that are found effective in controlling erosion, as well as some insect pests, are *Panicum virgatum, Tripsacum dactyloides, Parastrephia lepidophylla, Adesmia spinosissima, Lampaya medicinalis* (Los et al., 2001; Barrientos-Pérez et al., 2023).

Fig 2: Different agronomic practices

J. **Summer fallowing:** Summer fallowing is the concept of leaving soil without planting during the summer season. It is an efficient management technique to store rainwater and stores for later use. Fallowing after tillage has multi-fold benefits in agriculture increased infiltration rate, killing weed seeds and propagules, inoculum of pathogens, and eggs of insect pests (Nadeem et al., 2019).

K. **Windbreak and shelterbelts:** Windbreak and shelterbelts are the planting of trees, shrubs, hedges, or perennials in the boundary of the farmland. This plantation acts as an actual barrier against high wind and extreme run-off. It acts as a savior against wind and water erosion of soil. These barriers also benefit the cropland by creating a favorable microclimate for beneficial microbes. *Gliricidia sepium* and *Paspalum conjugatum are* reported to have a good barrier effect and can reduce soil loss due to erosion (Altieri and Nicholls, 2000). *Acacia tortilis, Eucalyptus camaldulensis, Dalbergia sissoo, Tecomella undulata* proved to enhance soil organic carbon and reduce daily air temperature in the shelter area (Prasad et al., 2013).

L. **Integrated nutrient management:** Integrated nutrient management, applying mineral fertilizers in addition to organic manure, boosts crop yield, increases the amount of organic carbon in the soil, and reduces soil loss. Integrated nutrient management enhanced soil health and soil organic carbon storage in the all-cropping and agro-forestry system (Bhattacharya et al. 2011).

Table 2: New agronomic interventions development for soil and water conservation

Interventions	Feature/s	Reference
Super-absorbent Polymer-hydrogel	Increase water retention and water holding capacity of soil by absorbing water by 400 times its volume and releases when scarcity of water in the soil.	El-Asmar et al., 2017: Dujeshwer et al., 2021
Planting basins	Planting basins involves digging holes and planting crops to concentrate nutrients and water close to the plant. Increase Water and nutrient use efficiency.	Kuyah et al., 2021
Precision Agriculture	Use of technology to optimize field-level management of soil,	Delgado et al. 2011; Nemade et

	water, and vegetation with GPS-guided equipment and remote sensing. These tools assist us in making better management decisions that concentrate management efforts at the most vulnerable fields within a given site and/or in watersheds.	al., 2023
Broad Bed and Furrow (BBF) system	Effective in maintaining water balance in the soil and acts as life-saving for crops, especially in dryland areas. Effective methods for water conservation in dryland areas.	Talukdar et al., 2023; Lakpale et al., 2012
Site-specific water management	Water is supplied to specific areas of demand; the land is divided into several sub-units and separate water treatment is applied to each unit. Also involves reducing water losses, recycling of wastewater, and in-situ rainwater conservation.	Kumar et al., 2014

Constraints and challenges in the adoption of soil and water conservation practices

As the soil and water conservation practices are found less effective on many levels cause a measurable impact on farmers' interest in adoption and mitigation. The gap in R&D and farmers' approaches and policymakers also does not succeed in creating awareness among the farmers. There are some key issues that highlight the challenges in soil and water conservation adoption-

i. Primary stakeholders are not involved in the planning of natural resources or the development of technologies through on-farm research for sustainable agricultural production.

ii. Soil water conservation technologies that were used alone resulted in little boost in agricultural productivity and failed to inspire and persuade farmers towards adoption.

iii. Indigenous technical knowledge technologies are not been utilized while preparing conservation measures, especially at a micro-management level for specific areas.

iv. For efficient soil and water conservation, mechanical and vegetative interventions are not designed and executed simultaneously, resulting in less effective single management.

v. Cost-effective engineering measures are a need of the present development for approaching small-scale farmers and for wider adoption.

Future prospects and recommendations

i. With traditional agronomic practices for soil and water conservation, new technologies like precision farming, and biochar, the inclusion of effective beneficial microbes should also be included by the farming communities for better outcomes.

ii. Financial assistance should be provided to the more sensitive land areas for the construction of engineering intervention to tackle both degraded and drought areas.

iii. More emphasis on upgrading the knowledge through awareness, workshops, field demonstrations, training, and other farmers' outreach programs should be undertaken regularly.

iv. Monitoring and site-specific management through new technologies like remote sensing, GIS, etc can be used for the most vulnerable areas.

Conclusion

Degraded lands do not support any form of vegetation, because they are deficient in soil nutrients and organic matter. When management and conservation measures are not implemented, the effects become irreversible and unproductive, leading to significant ecological and socioeconomic ramifications. Under the current challenges of climate and land degradation, a sustainable practice is the need of present and future generations. Encountering these challenges, a natural/biological practice needs to be given priority among the farming communities and stakeholders. Much research and literature reported that agronomic interventions are effective in minimizing land degradation problems with scientific adoption and management. The agronomic interventions are effective in a slope ranging from 3-6 %, but they can also be effective up to 15% slope with the integration of some engineering approaches for sustainable soil and water conservation techniques. A multidimensional approach comprising research, education, policy support, and stakeholders' cooperation while adopting cutting-edge technology is needed to overcome these issues.

References

Ailincai, C., Jitareanu, G., Bucur, D., Ailincai, D. and Zbant, M., (2009). Evaluation of the long-term effect of crop rotation on water runoff, soil, and nutrient losses in the Moldavian Plateau. *Cercetări Agronomice în Moldova, 42(1),* 19-29.

Altieri, M.A., and Nicholls, C., (2000). Agroecología: Teoría y práctica para una agricultura sustentable. Mexico City: UNEP.

Avasthe, R., Kumar, A., Saha, S., Singh, R., and Babu, S., (2021). Agronomic interventions for enhancing diverse agro-ecosystem service. *Indian Farming, 71*, 56-59.

Barrientos-Pérez, E., Carevic-Vergara, F. S., Rodriguez, J. P., Arenas-Charlín, J., and Delatorre-Herrera, J., (2023). Effect of native vegetative barriers to prevent wind erosion: a sustainable alternative for Quinoa (*Chenopodium quinoa* Willd.) Production. *Agriculture, 13(7),* 1432.

Bhattacharyya, R., Ghosh, B. N., Mishra, P. K., Mandal, B., Rao, C. S., Sarkar, D., Das, K., Anil, K. S., Lalitha, M., Hati, K. M., et al. (2015). Soil Degradation in India: Challenges and Potential Solutions. *Sustainability*, *7(4),* 3528-3570. https://doi.org/10.3390/su7043528.

Bhattacharyya, R., Kundu, S., Srivastava, A. K., Gupta, H. S., Prakash, V., and Bhatt, J. C., (2011). Long-term fertilization effects on soil organic carbon pools in the sandy loam soil of the Indian Himalayas. *Plant and Soil, 341*, 109–124.

Delgado, J. A., Khosla, R., and Mueller, T., (2011). Recent advances in precision (target) conservation. *Journal of Soil and Water Conservation, 66(6)*, 167A-170A. DOI: https://doi.org/10.2489/jswc.66.6.167A

Dujeshwer, K., Singh, R. K., Jatav, H. S., et al. (2021). Hydrogel-based *Trichoderma* formulations' effect on different varieties of rice under rainfed conditions of Indo-Gangetic Plains. *Environment, Development and Sustainability, 24,* 7035-7056. DOI: 10.1007/s10668-021-01738-w.

El-Asmar, J., Jaafar, H., Bashour, I., Farran, M. T., Saoud, I. P., (2017). Hydrogel banding improves plant growth, survival, and water use efficiency in two calcareous soils. *CLEAN–Soil, Air, Water, 45(7),* 1700251.

Farahani, S. S., Fard, S. F., Asoodar, M. A., (2016). Effects of contour farming on runoff and soil erosion reduction: A review study. *Elixir Agriculture, 101,* 44089-44093.

Hamdy, A. and Aly, A., (2014). Land degradation, agriculture productivity, and food security. Book of proceedings: Fifth International Scientific Agricultural Symposium "Agrosym 2014", Jahorina, Bosnia and Herzegovina, October 23-26, 2014.

Jinger, D., Kaushal, R., Kumar, R., Paramesh, V., et al., (2023). Degraded land rehabilitation through agroforestry in India: Achievements, current understanding, and future prospectives. *Frontier in Ecology and Evolution. 11,* 1088796. doi: 10.3389/fevo.2023.1088796.

Kumar, A., Choudhary, A. K., Suri, V. K., Bana, R. S., Pooniya, V., and Singh, U., (2014). Site-specific water management for sustainable agriculture. (In) Water Management in Agriculture, 327-336.

Kumawat, A., Yadav, D., Samadharmam, K., and Rashmi, I., (2021). 'Soil and Water Conservation Measures for Agricultural Sustainability'. Soil Moisture Importance. IntechOpen. doi:10.5772/intechopen.92895.

Kurrey, D., Singh, R. K., and Rajput, R., (2016). Status of soil erosion or land degradation in India. National Conference on Managing Soil Resources for Environmental Sustainability: Challenges and Perspectives, 09-10 December 2016.

Kuyah, S., Sileshi, G. W., Nkurunziza, L., et al., (2021). Innovative agronomic practices for sustainable intensification in sub-Saharan Africa. A review. *Agronomy for Sustainable Development, 41,* 16. https://doi.org/10.1007/s13593-021-00673-4.

Lakpale, R., and Tripathi, V. K., (2012). Broad-bed furrow and ridge and furrow method of sowing under different seed rates of soybean (*Glycine max* L.) for high rainfall areas of Chhattisgarh plains. *Soybean Research, 10,* 52-59.

Los, P., Anderson, S. H., and Gantzer, C. J., (2001). Vegetative Barriers for Erosion Control. Mu Extension, University of Missouri-Columbia, 1653.

Mbow, C., Smith, P., Skole, D., Duguma, L., and Bustamante, M., (2014). Achieving mitigation and adaptation to climate change through sustainable agroforestry practices in Africa. *Current Opinion in Environmental Sustainability, 6,* 8-14.

Mishra, P. K., and Tripathi, K. P., (2013). Soil and water conservation research for land management in India. *Indian Journal of Dryland Agricultural Research and Development, 28(1),* 1-18.

Mishra, P. K., Rai, A., and Rai, S. C., (2019). Agronomic Measures in Traditional Soil and Water Conservation Practices in the Sikkim Himalaya, India. *American Research Journal of Agriculture, 5(1),* 1-16.

Murthy, C. S., and Sesha Sai, M. V. R., (2010). Agricultural drought monitoring and assessment. Remote Sensing Application pub. ISRO-NRSC pp-303-329.

Nadeem, F., Nawaz, A., and Farooq, M., (2019). Crop Rotations, Fallowing, and Associated Environmental Benefits. *Oxford Research Encyclopedia of Environmental Science,* https://doi.org/10.1093/acrefore/9780199389414.013.197.

Nemade, S., Ninama, J., Kumar, S., et al., (2023). Advancements in Agronomic Practices for Sustainable Crop Production: A Review. *International Journal of Plant and Soil Science, 35,* 679-689.

Prasad, R., Mertia, R. S., Gajja, B. L., Singh, D., and Chandra, P., (2013). Morphological characteristics of tree shelterbelts and their impact on wind regimes, air temperature, and soil quality. *Indian Journal of Agroforestry, 15,* 74–81.

Reddy, S. R., (2019). Principles of Agronomy, sixth edition, Kalyani Pub. New Delhi. pp. 653-654.

Richet, J. B., Ouvry, J. F., and Saunier, M., (2017). The role of vegetative barriers such as fascines and dense shrub hedges in catchment management to reduce runoff and erosion effects: Experimental evidence of efficiency, and conditions of use. *Ecological Engineering, 103,* 455-469, https://doi.org/10.1016/j.ecoleng.2016.08.008.

Rutebuka, J., Munyeshuli, U. A., Nkundwakazi, O., Mbarushimana, K. D., Mbonigaba, J. J. M., Vermeir, P., and Verdoodt, A., (2021). Effectiveness of terracing techniques for controlling soil erosion by water in Rwanda. *Journal of Environmental Management, 277,* 111369. https://doi.org/10.1016/j.jenvman.2020.111369.

Ryan, J., Masri, S., Ibrikci, H., Singh, M., Pala, M., Harris, H., (2008). Implications of cereal-based crop rotations, nitrogen fertilization, and stubble grazing on soil organic matter in a Mediterranean-type environment. *Turkish Journal of Agriculture and Forestry, 32,* 289–297.

Sengupta, R., (2021). Land degradation in India hurts farmers and forest dwellers the most. *Down-To-Earth.* Retried on 06th March 2024 from Land degradation in India hurts farmers and forest dwellers the most (downtoearth.org.in).

Talukdar, N., Paul, A., Sarma, H. H., and Kakoti, M., (2023). Agronomic Practices for Soil Conservation. In the book: Frontiers in Agricultural Sustainability. Integrated Publications TM pub, New Delhi, India.

Triple Pundit., (2017). https://www.triplepundit.com/story/2017/how-climate-change-impacts-our-watersupply/20361. Accessed 13 May 2024.

Tyagi, V., Nagargade, M., Singh, R. K., (2020). Agronomic Interventions for Drought Management in Crops. In: New Frontiers in Stress Management for Durable Agriculture. Springer, Singapore. https://doi.org/10.1007/978-981-15-1322-0_24.

CHAPTER TEN

WATER MANAGEMENT AND CONSERVATION THROUGH AGROFORESTRY UNDER CHANGING ENVIRONMENTAL CONDITIONS

SARITA BISHT[1,*], AMIT KUMAR[2], ARADHANA[3] AND KARUNA GUSAIN[4]

[1]FOREST RESEARCH INSTITUTE DEHRADUN, 248006, INDIA

[2]CSIR-INSTITUTE OF HIMALAYAN BIORESOURCE TECHNOLOGY, PALAMPUR-176061

[3]DEPARTMENT OF BOTANY AND MICROBIOLOGY, GURUKUL KANGRI (DEEMED TO BE) UNIVERSITY, HARIDWAR, INDIA

[3]G.B. PANT NATIONAL INSTITUTE FOR HIMALAYAN ENVIRONMENT, KOSI-KATARMAL, ALMORA-263643

Abstract

Water management and conservation are critical aspects of sustainable agriculture, particularly in the face of changing environmental conditions. In this chapter we focus on the role of agroforestry systems in water management and conservation, highlighting their potential to mitigate water-related challenges while adapting to environmental changes. We also discuss the various ways that agroforestry contributes to water conservation and management. It examines specific agroforestry techniques such as alley cropping, riparian buffers, and silvopastoral systems, elucidating their roles in improving water availability, quality, and regulation within agricultural landscapes. Moreover, agroforestry systems facilitate groundwater recharge and regulate streamflow through the intricate network of tree root systems, thereby bolstering ecosystem stability and enhancing overall water security. It then delves into the specific mechanisms through which agroforestry can contribute to water management and conservation, including soil moisture retention, groundwater recharge, and erosion control. We address the challenges and opportunities associated with implementing agroforestry systems under changing environmental conditions, such as altered precipitation patterns and increased water stress. Drawing on case studies and research findings, the chapter highlights the potential of agroforestry to enhance water resilience in agricultural landscapes. Furthermore, we concluded with recommendations for promoting the adoption of agroforestry as a sustainable water management strategy and directions for future research in this field.

Keywords: Agroforestry, Water management, agroforestry techniques, Water resilience, agricultural landscapes

10.1 Introduction

Many ecosystem services and environmental advantages can be derived from agroforestry, which has been recognized for its importance (Dwivedi et al., 2023). Soil and water are the primary resources for agricultural production. The existence of damaged soil and contaminated water threatens human survival. The degradation of soil and water is directly related to overall environmental health. Climate change and water pollution are major global concerns. Addressing these concerns necessitates prioritizing long-term solutions for soil and water conservation, such as agroforestry. Agroforestry encompasses land-use systems where perennial woody plants are cultivated alongside herbaceous plants or livestock, either in a specific spatial layout, rotational scheme, or both. This approach serves dual purposes, combining productivity with various beneficial services. Within its productive aspect, the primary contributions come from the "3Fs": fuelwood, fodder, and fruit, while also providing resources such as construction wood, gums, resins, medicines, fibres, and a range of other economic assets, thereby enhancing food security (Kumar et al., 2020a&b: Kaushal et al., 2021). Agroforestry, an ancient land-use practice that integrates trees with crops and/or livestock, offers a promising solution to address water-related challenges in agricultural landscapes. By harnessing the synergies between trees, crops, and soil, agroforestry systems can enhance water availability, quality, and resilience while mitigating the impacts of environmental changes (Williams-Guillén, 2008; Jose, 2009). The basic concept of agroforestry is the anticipated contribution that both on- and off-farm tree productions will make to the management of natural resources and sustainable land use

(Nair, 2011). Water conservation is the process of using water efficiently and reducing waste. It involves making conscientious efforts to limit water consumption and protect this precious resource (Knox et al., 2012). Water is a vital resource for agriculture, playing a fundamental role in crop production, livestock farming, and ecosystem health. However, the availability and quality of water resources are increasingly threatened by climate change, land degradation, and human activities. As the global population grows and environmental pressures intensify, there is a pressing need to adopt sustainable approaches to water management and conservation in agriculture (Nair, 2011). There are various methods to conserve water, such as using water-saving technologies, altering agroforestry behaviours, and adopting sustainable water management techniques (Abdallah et al., 2021).

In this chapter, we explore the role of agroforestry in water management and conservation under changing environmental conditions. Drawing on a diverse range of studies and empirical evidence, we examine the mechanisms through which agroforestry contributes to sustainable water management, including soil moisture retention, groundwater recharge, erosion control, and regulation of streamflow. We also discuss the challenges and opportunities associated with implementing agroforestry in the context of altered precipitation patterns, increased water stress, and other environmental stressors. To provide a comprehensive overview, we present case studies and examples from different regions and agroecological contexts, highlighting the effectiveness of agroforestry practices in improving water resilience and enhancing ecosystem services. In this chapter, our objective is to emphasize the significance of incorporating agroforestry into both water management strategies and farming methods to secure the enduring sustainability of food production systems and environmental robustness. By utilizing the diverse advantages of agroforestry, we address water-related challenges, enhance ecosystem health, and promote climate resilience in agricultural environments.

10.2 Agroforestry Systems: Principles and Practices

Agroforestry is a land management approach that integrates trees with crops and/or livestock in agricultural systems. This multifunctional land-use strategy harnesses ecological interactions between different components to enhance productivity, biodiversity, and sustainability. Understanding the principles and practices of agroforestry is essential for implementing effective and resilient systems that contribute to water management and conservation. Agroforestry encompasses a diverse range of practices that involve the deliberate integration of trees with agricultural activities (Rao et al., 2007). At the core of agroforestry lies the integration of multiple components, including trees, crops, and livestock, within the same land unit (Tewari et al., 2024; Kumar et al., 2021a&b; Kumar et al., 2022). This diversity enables the exploitation of complementary ecological interactions, such as nutrient cycling, pest regulation, and microclimate modification. By combining different species, agroforestry systems can maximize resource utilization, minimize environmental impacts, and enhance overall resilience to biotic and abiotic stresses.

Principles

Agroforestry stands out for its multifaceted benefits, rooted in principles that encompass diversification, resource use efficiency, sustainability, and social and economic viability. Diversification lies at the core of agroforestry, as it integrates multiple plant species within the same area, thereby promoting biodiversity. This approach not only enhances ecosystem resilience but also reduces the risks associated with crop failures and insect outbreaks by producing a more balanced and robust agricultural environment (Nair, 2012). Moreover, by harnessing complementary interactions between trees, crops, and livestock, agroforestry optimizes resource use efficiency. Trees provide vital services such as shade, soil fertility improvement, and windbreaks, which increase overall production while reducing resource waste. Sustainability is a guiding principle in agroforestry systems, which are deliberately built to resemble natural ecosystems (Jose, 2009). By prioritizing soil health, and water conservation, and reducing reliance on external inputs like synthetic fertilizers and pesticides, agroforestry contributes significantly to environmental conservation and climate change mitigation efforts. Furthermore, agroforestry plays a pivotal role in enhancing rural livelihoods by providing additional sources of income, food security, and employment opportunities, particularly for smallholder farmers (Tewari et al., 2022; Kumar et al., 2018; Kumar et al., 2021c&d).

Agroforestry Practices

Agroforestry is a land-use management system that integrates trees and shrubs with crops and/or livestock, creating ecological and economic synergies. This practice has gained significant attention for its potential to address various environmental, social, and economic challenges faced by agriculture. Agroforestry systems vary widely based on climate, soil conditions, and local practices, but they generally fall into several categories:

Alley Cropping: In alley cropping, rows of trees or shrubs are planted alongside or between rows of annual crops. This configuration provides multiple benefits such as erosion control, nutrient cycling, and microclimate moderation for the crops. Well-known examples include the integration of leguminous trees like *Acacia* or *Leucaena* with maize or cassava in tropical regions (Garrett et al., 2021). Crop and grass yields are important. Tree species appropriate for hedgerows include *Leucaena leucocephala, Colliendra*, and *Sesbania*. Pigeon peas and castor crops thrive in the alleys of *Leucaena* (Sitaram, 2015).

Riparian Buffer Zones: Riparian buffer zones consist of vegetation strips, including trees, shrubs, and grasses, established along water bodies such as rivers, streams, and ponds. These buffer zones help filter pollutants, stabilize stream banks, and reduce sedimentation, thus improving water quality. Moreover, the dense root systems of riparian vegetation contribute to groundwater recharge and regulate water flow, mitigating floods and droughts (Schroth et al., 2004). Vegetation within riparian buffer zones helps to anchor soil along streambanks, reducing erosion caused by water flow. This stabilization prevents sedimentation in water bodies, maintaining water clarity and quality. The dense vegetation in riparian buffer zones acts as a natural filter, trapping and metabolizing pollutants from surface runoff and agricultural activities. Nutrients, sediments, pesticides, and other contaminants are removed or attenuated before reaching water bodies, thus safeguarding water quality (Lowrance et al., 1997). Riparian buffer zones absorb excess water during periods of heavy rainfall or flooding, reducing the risk of downstream flooding. The vegetation within these zones acts as a sponge, slowing down the flow of water and reducing its erosive force. Shade provided by riparian vegetation helps to regulate water temperature in streams and rivers, particularly in regions experiencing high temperatures (Allan, 2004). Maintaining optimal water temperatures is crucial for the survival and reproduction of aquatic organisms. Through preserving and restoring these transitional areas between land and water, stakeholders can enhance the resilience of aquatic ecosystems, mitigate the impacts of pollution and flooding, and support biodiversity. Implementing and maintaining riparian buffer zones should be a priority in integrated water resource management strategies, ensuring the sustainable use and protection of freshwater resources for present and future generations (Verma et al., 2023; Kumar et al., 2020; Kumar et al., 2021e).

Silvopastoral Systems: Silvopastoral systems combine trees or shrubs with livestock grazing. Trees provide shade, forage, and shelter for the animals, while the livestock contribute to nutrient cycling and soil fertility. Agroforestry systems such as silvopastoralism have been successfully implemented in various regions, including Mediterranean areas and parts of sub-Saharan Africa (Nair et al., 2021). Grasses possessing an extensive fibrous root system and dense foliage play a crucial role in mitigating water and soil erosion, making them the preferred choice for controlling runoff and soil loss in erosion-prone regions (Dagar and Gupta, 2016). Within silvopastoral systems, the foliage of these grasses enhances resistance against overflow, while their extensive root systems promote infiltration, thereby facilitating efficient water recharge (Kaushal et al., 2021). Sharda and Venkateswarlu (2007) reported a runoff of 5.8% and soil loss of 1.26 Mg ha^{-1} under the *Acacia* + *Dichanthium annulatum* silvopasture system, compared to 14.7% runoff and 3 Mg ha^{-1} of soil loss in agricultural catchments. Furthermore, in silvopasture settings, higher levels of infiltration were observed under *Leucaena* + *E. binata*, followed by eucalyptus or *Eulaliopsis binate*, in contrast to minimal infiltration in agricultural systems.

Forest Farming: Forest farming, a form of agroforestry, integrates tree cultivation with crops and/or livestock on the same land, offering not only traditional agricultural benefits but also a sustainable approach to water management and conservation. Forest farming improves water efficiency, reduces soil erosion, and increases biodiversity by emulating natural forest ecosystems (Jose, 2009). The multi-layered canopy characteristic of forest farming moderates temperatures, reduces evaporation, and enhances soil moisture retention, thus optimizing water management. Additionally, tree roots within forest farming systems improve soil structure, facilitating water infiltration and storage, leading to groundwater recharge, reducing surface runoff, and mitigating soil erosion (Schroth et al., 2013). The presence of trees also regulates microclimates, reducing water stress on crops and enhancing water use efficiency, particularly during dry periods. Moreover, forest farming promotes biodiversity, creating healthy ecosystems that contribute to improved water quality through the filtration of pollutants by vegetative cover and root systems, thus conserving water resources (Jose, 2009). According to Kumar and Nair (2011), Forest farming involves the intentional cultivation of crops beneath a forest canopy. This practice can include cultivating non-timber forest products (NTFPs) like mushrooms, medicinal herbs, fruits, and nuts. Forest farming helps to conserve biodiversity, enhance soil health, and provide economic opportunities for rural communities (Kumar et al., 2022; Pandey et al., 2018; Bisht et al., 2022).

Windbreaks and Riparian Buffers: Planting trees and shrubs along field perimeters or watercourses functions as windbreaks and riparian buffers, mitigating soil erosion, improving water quality, and fostering wildlife habitats. These linear agroforestry practices are prevalent in temperate regions and play a crucial role in safeguarding agricultural lands against wind and water erosion (Smith et al., 2008). The utilization of trees and other perennial woody plants for safeguarding agricultural fields is widespread in arid regions prone to wind erosion. Strong winds can cause detrimental effects on both soil structure and crop yield, resulting in the depletion of fertile topsoil in sandy soils and diminished crop productivity due to physical harm and moisture deficiencies. Wind erosion poses a significant challenge in arid regions across all continents, although the adoption of windbreaks is less common in tropical climates compared to temperate regions (Tewari et al., 2021; Kumar et al., 2021 Kumar et al., 2024).

The configuration of windbreaks is contingent upon site-specific environmental factors such as wind velocity, precipitation patterns, and soil characteristics, groundwater levels, alongside farming components including cultivated crops, livestock, and human requisites. Windbreaks may encompass single or multiple rows of trees, comprising various species and potentially incorporating secondary hedgerows. Establishing an interconnected network of windbreaks across multiple farms on a large scale can yield more substantial benefits in controlling soil erosion and enhancing crop yields compared to isolated windbreaks on individual farms. Similar to other agroforestry systems, the selection of perennial woody plants is crucial and is based on criteria such as growth patterns, seasonal fluctuations, multifunctionality, mutual benefit, and compatibility with cultivated crops (Kiepe and Rao, 1994).

10.3 Two different categories of windbreaks can be delineated

1. Small-scale windbreaks, established with diminutive trees such as *Gliricidia* and *Moringa oleifera*, compact bushes including *Balanites, Euphorbia, Zizyphus*, or even grass species like *Andropogon gayanus*.

2. Large-scale windbreaks, comprising substantial timber species such as eucalyptus and *Casuarina,* woody plants suitable for fuelwood like neem, *Cassia, Prosopis*, and fruit-bearing trees such as mango and cashew nut trees.

By strategically implementing windbreaks tailored to the specific conditions and agricultural systems prevalent in each locality, farmers can effectively mitigate soil erosion, shield crops from wind-induced damage, and bolster overall agricultural sustainability.

Home Gardens and Multistory Systems: Home gardens and multistory e systems involve planting a diverse range of crops, trees, and shrubs in small-scale agricultural settings. Arrangements of trees, shrubs, and herbaceous plants forming multiple layers vertically without a specific spatial pattern are referred to as multistory tree gardens. They are characterized by a dense canopy, a diverse array of species, substantial soil coverage, and leaf litter, and require intensive management. The upper layer typically comprises timber trees such as *Melia azadirach*, teak, *Terminalia* spp., shade trees like *Albizzia, Erythrina, Grevillea*, or palms such as coconut, areca nut, and peach palm. The middle layer consists of crops such as coffee, cocoa, mango, etc., while shade-tolerant annuals and other herbaceous plants form the understory (Nair, 1989; Bregman, 1991). These systems are primarily practised in humid tropical regions and on mountain slopes in the sub-humid tropics, where water availability is not a limiting factor. Their main objective is the efficient utilization of natural resources and the promotion of highly sustainable production (Kiepe and Rao, 1994)). These systems provide food security, diversify income sources, and promote resilience to climate change. Home gardens are prevalent in tropical regions and contribute significantly to household nutrition and livelihoods. This agroforestry system is widely used by small and marginal farmers in Kerala, including both tribal and non-tribal populations. This approach to agro-forestry is both simple to apply and highly rewarding (Sitaram, 2015).

Agroforestry practices offer numerous benefits, including enhanced biodiversity, improved soil fertility, increased carbon sequestration, climate change mitigation, and resilience to extreme weather events. Moreover, agroforestry can contribute to sustainable development goals by promoting food security, poverty alleviation, and rural livelihoods. Several organizations and research institutions actively promote agroforestry, including the World Agroforestry Centre (ICRAF), these organizations conduct research, provide technical assistance, and disseminate information to farmers and policymakers worldwide, facilitating the adoption of agroforestry practices.

10.4 Role of Agroforestry in Water Management and Conservation

Agroforestry plays a significant role in enhancing ecosystem services by improving the quality and availability of water, as well as by reducing the risk of flooding. The presence of surface litter, stems, and roots within agroforestry systems effectively slows down water flow, leading to increased sedimentation and nutrient retention, as highlighted by Schultz et al. (2009). Moreover, agroforestry practices serve as a proven method for ensuring access to clean water. In contrast to conventional agricultural systems where less than half of the applied nitrogen and phosphorus fertilizers are utilized by crops, agroforestry systems help mitigate this issue. In conventional agriculture, excess fertilizer often finds its way into water bodies through surface runoff or leaching, resulting in contamination and decreased water quality (Cassman, 1999). For instance, agricultural runoff contributes to the excessive delivery of sediment, nutrients, and pesticides to water bodies, thereby exacerbating issues such as eutrophication in regions like the Gulf of Mexico (Jose, 2009).

I. **Soil Moisture Retention:** Agroforestry systems help in enhancing soil moisture retention. The presence of trees and shrubs in agroforestry systems creates a microclimate that reduces soil evaporation and transpiration, thus minimizing water loss from the soil. The roots of trees and shrubs also help improve soil structure, increasing water infiltration, and reducing surface runoff, which ultimately aids in maintaining soil moisture levels for crops (Kaushal et al., 2021). Mulching is the practice of applying organic materials like straw, leaves, or wood chips to cover the soil surface. Its primary purpose is to retain soil moisture by minimizing evaporation. This method not only conserves water but also enhances soil fertility and suppresses weed growth. A study carried out in Kenya showcased the benefits of combining agroforestry with mulching. Farmers who adopted this approach reported notable enhancements in water retention, resulting in higher crop yields and reduced water demands (Patil Shirish et al., 2013).

II. **Groundwater Recharge:** In the upcoming decades, the primary problem for India's agricultural and rural progress will be groundwater depletion. Hence, scrutiny of watershed management initiatives' effect on replenishing groundwater is conducted through both extractive and participatory methods (Sitaram, 2015). Agroforestry practices contribute significantly to groundwater recharge. Trees and shrubs in agroforestry systems intercept rainfall, allowing it to infiltrate into the soil and replenish groundwater reserves. Moreover, the deep root systems of trees facilitate water percolation into the ground, promoting groundwater recharge. By reducing surface runoff and enhancing infiltration rates, agroforestry helps sustain groundwater levels, which are essential for agricultural and domestic water supply (Kaushal et al., 2021). Agroforestry buffers improve the quality of shallow groundwater by root uptake and enhance conditions for denitrification since the retention time is sufficient for tree roots and bacteria to interact with groundwater. Reductions ranging from 65% to 90% for N and 24% to 81% for P in groundwater have been observed (Udawatta and Gantzer, 2022).

III. **Erosion Control and Runoff:** Soil erosion and runoff pose significant challenges within agroforestry systems, resulting in the depletion of essential topsoil and adding to water contamination concerns. Effective soil management strategies, including the maintenance of vegetative cover, contour ploughing, and mulching, offer viable solutions to mitigate erosion and runoff. By ensuring continuous vegetation or mulch cover over the soil, the erosive effects of rainfall are diminished, thereby preventing the displacement of soil particles and their subsequent transport by runoff. This approach not only conserves soil fertility but also serves to safeguard water bodies against sediment accumulation and pollution (Pavlidis and Tsihrintzis, 2018).

Aagroforestry serves as an effective measure for erosion control. Planting trees on erosion control structures proves highly efficient in stabilizing various types of land. The presence of vegetation, particularly trees and shrubs, helps in stabilizing soil and preventing erosion caused by wind and water. Tree roots bind soil particles together, reducing soil erosion by providing physical support and anchorage. Moreover, the canopy cover of trees mitigates the impact of rainfall, minimizing soil splashing and surface runoff. These erosion control mechanisms in agroforestry systems contribute to the conservation of soil and water resources, safeguarding agricultural productivity and ecological integrity. This practice not only stabilizes the land but also provides valuable resources like fuel wood, fodder, poles, and small timber., locally available tree species are chosen for planting on erosion-control structures, tailored to the specific priorities and needs of farmers (Kaushal et al., 2021).

Grassed watercourses serve as effective filters for sediment removal. Depending on soil and climate conditions, a combination of fast-growing grasses for temporary control and resilient perennial grasses for long-term management is selected to establish vegetation in the waterway. According to Rao et al. (2015), grass filter strips transition supercritical flows to subcritical ones, significantly reducing sediment concentration in runoff water from 4.2 to 0.65 g L−1 and decreasing runoff velocity from 1.06 to 0.47 m s^{-1}.

Brushwood check dams are utilized for runoff control in small and medium-sized gullies. To construct them, small branches, preferably from coppiceable species, are arranged in two parallel rows across the gully and filled with brushwood between these vertical stakes. Various tree or shrub species like alder, pine, bamboo, willow, or poplar can serve as posts. Flexible branches from trees such as willow or poplar, along with flexible stems from shrubs like *Tamarix* or *Arundinaria,* as well as bamboo stem strips, may be employed as interlink materials. These structures trap a significant amount of soil, which, combined with enhanced moisture availability, facilitates the growth of economically beneficial vegetation. The sediment deposited behind these structures aids in stabilizing gully beds (Dhyani et al 1988; Kaushal et al., 2021).

IV. Regulation of Stream Flow: Agroforestry practices play a crucial role in regulating stream flow patterns. The vegetation in agroforestry systems intercepts rainfall and reduces peak flows by slowing down the movement of water over the land surface. This regulation of stream flow helps prevent floods downstream and maintains stable stream channels. Furthermore, agroforestry buffers along water bodies act as riparian zones, filtering pollutants, trapping sediments, and providing habitat for aquatic organisms. These riparian buffers contribute to water quality improvement and ecosystem resilience (Kuyah et al., 2017). Agroforestry systems help clean runoff water by enhancing infiltration, sediment deposition, nutrient retention, and improving groundwater quality through nutrient uptake (Kumar et al., 2020a). Agroforestry offers multifaceted benefits for water management and conservation, including soil moisture retention, groundwater recharge, erosion control, and regulation of stream flow. By integrating trees and shrubs into agricultural landscapes, agroforestry enhances water efficiency, preserves soil quality, mitigates environmental degradation, and promotes sustainable land use practices (Marques et al., 2016).

10.5 Integrating technology for enhanced water management in agroforestry

Integrating technology for enhanced water management in agroforestry represents a promising approach to optimize water use efficiency, conserve water resources, and improve agricultural productivity sustainably. By leveraging advancements in various technological domains, such as remote sensing, precision agriculture, and irrigation systems, agroforestry practitioners can make informed decisions to better manage water within their systems (Baca et al., 2014).

- **Remote Sensing and GIS (Geographic Information Systems):** Remote sensing technologies, including satellite imagery and drones, offer valuable insights into vegetation health, soil moisture levels, and land use patterns in agroforestry systems. GIS tools help in mapping and monitoring water resources, identifying suitable sites for agroforestry, and assessing the spatial distribution of tree and crop species (Sharma et al., 2023).

- **Precision Agriculture:** Precision agriculture techniques, such as variable rate irrigation and fertilization, enable targeted application of water and nutrients based on site-specific conditions and crop requirements. Sensor technologies, including soil moisture sensors and weather stations, provide real-time data on soil moisture levels, weather patterns, and crop water demand, allowing for timely irrigation scheduling and water conservation (Pedersen and Lind, 2017).

- **Drip Irrigation Systems:** Drip irrigation systems have emerged as a transformative tool in water management, offering efficient water delivery directly to the root zone of plants, thus minimizing water waste and maximizing crop yields, thereby revolutionizing agricultural practices worldwide (Kumar and Kumar, 2018). Unlike traditional methods such as flood or sprinkler systems, drip irrigation delivers water precisely to the roots, reducing evaporation and runoff, and ensuring optimal water use efficiency. This precise irrigation technique promotes healthier plant growth and higher crop yields by preventing water stress and providing nutrients directly to the root zone throughout the growing season (Sidhu et al., 2021). Moreover, drip irrigation systems conserve not only water but also energy and fertilizers, reducing the need for pumping and treating water while minimizing nutrient leaching into the soil, thereby conserving both water and soil quality. Furthermore, the adaptability of drip irrigation systems to various crops, soils, and topographies, from large-scale agriculture to home gardening, underscores their versatility and effectiveness in meeting diverse agricultural needs (Baca et al., 2014). Additionally, by minimizing water usage and reducing the need for chemical inputs, drip irrigation contributes to environmental sustainability by mitigating soil erosion, reducing water pollution from agricultural runoff, and conserving natural resources for future generations.

- **Agroforestry Modeling and Decision Support Systems:** Agroforestry modelling tools and decision support systems integrate data on soil properties, climate conditions, vegetation dynamics, and hydrological processes to simulate water availability and productivity outcomes. These modelling approaches help in designing and optimizing agroforestry systems for specific environmental and socio-economic objectives, such as maximizing water use efficiency or enhancing ecosystem services (Rahman et al., 2023).

Integrating these technologies into agroforestry practices requires capacity building, infrastructure development, and institutional support. Additionally, interdisciplinary collaboration among researchers, extension agents, policymakers, and farmers is essential to ensure the effective adoption and scaling of technology-enabled water management solutions in agroforestry.

10.6 Challenges and Opportunities

Agroforestry encounters various socioeconomic obstacles in India that hinder its complete potential for conserving biodiversity. These challenges encompass factors such as the limited size of landholdings, which restricts the implementation of agroforestry practices, and the absence of access to quality inputs, including high-yielding and disease-resistant tree species. Furthermore, inadequate market accessibility for agroforestry products undermines the economic feasibility of these systems. Additionally, there exists a prevalent lack of awareness among farmers regarding the advantages of agroforestry, compounded by cultural preferences favouring traditional agricultural methods. Financial constraints, including the initial investment necessary for transitioning to agroforestry systems and the extended period required to yield returns from tree crops compared to annual crops, represent significant barriers. Moreover, insecurity regarding land tenure can discourage investment in long-term agroforestry practices essential for biodiversity conservation (Wato and Amare, 2020).

Agroforestry offers opportunities to conserve water by improving soil structure and reducing evaporation through tree cover. Trees can also act as natural barriers to wind, reducing water loss from soil surfaces (Wallace et al., 2005). Tree species enhance the soil's ability to retain moisture and improve both its hydraulic conductivity and moisture storage capacity throughout the dry and rainy seasons compared to areas without trees. Agroforestry techniques such as windbreaks and shelterbelts mitigate wind erosion by reducing wind speed. Excessive fertilizers often get washed away from agricultural lands through surface runoff or leaching into the subsurface, leading to water source contamination and degraded water quality (Cassman, 1999). Agroforestry systems aid in purifying runoff water by decreasing runoff speed, facilitating infiltration, sediment accumulation, and nutrient retention, and enhancing groundwater quality through nutrient capture (Kumar et al., 2020). Agroforestry enhances ecosystem services such as water purification, soil conservation, and climate regulation. Multi-functional agroforestry landscapes deliver multiple benefits, including improved water quality, reduced flood risk, and enhanced biodiversity conservation (Kay, 2018). It also can strengthen community resilience to climate change by improving food security, water availability, and livelihood opportunities. Community-based agroforestry initiatives can empower local stakeholders to manage natural resources sustainably and adapt to changing environmental conditions. Organic matter from vegetation serves as a long-term reservoir for nutrients. Improved nutrient absorption by agroforestry vegetation retains nutrients within the system, making them available for uptake by other plants, consequently lowering nutrient loss and mitigating water pollution. The effectiveness of riparian buffers in conserving resources is extensively documented in temperate regions (Udawatta and Gantzer, 2022). Naiman et al. (2005) found that established riparian buffers accumulate significant nutrient quantities within their biomass, with estimates reaching 10 Mg P ha^{-1} y^{-1} (equivalent to 4.46 to P ac^{-1} yr^{-1}).

10.7 Policy, Education, and Community Involvement

Agroforestry stands as a sustainable approach to water management and conservation, especially crucial under changing environmental conditions such as climate change and increasing water scarcity. Adopting and scaling up agro-forestry practices require a supportive policy environment. This act is highlighted by the complexities involved in planning and distributing planting material for trees, which require nurseries. Agro-forestry approaches also require combining trees and crops, with returns often created over several years. Therefore, agroforestry practices require rigorous and supportive regulations for their growth and wider acceptance (Mugendi et al., 2007). Policy can play an important role in supporting or suppressing agroforestry sustainability and environmental soundness.

10.8 The Importance of Supportive Policy Frameworks

Establishing supportive policy frameworks is crucial for shaping and directing water conservation endeavours within agriculture. These frameworks encompass a combination of government initiatives and international collaborations. The Government of India has introduced a range of initiatives and regulations aimed at promoting water conservation in agriculture. Policies like the National Water Policy, Pradhan Mantri Krishi Sinchayee Yojana (PMKSY), and the Atal Bhujal Yojana are designed to enhance water use efficiency, encourage the adoption of micro-irrigation techniques, and replenish groundwater levels. The overarching goal of these initiatives is to tackle the challenges posed by water scarcity, inefficient water utilization, and the deterioration of water resources (Gany et al., 2019). Furthermore, international cooperation and assistance play a significant role in bolstering India's water conservation endeavours in agriculture. Collaborative efforts with organizations such as the World Bank, the International Water Management Institute (IWMI), and the Food and Agriculture Organization (FAO) have facilitated the exchange of knowledge, technology, and resources. These partnerships have supported various projects and research initiatives aimed at improving water management practices in India, thereby contributing to the sustainability of agricultural water use (Rastogi et al., 2024)

10.9 Community-Based Water Management Strategies.

Engaging local communities is essential for ensuring the effectiveness and long-term viability of water management strategies. Community-based approaches ensure that solutions are tailored to local needs and are embraced by the wider population. Farmer cooperatives and water user associations play a crucial role in promoting participatory water management. These organizations empower farmers to collectively oversee water resources, participate in decision-making processes related to water allocation, and adopt water-saving practices (Hu et al., 2014). The establishment of such groups has been actively encouraged through various government initiatives, fostering a sense of ownership and stewardship among farmers toward water resources. NGOs and local grassroots organizations also play a pivotal role in driving community-led water conservation efforts and fostering a shift in mindset. Entities like the Foundation for Ecological Security (FES) and the Watershed Organization Trust (WOTR) collaborate directly with rural communities, facilitating the implementation of water conservation projects, providing capacity-building support, and disseminating knowledge (Rastogi et al., 2024).

10.10 Education and Awareness

Promoting a culture of water conservation and sustainable agricultural practices heavily relies on education and awareness. Training initiatives targeted at farmers play a pivotal role in equipping them with the necessary knowledge and skills for effective water management. These programs, commonly facilitated by government agencies, research institutions, and non-governmental organizations (NGOs), encompass a diverse range of topics such as efficient irrigation methods, soil health preservation, and the adoption of climate-smart agricultural techniques. Public awareness campaigns and educational outreach endeavours, often conducted through Farmer Field Schools, serve to broaden understanding of the significance of water conservation within communities. Government-led initiatives like the 'Jal Shakti Abhiyan' and the 'Catch the Rain' campaign in India are designed to actively involve citizens in water conservation endeavours. These campaigns underscore the individual responsibility of each citizen in safeguarding water resources, fostering a collective sense of ownership and commitment to sustainable water management practices (Biswas, 2019; Rastogi et al., 2024).

10.11 Future Directions and Research Needs

As the impacts of climate change continue to unfold, the role of agroforestry in water management and conservation becomes increasingly significant. To address the challenges posed by changing environmental conditions and ensure the effectiveness of agroforestry practices, several future directions and research needs must be considered.

The adoption of soil conservation technologies in agroforestry ultimately hinges on the economic incentives for farmers. Although the potential of agroforestry is widely acknowledged, the scientific community has yet to provide concrete evidence demonstrating that agroforestry technologies are indeed more productive, cost-effective, and sustainable compared to traditional methods. Existing soil conservation research has primarily focused on quantifying reductions in soil and water loss, neglecting to measure improvements in soil fertility and economic benefits. It is crucial to distinguish between soil fertility improvements resulting from soil conservation practices and those stemming from other factors such as nitrogen fixation, nutrient cycling, and organic matter addition by trees. Additionally, understanding the long-term effects of agroforestry technologies on the resource base is imperative (Hurni et al., 2008).

Agroforestry systems are inherently complex due to the presence of multiple species and numerous variables influencing their performance. Further research is warranted on soil conservation technologies tailored specifically for agroforestry, such as contour hedgerow intercropping and windbreaks. This research should focus on selecting suitable species and management practices to minimize competition with crops. Additionally, investigations into the components of the water balance are essential for comprehending the sustainability of these systems, particularly in fragile environments. Emphasizing research on reclamation forestry and its applicability to delicate ecosystems is justified (Kiepe and Rao, 1994).

The role of agroforestry in watershed development and the criteria for selecting technologies for various purposes require attention. Conducting on-farm research is essential to understand farmer constraints, identify opportunities for adapting technologies to local conditions, and grasp social and policy constraints affecting adoption rates. Soil conservation benefits can be observed at different levels, including the farm, watershed, regional, or global scales (Kassam et al., 2014). While quantifying benefits at the farm level is relatively straightforward, evaluating social and environmental improvements at higher scales poses challenges. It is imperative to quantify the economic, social, and ecological impacts of conservation technologies across various scales using appropriate criteria to emphasize their value to both planners and practitioners of agroforestry (Nair and Toth 2016).

Conclusion

In conclusion, water management and conservation through agroforestry offer promising solutions to address the challenges posed by changing environmental conditions. Agroforestry practices, which integrate trees and shrubs into agricultural landscapes, play a crucial role in enhancing soil moisture retention, mitigating erosion, and promoting sustainable water use. As climate change intensifies, the importance of agroforestry in water management becomes increasingly evident, highlighting the need for adaptive strategies that can withstand shifting climatic patterns. However, to fully realize the potential of agroforestry for water management and conservation, concerted efforts are required. Future research should focus on developing climate-resilient agroforestry systems, integrating agroforestry into broader water resource management strategies, and quantifying the water benefits of agroforestry practices. Additionally, there is a need for policy support, capacity building, and interdisciplinary collaboration to promote the adoption of agroforestry-based water management practices. Overall, water management and conservation through agroforestry hold promise for enhancing agricultural resilience, improving ecosystem health, and ensuring the sustainable use of water resources in the face of changing environmental conditions. By harnessing the synergies between trees, crops, and soil, agroforestry has the potential to contribute significantly to water security and environmental sustainability now and in the future.

References

Abdallah, A. M., Jat, H. S., Choudhary, M., Abdelaty, E. F., Sharma, P. C., & Jat, M. L. 2021. Conservation agriculture effects on soil water holding capacity and water-saving varied with management practices and agroecological conditions: A Review. *Agronomy*, *11*(9), 1681.

Allan, J. D. 2004. Landscapes and riverscapes: the influence of land use on stream ecosystems. *Annu. Rev. Ecol. Evol. Syst.*, *35*, 257-284.

Baca, M., Läderach, P., Haggar, J., Schroth, G., & Ovalle, O. 2014. An integrated framework for assessing vulnerability to climate change and developing adaptation strategies for coffee-growing families in Mesoamerica. *PloS one*, *9*(2), e88463.

Bisht, S., Kumar, A., Kumar, N., Singh, H. and Kumar, P. 2022. Biofuel Production by Using Biomass and Its Application. In book: Renewable Energy and Green Technology in CRC Press.

Biswas, A. 2019. Catalyzing Peoples' Participation in Groundwater Management. *Groundwater Development and Management: Issues and Challenges in South Asia*, 505-528.

Cassman, K. G. 1999. Ecological intensification of cereal production systems: yield potential, soil quality, and precision agriculture. *Proceedings of the National Academy of Sciences*, *96*(11), 5952-5959.

Dagar, J. C., & Gupta, S. 2016. Agroforestry: Potentials for rehabilitation of degraded lands, constraints and the way forward. In J. C. Dagar, & J. C. Tewari (Eds.), *Agroforestry Research Developments* (pp. 47–98). New York: Nova Publishers.

Dhyani, S. K., Dadhwal, K. S., & Katiyar, V. S. 1988. Biological measures or rehabilitation of abandoned mine area – preliminary studies. *Indian Farming*, *38*(7), 7–10.

Gany, A. H. A., Sharma, P., & Singh, S. 2019. Global review of institutional reforms in the irrigation sector for sustainable agricultural water management, including water users' associations. *Irrigation and drainage*, *68*(1), 84-97.

Garrett, H. E. G., Wolz, K. J., Walter, W. D., Godsey, L. D., & McGraw, R. L. 2021. Alley cropping practices. *North American agroforestry*, 163-204.

Hu, X. J., Xiong, Y. C., Li, Y. J., Wang, J. X., Li, F. M., Wang, H. Y., & Li, L. L. 2014. Integrated water resources management and water users' associations in the arid region of northwest China: A case study of farmers' perceptions. *Journal of Environmental Management*, *145*, 162-169.

Hurni, H., Herweg, K., Portner, B., & Liniger, H. 2008. Soil erosion and conservation in global agriculture. *Land use and soil resources*, 41-71.

Jose, S. 2009. Agroforestry for ecosystem services and environmental benefits: an overview. *Agroforestry systems*, *76*, 1-10.

Kassam, A., Derpsch, R., & Friedrich, T. 2014. Global achievements in soil and water conservation: The case of Conservation Agriculture. *International Soil and Water Conservation Research*, *2*(1), 5-13.

Kaushal R, Tewari S., Thapliyal SD., Kumar, A., Roy, T., Islam, S., Lepcha, STS., Durai, J. 2021. Build-up of Labile, and non-labile carbon fractions under fourteen-year-old bamboo plantations in the Himalayan foothills. Heliyon 7(3):e07850

Kaushal, R., Mandal, D., Panwar, P., Kumar, P., Tomar, J. M. S., & Mehta, H. 2021. Soil and water conservation benefits of agroforestry. In *Forest resources resilience and conflicts* (pp. 259-275). Elsevier.

Kay, S. 2018. *Assessment of ecosystem services provided by agroforestry systems at the landscape scale* (Doctoral dissertation, University of Zurich).

Kiepe, P., & Rao, M. R. 1994. Management of agroforestry for the conservation and utilization of land and water resources. *Outlook on Agriculture*, *23*(1), 17-25.

Knox, J. W., Kay, M. G., & Weatherhead, E. K. 2012. Water regulation, crop production, and agricultural water management—Understanding farmer perspectives on irrigation efficiency. *Agricultural water management*, *108*, 3-8.

Kumar, A., Dwivedi, G. K., Tewari, S., Sah, V. K., Singh, H., Kumar, P., Kumar, N., Kaushal, R. 2020a. Soil organic carbon pools under *Terminalia chebula* Retz. based agroforestry system in the Himalayan foothills, India. Current Science (00113891), 118(7).

Kumar, A., Dwivedi, G.K., Tewari, S., Paul, J., Anand, R., Kumar, N., Kumar, P., Singh, H. and Kaushal, R. 2020b. Carbon Mineralization and Inorganic Nitrogen Pools under *Terminalia chebula* Retz.-Based Agroforestry System in Himalayan Foothills, India. Forest Science. 66(5), 634-643

Kumar, A., Hasanain, M., Singh, R., Verma, G., Kumar, D., & Mishra, R. 2020. Role of agroforestry measures for soil and water conservation. *Food and Scientific Reports*, *1*(March), 49-52.

Kumar, A., Kumar, P., Singh, H. and Kumar, N., 2021a. Impact of plant functional traits on near saturated hydraulic conductivity of soil under different forests of Kempty watershed in Garhwal Himalaya. Indian Journal of Soil Conservation, 49(1), pp.38-44.

Kumar, A., Kumar, P., Singh, H., Kumar, N. 2021b. Adaptation and mitigation potential of roadside trees with bio-extraction of heavy metals under vehicle emissions and their impact on physiological traits during seasonal regimes. Urban Forestry & Urban Greening, 58, p.126900.

Kumar, A., Kumar, P., Singh, H., Kumar, N. 2021c. Modulation of Plant Functional Traits Under Essential Plant Nutrients During Seasonal Regime in Natural Forests of Garhwal Himalayas. Plant and Soil 465(1):1-16

Kumar, A., Kumar, P., Singh, H., Kumar, N. 2021d. Relationship of physiological plant functional traits with soil carbon stock in the temperate forest of Garhwal Himalaya. Current Science 120(7): 1064-1068.

Kumar, A., Singh, H., Kumari, G., Bisht, S., Malik, A., Kumar, N., Singh, M., Raturi, A., Barthwal, S., Thakur, A. and Kaushal, R. 2022.

Adaptive resilience of roadside trees to vehicular emissions via leaf enzymatic, physiological, and anatomical trait modulations. Environmental Pollution, p.120191.

Kumar, A., Tewari, S., Singh, H., Kumar, P., Bisht, S. 2021e. Biomass accumulation and carbon stock in different agroforestry systems prevalent in the Himalayan foothills. Current Science 120(6), 1083-1088

Kumar, A., Tewari, S., Singh, Indra., Panday, R., Kumar, D. and Anand, R. 2018. Effect of nutrient sources on growth, yield and quality of turmeric under harad (*Terminalia chebula*) based agroforestry system Indian J. of Agroforestry, 20(2). 1-6

Kumar, M., & Kumar, R. 2018. Hydraulics of water and nutrient application through drip irrigation-A review. *Journal of Soil and Water Conservation*, *17*(1), 65-74.

Kumar, N., Jeena, N., Kumar, A., Khwairakpam, R., Singh, H. 2021. Comparative response of rice cultivars to elevated air temperature in Bhabar region of Indian Himalaya: Status on Yield attributes. Heliyon 7(6):e07474

Kumar, N., Kumar, A., Jeena, N., Singh, R., and Singh, H. 2020. Factors Influencing Soil Ecosystem and Agricultural Productivity at Higher Altitudes. Microbiological Advancements for

Kumar, N., Singh, H., Giri, K., Kumar, A., Joshi, A., Yadav, S., Singh, R., Bisht, S., Kumari, R., Jeena, N., Khairakpam, R., and Mishra, G. 2024. Physiological and molecular insights into the allelopathic effects on agroecosystems under changing environmental conditions. Physiology and Molecular Biology of Plants. 248. 119255.

Kumar, N., Verma, S., Kumar, A., Dashila, H., Suyal, D.C., Kumari, G., Rawat, S., Jeena, N., Singh, M., Kumar, A. 2022. Molecular insights into stress-responsive genes in the mitigation of environmental stresses. Unravelling Plant-Microbe Synergy, Edition: 1, Publisher: Elsevier

Kuyah, S., Öborn, I., & Jonsson, M. 2017. Regulating ecosystem services delivered in agroforestry systems. *Agroforestry: Anecdotal to modern science*, 797-815.

Lowrance, R., Altier, L. S., Newbold, J. D., Schnabel, R. R., Groffman, P. M., Denver, J. M., ... & Robinson, J. L. 1997. Water quality functions of riparian forest buffers in Chesapeake Bay watersheds. *Environmental management*.

Marques, M. J., Schwilch, G., Lauterburg, N., Crittenden, S., Tesfai, M., Stolte, J., & Doko, A. 2016. Multifaceted impacts of sustainable land management in drylands: a review. *Sustainability*, *8* (2), 177.

Mugendi, D. N., Waswa, B., Mucheru-Muna, M., & Mugwe, J. N. 2007. Agroforestry for Land and Water Management in Kenya. *Environment and Sustainable Development: A Guide for Tertiary Education in Kenya; Waswa, F., Otor, S., Olukoye, G., Mugendi, D., Eds*, 122-138.

Nair, P. R., & Toth, G. G. 2016. Measuring agricultural sustainability in agroforestry systems. *Climate Change and Multi-Dimensional Sustainability in African Agriculture: Climate Change and Sustainability in Agriculture*, 365-394.

Nair, P. R. 2011. Agroforestry systems and environmental quality: introduction. *Journal of environmental quality*, *40*(3), 784-790.

Nair, P. R., Kumar, B. M., Nair, V. D., Nair, P. R., Kumar, B. M., & Nair, V. D. 2021. Silvopastoral Systems (SPS) in the Tropics and Subtropics. *An Introduction to Agroforestry: Four Decades of Scientific Developments*, 169-193.

Pandey R., Dubey S., Tewari L, Kumar A., Kumar D., Sharma N. and Tewari S.K. 2018. Performance of soybean under poplar based agroforestry system in Terai region. Indian J. of Agroforestry, 20(1), 31-34.

Patil Shirish, S., Kelkar Tushar, S., & Bhalerao Satish, A. 2013. Mulching: A soil and water conservation practice. *Res. J. Agric. For. Sci*, *2320*, 6063.

Pavlidis, G., & Tsihrintzis, V. A. 2018. Environmental benefits and control of pollution to surface water and groundwater by agroforestry systems: a review. *Water Resources Management*, *32*, 1-29.

Pedersen, S. M., & Lind, K. M. 2017. Precision agriculture–from mapping to site-specific application. *Precision agriculture: Technology and economic perspectives*, 1-20.

Rahman, M. H. U., Ahrends, H. E., Raza, A., & Gaiser, T. 2023. Current approaches for modeling ecosystem services and biodiversity in agroforestry systems: Challenges and ways forward. *Frontiers in Forests and Global Change*, *5*, 1032442.

Rao, K. P. C., Verchot, L. V., & Laarman, J. 2007. Adaptation to climate change through sustainable management and development of agroforestry systems. *Journal of SAT agricultural research*, *4*(1), 1-30.

Rastogi, M., Kolur, S. M., Burud, A., Sadineni, T., Sekhar, M., Kumar, R., & Rajput, A. 2024. Advancing Water Conservation Techniques in Agriculture for Sustainable Resource Management: A review. *Journal of Geography, Environment and Earth Science International*, *28*(3), 41-53.

Schroth, G., da Fonseca, G. A., Harvey, C. A., Gascon, C., Vasconcelos, H. L., & Izac, A. M. N., (Eds.). 2013. *Agroforestry and biodiversity conservation in tropical landscapes*. Island press.

Sharda, V. N., & Venkateswarlu, B. 2007. Crop diversification and alternate land use systems in atershed Management. In Best–bet options for integrated watershed management. Proceedings of the comprehensive assessment of watershed programs in India (pp. 111–128), July 25–27. Patancheru, Andhra Pradesh, India: ICRISAT.

Sharma, P., Bhardwaj, D. R., Singh, M. K., Nigam, R., Pala, N. A., Kumar, A., & Thakur, P. 2023. Geospatial technology in agroforestry: status, prospects, and constraints. *Environmental Science and Pollution Research*, *30*(55), 116459-116487.

Sidhu, R. K., Kumar, R., Rana, P. S., & Jat, M. L. 2021. Automation in drip irrigation for enhancing water use efficiency in cereal systems of South Asia: Status and prospects. *Advances in agronomy*, *167*, 247-300.

Sitaram, S. S. 2015. Role of Agroforestry in Watershed Management.

Smith, P., Martino, D., Cai, Z., Gwary, D., Janzen, H., Kumar, P., & Smith, J. 2008. Greenhouse gas mitigation in agriculture. *Philosophical Transactions of the Royal Society B: Biological Sciences*, *363*(1492), 789-813.

Tewari, S., Tewari, L., Dubey, A., Kumar, A., Kumar, N., Kaushal, R. 2022. Use of the RAPD marker to determine the genetic diversity of various *Dalbergia sissoo* Roxb. (Shisham) genotypes and their evolutionary relationship in Indian subcontinents. Vegetos,

Tewari, S., Tewari, L., Kumar, A., Bisht, S, Kaushal, R. and Kumar, A. 2024. Germination Behaviour of Teak Genotypes and Effect of Different Treatments on Seed Germination Characteristics in Himalayan Foothills, India. Indian Forestry, 150 (3): 244-254

Tewari, S., Tewari, L., Kumar, A., Kaushal, R. 2021. Performance of *Glycine max* L. varieties under *Tectona grandis* L. and *Dalbergia sissoo* Roxb. based agroforestry system Indian J. of Agroforestry, 20(2) 1-6

Udawatta, R. P., & Gantzer, C. J. 2022. Soil and water ecosystem services of agroforestry. *Journal of Soil and Water Conservation*, *77*(1), 5A-11A.

Verma, T., Bhardwaj, D.R., Sharma, U., Sharma, P., Kumar, D., Kumar, A. and Kumar, A. 2023. Agroforestry systems in the mid-hills of the north-western Himalaya: A sustainable pathway to improved soil health and climate resilience. Journal of Environmental Management. 348, 119264.

Wallace, J. S., Young, A., & Ong, C. K. 2005. 26 The potential of agroforestry for sustainable land and water management. *This page was intentionally left blank*, 652.

Wato, T., & Amare, M. 2020. Opportunities and challenges of scaling up agroforestry practices in Sub-Saharan Africa: a review. *Agricultural Reviews, 41(3), 216-226.*

CHAPTER ELEVEN

POTENTIALS OF AGROFORESTRY FOR LIVELIHOOD IMPROVEMENT

KHWAIRAKPAM ROLINCE, KHULAKPAM APSHAHANA AND GOPAL SHUKLA

DEPARTMENT OF FORESTRY
NORTH EASTERN HILL UNIVERSITY,
TURA CAMPUS, MEGHALAYA

Abstract

A livelihood is considered sustainable if it can endure and bounce back from stresses and shocks, and maintain or enhance its capabilities and assets both now and in the future, without depleting the natural resource base. With the rapid growth of population, there has been an increase in overexploitation and degradation of forest lands. Many forest areas have been lost to urbanization and conversion into agricultural land Agroforestry is one of the most sustainable land management systems globally, offering significant socioeconomic benefits to farmers. If one crop fails, other crops can compensate for the loss. Thus, agroforestry systems are developed with sustainability in mind, emphasizing resilience, diversity, and the prevention of negative side effects. Agroforestry has been recognized for its potential to offer alternative sources of income and employment for the rural poor. It enhances sustainable livelihoods through four key pathways: food production, health and nutrition, provision of wood-based energy, and income generation. By encouraging farmers to plant trees on their cropland, agroforestry increases species variety, enhances productivity and income, and helps poor farmers sustain their livelihood.

11.1 Introduction

Livelihood refers to how people make money for themselves. A person's livelihood is defined by their means of subsistence, which include food, income, property, and skills (Chambers and Conway, 1991). A livelihood is deemed sustainable when it can endure and recover from stresses and shocks, as well as preserve or increase its capabilities and assets in the present and future, without depleting its supply of natural resources (FAO 2019). Livelihoods can include a variety of on-farm and off-farm activities that provide alternative approaches to acquiring food and money (Kumar et al., 2020a&b: Kaushal et al., 2021). Consequently, each household can have multiple sources of entitlement that form its livelihood. Since the millennial, livelihood in the rural areas is mainly sustained by dwelling inside the forest and from the crops. With the rapid growth of population, there has been an increase in overexploitation and degradation of forest lands. Many forest areas have been lost to urbanization and conversion into agricultural land. Furthermore, rapid urbanization and industrialization have led to a significant decrease in agricultural land. Cultivable land declined by 0.14% from 1976 to 2000 and by 0.73% from 2000 to 2010, with this trend expected to continue worsening (Hasan et al., 2013). Additionally, the current land use system, which allocates separate areas for agriculture and forestry, will not be sufficient to meet the needs of rural communities shortly (Hanif and Bari, 2013). Agroforestry has the potential to help mitigate the loss of agricultural and forest land.

Agroforestry, though relatively a new term, unbeknownst to humans, has been practised by mankind since time immemorial. Ancient civilizations, such as the Mayans and Greeks, integrated trees with crops and livestock to enhance soil fertility and yield stability (Tewari et al., 2024; Kumar et al., 2021a&b; Kumar et al., 2022). In India, the Indian Council of Agricultural Research, New Delhi initiated the All India Coordinated Research Project (AICRP) on Agroforestry in 1983, marking the start of agroforestry research during the eighth decade of the previous century. With this, the paradigm shift towards viewing agroforestry as a “sustainable land use system” contrasts with traditional approaches like monocropping or large-scale plantations (Russell and Franzel, 2004). Many authors have tried defining agroforestry in many ways. The commonly used definition is given by ICRAF, “Agroforestry is a collective name for land-use systems and technologies where woody perennials (trees, shrubs, palms, bamboos, etc.) are deliberately used on the same land-management units as crops and/or animals, in some form of spatial arrangement or temporal sequence.” In agroforestry systems, there are both ecological and economic interactions between the different components (Lundgren and Raintree, 1982). The agroforestry system supplies ample food, timber, fodder, fruit, fuelwood, construction materials, raw materials, and other products for small-scale forest-based enterprises and cottage industries (Abedin *et al*, 1987; Hasan *et al*, 1997; Rahman, 2011). Among the most commercially valuable products from cropland trees are poles and pulpwood, as these trees are often short-rotation species (Ghosh, 2011). Trees within crop fields serve as insurance against sudden crop failures, help protect crops from environmental hazards, and provide additional income (Chakraborty *et al.*, 2015). Additionally, if one crop fails, other crops can compensate for the loss. Thus, agroforestry systems are developed with sustainability in mind, emphasizing resilience, diversity, and the prevention of negative side effects (Brooks, 1995). Although agroforestry is increasingly recognized as an environmentally and economically viable land use, its economic potential as a competing land use from both private and social perspectives has not been thoroughly examined (Hazell and Wood, 2000; Molua, 2005; Franzel et al., 2001; Franzel and Scherr, 2002; Swinkels and Scherr, 1992). To enhance the role of agroforestry systems and innovations in poverty alleviation, it's crucial to delve into, support, and broaden market access and value-added opportunities for small-scale farmers, particularly in specialized markets and high-value products that go beyond basic staple crops (Tewari et al., 2022; Kumar et al., 2018; Kumar et al., 2021c&d). Well-designed agroforestry systems have the potential to transform bio-production systems sustainably, enhancing both the quality and quantity of food grains, fruits, fodder, industrial raw materials, fish, eggs, chickens, red meat, milk, silk, lace, and various other animal products. Additionally, they can increase the production of non-wood forest products (NWFPs), supporting the resurgence of sustainable livelihoods and having a significant, positive impact on ecosystem health (Quli *et al.*, 2017; Bossio *et al.*, 2007).

11.2 Agroforestry systems practised in different ecological zones of India

Various patterns of agroforestry systems are practised across different agroecological regions of India, reflecting the biophysical and social variations. Trees are planted along the borders or within fields, either systematically or at irregular intervals, often alongside crops like rice, wheat, pulses, jute, oilseeds, sugarcane, vegetables, and others. Additionally, farmers also grow shade-tolerant crops such as turmeric, ginger, and aroids under trees with high canopy coverage, such as jackfruit and mahogany (Kumar et al., 2022; Pandey et al., 2018; Bisht et al., 2022; Miah *et al*., 2002).

Table. Prominent agroforestry systems in different regions of India.

Agro-climatic zone	Agroforestry Systems	Tree Components	Crop/Grass
Western Himalayas	Silvopasture (RF)	*Grewia option*	*Setaria spp*
		Morus alba	*Setaria spp*
	Agrihorticulture	*Malus pumila*	Millets, Wheat
		Prunus persica	Maize, Soyabean
Eastern Himalayas	Agrisilviculture	*Anthocephalus cadamba*	Rice
	Agrihorticulture	*Alnus nepalensis*	Large Cardamom/ Coffee
	Silvipasture	*Parkia timoriana, Morus alba*	Bamboos
		Bauhinia variegate, Ficus spp, Morus alba	Napier
Lower Gangetic Plains	Agrisilviculture (Irri)	*Eucalyptus spp, Albizia lebbeck*	Rice
	Agrihorticulture (Irri)	*Mangifera indica, Musa spp, Litchi chinensis*	Wheat, Paddy, Maize
	Silvipasture	*Morus alba, Albizia lebbeck*	*Dicanthium pennisetum*
Middle Gangetic Plains	Agrisilviculture (Irri)	*Populus deltoids*	Sugarcane-Wheat
		Eucalyptus spp	Rice-Wheat
	Agrisilviculture	*Dalbergia sissoo*	Sesamum
	Agrihorticulture (Irri)	*Mangifera indica, Citrus spp*	Rice-Wheat
	Silvopasture	*Albizia lebbeck*	*Chrysopogon, Dicanthium*
Trans Gangetic Plains	Agrihorticulture (Irri)	*Phyllanthus emblica*	Black gram/ Green gram
	Agrisilviculture	*Azadirachta indica*	Black gram-Wheat/Mustard

	Silvopasture	*Bauhinia variegate, Albizia lebbeck*	*Cenchrus, Pennisetum*
Upper Gangetic Plains	Agrisilviculture (Irri)	*Populus deltoids*	Wheat, Bajra fodder
		Eucalyptus spp	Rice-Wheat
	Silvopasture	*Bauhinia variegate, Albizia lebbeck*	*Chrysopogon, Poa*
Eastern Plateau and Hills	Agrisilviculture	*Gmelina arborea*	Rice, Linseed
		Acacia nilotica	Rice
	Silvipasture	*Acacia mangium, A. nilotica,*	Bamboos
		Luecaena leucocephala	*Chrysopogon, Pennisetum, Dicanthium*
Central Plateau and Hills	Agrihorticulture (Irri)	*Psidium guajava*	Bengal gram/ Groundnut
	Agrihorticulture (RF)	*Phyllanthus emblica*	Black gram/ Green gram
	Agrisilviculture	*Acacia nilotica/ Luecaena leucocephala/ Azadirachta indica/ Albizia lebbeck*	Soyabean, Black gram-Mustard/Wheat
	Silvopasture (RF- and degraded lands)	*Albizia amara, Leucaena leucocephala, Dichrostycus cinera*	*Chrysopogon, Stylosanthes hamata, S. Scabra*
	TBOs (RF)	*Jatropha curcas*	-
Western Plateau and Hills	Agrisilvihorticulture (Irri)	*Tectona grandis, Achrus zapot*	Rice, Maize
	Agrihorticulture	*Areca catechu*	Black pepper, Cardamom
	Silvipasture	*Acacia mangium, Albizia amara*	*Cenchrus*
Southern Plateau and Hills	Agrisilviculture (RF)	*Eucalyptus spp, Casuarina equisetifolia, Ailanthus excels*	Cotton, Groundnut

	Agrisilviculture (Irri)	*Eucalyptus tetricornis*, *Melia dubia*	Chilli
	Agrihorticulture	*Tamarindus indica*	Chilli
	TBOs	*Milettia pinata*	-
East Coast Plains and Hills	Agrisilviculture (RF)	*Ailanthus excelsa*, *Acacia leucophloea*	Cow pea
	TBOs	*Milettia pinnata*	-
	Silvopasture	*Artocarpus spp.*	*Chrysopogon*, *Napier*, *Cenchrus*
West Coast Plains and Hills	Agrisilviculture (RF)	*Acacia auriculiformis*	Black pepper
			Rice
	Agrihorticulture (RF)	*Artocarpus heterophyllus*	Black pepper
	Agrihorticulture	*Cocos nucifera*/ *Areca catechu*	Rice
	Agrisilviculture	*Casuarina equisetifolia*	Rice
	Silvipasture	*Hardwickiana binnata*	*Cenchrus*
Gujarat Coast Plains and Hills	Agrisilviculture	*Azadirachta indica*, *Ailanthus excelsa*	Cow pea, Green gram
	Silvopasture	*Prosopis juliflora*, *Acacia nilotica*	*Cenchrus*, *Setaria*
Western Dry Region	Agrisilviculture	*Prosopis cineraria*, *Tecomella indica*, *Acacia nilotica*, *Azadirachta indica*	Pearl millet
	TBOs	*Jatropha curcas*	-
	Silvopasture	*Albizia lebbeck*, *Hardwickiana binnata*	Cenchrus
All Islands	Agrihorticulture	*Cocos nucifera*	Rice
	Silvipasture	*Bauhinia spp*, *Erythrina indica*, *Leucaena leucochephala*	*Cenchrus*, *Pennisetum*

Irri: irrigated; **RF:** rainfed; **TBOs:** tree-borne oil seed ***(Source: Dhyani et al., 2009)***

11.3 Significance of Agroforestry in improving livelihood

Agroforestry serves as a key strategy to involve forest inhabitants in the rehabilitation of degraded forest lands. It is a dynamic, ecologically sound natural resource management system that integrates trees and woody perennials into agricultural and grazing lands, enhancing, diversifying and sustaining production for increased social, economic, and environmental benefits (Verma et al., 2023; Kumar et al., 2020; Kumar et al., 2021e). The farms-industries connections have also contributed to making the agroforestry systems more sustainable than traditional cropping practices (Kareemulla *et al*., 2005; Saxena, 2000).

Agroforestry has been recognized for its potential to offer alternative sources of income and employment for the rural poor. It enhances sustainable livelihoods through four key pathways: food production, health and nutrition, provision of wood-based energy, and income generation (Kuyah *et al*., 2020). It bolsters household livelihood security by producing a variety of goods such as food and food additives, fuelwood, fibre, medicines, gums and resins, oils and fragrances, and livestock fodder. Agroforestry also supports the production of food from crops and livestock, including meat, milk, and honey. These products can be used directly for household consumption or sold in markets to generate income. This income can then cover expenses related to housing, education, sanitation, and social integration, thereby improving overall livelihood security (Tewari et al., 2021; Kumar et al., 2021 Kumar et al., 2024; Kuyah *et al*., 2019).

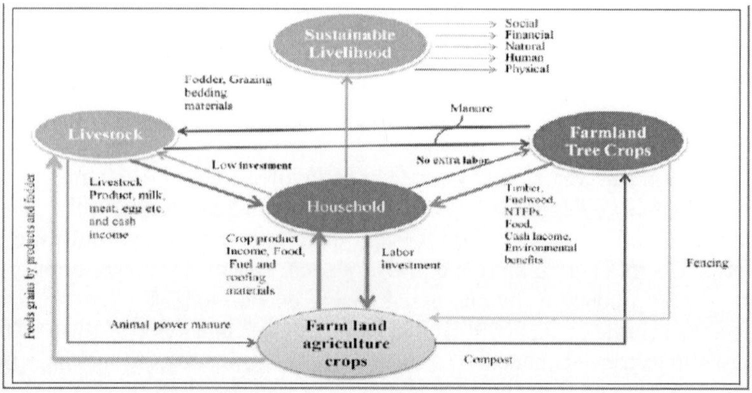

Fig 1. Contribution of Agroforestry on Sustainable Livelihood (Chakraborty et. al., 2015)

Food Production: By promoting crop productivity and supplying edible tree and livestock products, agroforestry improves food and nutritional security (Dewees *et al.* 2011; Jamnadass *et al.* 2013; Franzel *et al.* 2014). Trees improve the growth and yield of crops when appropriate species are planted at optimal densities and suitable pruning regimes are applied (Bayala *et al.*, 2002, 2015). Fruits and vegetables common in home gardens enhance food security by maturing at various times throughout the year.

Health and Nutrition: Unlike the dietary simplification that comes with conventional agricultural intensification centred on cereals (Fanzo *et al.*, 2013), agroforestry enhances dietary diversity by providing a wider range of food options (Jamnadass *et al.*, 2013; Kehlenbeck *et al.*, 2013). The variety of products (fruits, vegetables, spices, etc.) available throughout the year in systems like home gardens not only supports food security during "lean" seasons but also ensures dietary diversity. These products are rich in mineral nutrients, which enhance household nutritional security, particularly for vulnerable groups such as women and children. Fruits, nuts, kernels, oils, condiments, vegetables, and medicinal and aromatic plants are key sources of dietary minerals that contribute to nutritional security (Kalinganire *et al.*, 2007; Kehlenbeck *et al.*, 2013). Many of these products have higher levels of micronutrients, vitamins, fibre, and protein compared to staple crops (Whitney et al., 2017). Agroforestry enables communities to access products by domesticating indigenous shrub and tree species, ensuring households have a year-round supply of vitamin-rich fruits and vegetables (Akinnifesi *et al.*, 2007; Ofori *et al.*, 2014; Fanzo *et al.*, 2013).

Wood-based Energy: Agroforestry fuelwood production offers more sustainable alternatives compared to forest and woodland sources (Iiyama *et al.*, 2014). For instance, firewood from agroforestry is often harvested by selecting branches or collecting deadwood from trees, and woodlots are sometimes established on farms for charcoal production. Having fuelwood on the farm is crucial for women and children, who are typically responsible for fuelwood collection for cooking in most rural areas. Fuelwood is often reported as a priority use of on-farm trees in many surveys of on-farm agrobiodiversity or the livelihood benefits of trees (Tabuti, 2012; Reppin *et al.*, 2019).

Income: Agroforestry enhances household income by providing surplus staples or vegetables, livestock or livestock products, and various tree products for sale. These marketable tree products include fruits, nuts, kernels, edible leaves, oils, condiments, gums, resins, building poles, stakes for climbing beans, timber, fuelwood, fodder, and medicines (Kalinganire *et al.*, 2007; Dewees et al., 2011; Place et al., 2016). Non-timber forest products provide farmers with a consistent income, though the value of these products varies significantly by species and region (Place et al., 2016). Agroforestry also offers households opportunities to generate income through processing and adding value to tree products, as well as establishing tree nurseries (Asaah et al., 2011). The production of wood on farms, including timber and poles, for local markets is on the rise (Place et al., 2016; Sharma et al., 2016), significantly contributing to the improvement of rural livelihoods through income generation (Kiplagat et al., 2011). Economic studies evaluating farm returns to land and labour, using net present values (NPV) and benefit-cost ratios (BCR), show that fertilizer trees perform as well as or better than inorganic fertilizers (Ajayi et al., 2009; Kamanga et al., 2010).

Agroforestry is an effective way to ensure livelihood security through its various benefits and services. This impressive land-use technique offers a solution to the issue of low land productivity faced by resource-limited smallholder farmers (Puri & Nair, 2004). Agroforestry products, including food crops and tree resources, meet the diverse needs of rural communities. By encouraging farmers to plant trees on their cropland, agroforestry increases species variety, enhances productivity and income, and helps poor farmers sustain their livelihoods (Rahman et al., 2012).

11.4 Conclusion

The potential of agroforestry to improve livelihoods is vast and diverse. Agroforestry systems offer sustainable solutions to many problems faced by rural communities by integrating trees and shrubs into agricultural land. These systems make farming more productive, provide multiple sources of income, and increase resilience to environmental stress. Soil health helps increase crop yields, and nitrogen fixation increases water retention. These benefits can lead to sustainable and profitable farming, improving household food security and income. , medicinal plants, and non-timber forest products. Diversification of farming activities reduces financial risk and provides additional income throughout the year. In areas where economic activity is limited, agroforestry can play an important role in meeting local needs and reducing dependence on external inputs. Agroforestry systems sequester carbon, help mitigate climate change, and increase biodiversity by providing habitat for a variety of plant and animal species. These ecological services are important for maintaining the overall health of rural areas and ensuring the sustainability of agriculture. When local communities are involved in the planning and management of agroforestry programs, these measures will be successful and sustainable. This partnership fosters the sense of ownership and responsibility necessary for long-term sustainability. Agroforestry can contribute to the economy and health of rural communities by providing agricultural production, income diversification and important environmental services. Therefore, agroforestry practices need to be adopted and supported by appropriate policies, research and community participation to realise their potential to improve livelihoods and promote growth.

References

Abedin, Z., Aktar, S., Haque, F. and Alam, S., 1987, November. Uses of multipurpose trees on the small farms of the low-rainfall Ganges floodplain soils of Bangladesh. In *Proceedings of an international workshop held in Pattaya, Thailand* (pp. 31-47).

Ahmad, S., Caihong, Z. and Ekanayake, E.M.B.P., 2021. Livelihood improvement through agroforestry compared to conventional farming system: Evidence from Northern Irrigated Plain, Pakistan. *Land*, *10*(6), p.645.

Ajayi, O.C., Akinnifesi, F.K., Sileshi, G. and Kanjipite, W., 2009. Labour inputs and financial profitability of conventional and agroforestry-based soil fertility management practices in Zambia. *Agrekon*, *48*(3), pp.276-292.

Akinnifesi, F.K., Ajayi, O.C., Sileshi, G., Matakala, P., Kwesiga, F.R., Ham, C., Kadzere, I., Mhango, J., Mng'Omba, S.A., Chilanga, T. and Mkonda, A., 2007. Creating opportunities for domesticating and commercializing miombo indigenous fruit trees in Southern Africa. In *Indigenous fruit trees in the tropics: domestication, utilization and commercialization* (pp. 137-170). Wallingford UK: CABI.

Asaah, E.K., Tchoundjeu, Z., Leakey, R.R., Takousting, B., Njong, J. and Edang, I., 2012. Trees, agroforestry and multifunctional agriculture in Cameroon. In *Sustainable Intensification* (pp. 110-119). Routledge.

Bayala, J., Sanou, J., Teklehaimanot, Z., Ouedraogo, S.J., Kalinganire, A., Coe, R. and van Noordwijk, M., 2015. Advances in knowledge of processes in soil–tree–crop interactions in parkland systems in the West African Sahel: A review. *Agriculture, Ecosystems & Environment*, *205*, pp.25-35.

Bayala, J., Teklehaimanot, Z. and Ouedraogo, S.J., 2002. Millet production under pruned tree crowns in a parkland system in Burkina Faso. *Agroforestry systems*, *54*, pp.203-214.

Bisht, S., Kumar, A., Kumar, N., Singh, H. and Kumar, P. 2022. Biofuel Production by Using Biomass and Its Application. In book: Renewable Energy and Green Technology in CRC Press.

Bossio, D., Critchley, W., Geheb, K., van Lynden, G., Mati, B., Bhushan, B., Hellin, J., Jacks, G., Kolff, A., Nachtergaele, F. and Neely, C., 2007. Conserving land–protecting water. *Water for Food, Water for Life: a Comprehensive Assessment of Water Management in Agriculture. Earthscan, London and International Water Management Institute, Colombo, Sri Lanka*, pp.551-583.

Brooks, K.N., Gregersen, H.M. and Ffolliott, P.F., 1995. Role of Agroforestry in Sustainable Land-Use Systems[1]. In Agroforestry and Sustainable Systems: Symposium proceedings (Vol. 261, p. 199). US Department of Agriculture, Forest Service, Rocky Mountain Forest and Range Experiment Station.

Chakraborty, M., Haider, M.Z. and Rahaman, M.M., 2015. Socio-economic impact of cropland agroforestry: evidence from Jessore district of Bangladesh. *International Journal of Research in Agriculture and Forestry*, *2*(1), pp.11-20.

Chambers, R. and Conway, G., 1991. Sustainable livelihood.

Dewees, P., Place, F., Scheer, S.J. and Buss, C., 2011. *Investing in trees and landscape restoration in Africa: what, where, and how*. World Bank.

Dhyani, S.K., Kareemulla, K., Ajit, A. and Handa, A.K., 2009. Agroforestry potential and scope for development across agro-climatic zones in India.

Fanzo, J., Hunter, D., Borelli, T. and Mattei, F., 2013. *Diversifying food and diets*. London: Routledge.

Franzel, S., Carsan, S., Lukuyu, B., Sinja, J. and Wambugu, C., 2014. Fodder trees for improving livestock productivity and smallholder livelihoods in Africa. *Current opinion in environmental sustainability*, *6*, pp.98-103.

Franzel, S., Coe, R., Cooper, P., Place, F. and Scherr, S.J., 2001. Assessing the adoption potential of agroforestry practices in sub-Saharan Africa. *Agricultural systems*, *69*(1-2), pp.37-62.

Franzel, S.C. and Scherr, S.J. eds., 2002. *Trees on the farm: assessing the adoption potential of agroforestry practices in Africa*. CABI.

Ghosh, S.R., Wadud, M.A., Mondol, M.A. and Rahman, G.M.M., 2011. Optimization of plant density of Akashmoni (Acacia auriculiformis) for production of fuel wood in the bunds of cropland. *Journal of Agroforestry and Environment*, *5*(2), pp.1-6.

Handa, A.K., Toky, O.P., Dhyani, S.K., Chavan, S.B. and Toky, I.D., 2016. Innovative agroforestry for livelihood security in India. *World Agriculture*, *7*, pp.7-16.

Hanif, M.A. and Bari, M.S., 2013. Potentiality of potato-based agrisilvicultural land use system in the northern part of Bangladesh. *Sci Secure J Biotechnol*, *2*, pp.61-65.

Hanif, M.A., Roy, R.M., Bari, M.S., Ray, P.C., Rahman, M.S. and Hasan, M.F., 2018. Livelihood improvements through agroforestry: Evidence from Northern Bangladesh. *Small-scale Forestry*, *17*, pp.505-522.

Hasan, M.K., Satter, M.A., Asaduzzaman, S.M., Hussain, M.A. and Haque, M.F., 1997. Agroforestry systems in the crop field and homestead at Bagherpara FSR Site, Jessore. In *Proc. National Workshop on Agroforestry Research, Dhaka, Bangladesh* (pp. 55-62).

Hasan, M.N., Hossain, M.S., Islam, M.R., Bari, M.A., Karim, D. and Rahman, M.Z., 2013. Trends in the availability of agricultural land in Bangladesh. *Soil Resource Development Institute (SERDI), Ministry of Agriculture, Bangladesh, Dhaka.*

Hazell, P. and Wood, S., 2000. From science to technology adoption: the role of policy research in improving natural resource management. *Agriculture, ecosystems & environment*, *82*(1-3), pp.385-393.

Iiyama, M., Neufeldt, H., Dobie, P., Njenga, M., Ndegwa, G. and Jamnadass, R., 2014. The potential of agroforestry in the provision of sustainable wood fuel in sub-Saharan Africa. *Current Opinion in Environmental Sustainability*, *6*, pp.138-147.

Jamnadass, R., Place, F., Torquebiau, E., Malézieux, E., Liyama, M., Sileshi, G., Kehlenbeck, K., Masters, E., McMullin, S. and Dawson, I., 2013. Agroforestry, food and nutritional security.

Kalinganire, A., Weber, J.C., Uwamariya, A. and Kone, B., 2007. Improving rural livelihoods through domestication of indigenous fruit trees in the parklands of the Sahel. In *Indigenous fruit trees in the tropics: domestication, utilization and commercialization* (pp. 186-203). Wallingford UK: CABI.

Kamanga, B.C.G., Waddington, S.R., Robertson, M.J. and Giller, K.E., 2010. Risk analysis of maize-legume crop combinations with smallholder farmers varying in resource endowment in central Malawi. *Experimental agriculture*, *46*(1), pp.1-21.

Kareemulla, K., Rizvi, R.H., Kumar, K., Dwivedi, R.P. and Singh, R., 2005. Poplar agroforestry systems of western Uttar Pradesh in northern India: a socioeconomic analysis. *Forests, trees and livelihoods*, *15*(4), pp.375-381.

Kaushal R, Tewari S., Thapliyal SD., Kumar, A., Roy, T., Islam, S., Lepcha, STS., Durai, J. 2021. Build-up of Labile, and non-labile carbon fractions under fourteen-year-old bamboo plantations in the Himalayan foothills. Heliyon 7(3):e07850

Kehlenbeck, K., Asaah, E. and Jamnadass, R., 2013. Case study 3: Diversity of indigenous fruit trees and their contribution to nutrition and livelihoods in sub-Saharan Africa: examples from Kenya and Cameroon. In *Diversifying Food and Diets* (pp. 257-269).

Kiplagat, J.K., Wang, R.Z. and Li, T.X., 2011. Renewable energy in Kenya: Resource potential and status of exploitation. *Renewable and sustainable energy reviews*, *15*(6), pp.2960-2973.

Kumar, A., Dwivedi, G. K., Tewari, S., Sah, V. K., Singh, H., Kumar, P., Kumar, N., Kaushal, R. 2020a. Soil organic carbon pools under *Terminalia chebula* Retz. based agroforestry system in the Himalayan foothills, India. Current Science (00113891), 118(7).

Kumar, A., Dwivedi, G.K., Tewari, S., Paul, J., Anand, R., Kumar, N., Kumar, P., Singh, H. and Kaushal, R. 2020b. Carbon Mineralization and Inorganic Nitrogen Pools under *Terminalia chebula* Retz.-Based Agroforestry System in Himalayan Foothills, India. Forest Science. 66(5), 634-643

Kumar, A., Kumar, P., Singh, H. and Kumar, N., 2021a. Impact of plant functional traits on near-saturated hydraulic conductivity of soil under different forests of Kempty watershed in Garhwal Himalaya. Indian Journal of Soil Conservation, 49(1), pp.38-44.

Kumar, A., Kumar, P., Singh, H., Kumar, N., 2021b. Adaptation and mitigation potential of roadside trees with bio-extraction of heavy metals under vehicle emissions and their impact on physiological traits during seasonal regimes. Urban Forestry & Urban Greening, 58, p.126900.

Kumar, A., Kumar, P., Singh, H., Kumar, N., 2021c. Modulation of Plant Functional Traits Under Essential Plant Nutrients During Seasonal Regime in Natural Forests of Garhwal Himalayas. Plant and Soil 465(1):1-16

Kumar, A., Kumar, P., Singh, H., Kumar, N., 2021d. Relationship of physiological plant functional traits with soil carbon stock in the temperate forest of Garhwal Himalaya. Current Science 120(7): 1064-1068.

Kumar, A., Singh, H., Kumari, G., Bisht, S., Malik, A., Kumar, N., Singh, M., Raturi, A., Barthwal, S., Thakur, A. and Kaushal, R., 2022. Adaptive resilience of roadside trees to vehicular emissions via leaf enzymatic, physiological, and anatomical trait modulations. Environmental Pollution, p.120191.

Kumar, A., Tewari, S., Singh, H., Kumar, P., Bisht, S., 2021e. Biomass accumulation and carbon stock in different agroforestry systems prevalent in the Himalayan foothills. Current Science 120(6), 1083-1088

Kumar, A., Tewari, S., Singh, Indra., Panday, R., Kumar, D. and Anand, R. 2018. Effect of nutrient sources on growth, yield and quality of turmeric under harad (*Terminalia chebula*) based agroforestry system Indian J. of Agroforestry, 20(2). 1-6

Kumar, N., Jeena, N., Kumar, A., Khwairakpam, R., Singh, H. 2021. Comparative response of rice cultivars to elevated air temperature in Bhabar region of Indian Himalaya: Status on Yield attributes. Heliyon 7(6):e07474

Kumar, N., Kumar, A., Jeena, N., Singh, R., and Singh, H. 2020. Factors Influencing Soil Ecosystem and Agricultural Productivity at Higher Altitudes. Microbiological Advancements for Higher Altitude Agro-Ecosystems & Sustainability. Springer. p. 55- 71.

Kumar, N., Singh, H., Giri, K., Kumar, A., Joshi, A., Yadav, S., Singh, R., Bisht, S., Kumari, R., Jeena, N., Khairakpam, R., and Mishra, G. 2024. Physiological and molecular insights into the allelopathic effects on agroecosystems under changing environmental conditions. Physiology and Molecular Biology of Plants. 248. 119255.

Kumar, N., Verma, S., Kumar, A., Dashila, H., Suyal, D.C., Kumari, G., Rawat, S., Jeena, N., Singh, M., Kumar, A. 2022. Molecular insights into stress-responsive genes in the mitigation of environmental stresses. Unravelling Plant-Microbe Synergy, Edition: 1, Publisher: Elsevier

Kuyah, S., Sileshi, G.W., Luedeling, E., Akinnifesi, F.K., Whitney, C.W., Bayala, J., Kuntashula, E., Dimobe, K. and Mafongoya, P.L., 2020. The potential of agroforestry to enhance livelihood security in Africa. *Agroforestry for Degraded Landscapes: Recent Advances and Emerging Challenges-Vol. 1*, pp.135-167.

Kuyah, S., Whitney, C., Jonsson, M., Sileshi, G., Öborn, I., Muthuri, C. and Luedeling, E., 2019. Can agroforestry enhance ecosystem services provision without reducing productivity?

Lundgren, B. and Raintree, J.B. 1983. Sustained agroforestry. In B. Nestel (ed). Agricultural Research for Development: Potentials and Challenges in Asia. ISNAR. The Hague

Miah, M.G., Ahmed, F.U., Ahmed, M.M., Alam, M.N., Choudhury, N.H. and Hamid, M.A., 2002, November. Agroforestry in Bangladesh: Potentials and opportunities. In *South Asia regional agroforestry consultation workshop was held on* (pp. 23-25).

Molua, E.L., 2005. The economics of tropical agroforestry systems: the case of agroforestry farms in Cameroon. *Forest policy and economics*, *7*(2), pp.199-211.

Nair PKR (1993) An introduction to agroforestry. Kluwer, Dordrecht

Ofori, D.A., Gyau, A., Dawson, I.K., Asaah, E., Tchoundjeu, Z. and Jamnadass, R., 2014. Developing more productive African agroforestry systems and improving food and nutritional security through tree domestication. *Current Opinion in Environmental Sustainability*, *6*, pp.123-127.

Pandey R., Dubey S., Tewari L., Kumar A., Kumar D., Sharma N. and Tewari S.K. 2018. Performance of soybean under poplar-based agroforestry system in Terai region. Indian J. of Agroforestry, 20(1), 31-34.

Place, F., Garrity, D., Mohan, S. and Agostini, P., 2016. *Tree-based production Systems for Africa's Drylands*. World Bank Publications.

Puri, S. and Nair, P.K.R., 2004. Agroforestry research for development in India: 25 years of experiences of a national program. *Agroforestry Systems*, *61*, pp.437-452.

Quli, S.M.S., Islam, M.A. and Singh, P.K., 2017. Mitigating livelihood crisis through agroforestry interventions in rural India. *Jharkhand Journal of Development and Management Studies*, *15*(1), pp.7159-7178.

Rahman, S.A., 2011. Cost-benefit and livelihood impacts of agroforestry in Bangladesh: Impacts of agroforestry in Bangladesh.

Rahman, S.A., Imam, M.H., Snelder, D.J. and Sunderland, T., 2012. Agroforestry for livelihood security in Agrarian landscapes of the Padma floodplain in Bangladesh. *Small-scale Forestry*, *11*, pp.529-538.

Rasul, G. and Thapa, G.B., 2006. Financial and economic suitability of agroforestry as an alternative to shifting cultivation: The case of the Chittagong Hill Tracts, Bangladesh. *Agricultural Systems*, *91*(1-2), pp.29-50.

Reppin, S., Kuyah, S., de Neergaard, A., Oelofse, M. and Rosenstock, T.S., 2020. Contribution of agroforestry to climate change mitigation and livelihoods in Western Kenya. *Agroforestry Systems*, *94*, pp.203-220.

Russell, D. and Franzel, S., 2004. Trees of prosperity: Agroforestry, markets and the African smallholder. Agroforestry systems, 61, pp.345-355.

Saxena, N.C., 2000. Farm and agroforestry in India-Policy and legal issues. *Planning Commission. Government of India. 50p*.

Swinkels, R.A. and Scherr, S.J., 1992. Economic analysis of agroforestry technologies. An annotated bibliography.

Tabuti, J.R., 2012. Important woody plant species, their management and conservation status in Balawoli sub-county, Uganda.

Tewari, S., Tewari, L., Dubey, A., Kumar, A., Kumar, N., Kaushal, R. 2022. Use of the RAPD marker to determine the genetic diversity of various *Dalbergia sissoo* Roxb. (Shisham) genotypes and their evolutionary relationship in Indian subcontinents. Vegetos,

Tewari, S., Tewari, L., Kumar, A., Bisht, S, Kaushal, R. and Kumar, A. 2024. Germination Behaviour of Teak Genotypes and Effect of Different Treatments on Seed Germination Characteristics in Himalayan Foothills, India. Indian Forestry, 150 (3): 244-254

Tewari, S., Tewari, L., Kumar, A., Kaushal, R. 2021. Performance of *Glycine max* L. varieties under *Tectona grandis* L. and *Dalbergia sissoo* Roxb. based agroforestry system Indian J. of Agroforestry, 20(2) 1-6

Tiwari, P., Kumar, R., Thakur, L., Salve, A. and Parmar, Y., 2017. Agroforestry for sustainable rural livelihood: a review. *International Journal of Pure and Applied Bioscience*, *5*(1), pp.299-309.

Verma, T., Bhardwaj, D.R., Sharma, U., Sharma, P., Kumar, D., Kumar, A. and Kumar, A. 2023. Agroforestry systems in the mid-hills of the north-western Himalaya: A sustainable pathway to improved soil health and climate resilience. Journal of Environmental Management. 348, 119264.

Whitney, C.W., Tabuti, J.R., Hensel, O., Yeh, C.H., Gebauer, J. and Luedeling, E., 2017. Home gardens and the future of food and nutrition security in southwest Uganda. *Agricultural Systems*, *154*, pp.133-144.

CHAPTER TWELVE

INNOVATIONS IN AGROFORESTRY: HARNESSING TECHNOLOGY FOR SUSTAINABILITY

ABHAY KUMAR[1*], NARESH SINGH[5], RIKESH KUMAR[2], TANU SHREE LAKRA[3], MUNESHWAR PRASAD[1], DINESH MAHTO[1], VARSHA KUMARI[1], RASHMI PRIYADARSHI[4] AND RAM GOPAL[6]

[1]KVK, JEHANABAD, BIHAR AGRICULTURAL UNIVERSITY, SABOUR, BHAGALPUR, BIHAR, INDIA-813210.
[2]FACULTY OF FORESTRY, BIRSA AGRICULTURAL UNIVERSITY, RANCHI, JHARKHAND, INDIA-834006.
[3] INSTITUTE OF FOREST PRODUCTIVITY, RANCHI, JHARKHAND, INDIA-835303.
[4]KVK, PURNEA, BIHAR AGRICULTURAL UNIVERSITY, SABOUR, BHAGALPUR, BIHAR, INDIA-813210.
[5] COLLEGE OF FORESTRY, RANICHAURI, VCSG UUHF, BHARSAR, PAURI GARHWAL, UTTRAKHAND, INDIA-246123
[6]NORTH EASTERN HILL UNIVERSITY, TURA CAMPUS, MEGHALAYA

Abstract

Agroforestry is a sustainable land use practice has enormous potential to achieve net zero carbon, and many sustainable development goals (SDGs) like poverty, hunger, sustainable economic growth, women empowerment, afforestation and deforestation with mitigation of climate change. Innovative agroforestry involvements integrating tree, crop, animal production, horticulture etc. into cropping systems exist that can help farmers increase yields and provide flexibility for supporting livelihoods, particularly among rural communities.

In the current scenario, innovative agroforestry involvements have the huge potential to conserve natural resources, biodiversity, and landscape, improve productivity and provide resilience to climate change. Thus, this chapter explores the potential of innovation in agroforestry and technologies like Agroforestry towards net zero carbon, Remote Sensing and GIS in Agroforestry, Land Use/Land Cover Mapping in Agroforestry, Climate Smart Agroforestry, Estimation of Carbon sequestration&Carbon Credit, Agroforestry encourages jobs and green growth, Economics of agroforestry system, Agroforestry landscape, Urban agroforestry for food and nutritional security, Agroforestry: pests, diseases and weeds, Roots dynamics in agroforestry, Agroforestry modelling etc. for the sustainable management of the natural resource, land, water and food security among climate change.

Keywords: Agroforestry, innovation, economics, modelling, net zero carbon, climate change

12.1 Introduction

Globally, modern innovations have become an important substance following the increase of sustainability farming with agroforestry, technological platforms for digital sensor technology, artificial intelligence (AI), remote sensing, sensor designs technology and data analytics are just a few of the many methods that are included in these developments (Kumar et al., 2020a&b: Kaushal et al., 2021). They possess the huge potential to change conventional techniques for agriculture sciences, maximize usage of resources, lessen their negative effects on the environment and enhance the sustainable development of agriculture. Developing countries have distinctive and complicated obstacles to achieving sustainable agriculture. Key obstacles include irregularity of innovation and adoption, insufficient infrastructure, scarce access to current farming techniques and lack of funding. In addition, macroeconomic factors like poverty, insufficient levels of education and social injustices might make it more difficult to hold and implement technical solutions. Climatic effects can affect various areas of contemporary agriculture, including profitability, prices, markets, finance and societal variables. Climate change and environmental degradation can harm the buffering effect of trees and forests during times of food insecurity. Sustainable farming aims at boosting resilience, livelihoods and food security, while also addressing adaptation and mitigation priorities (Tewari et al., 2024; Kumar et al., 2021a&b; Kumar et al., 2022). Agroforestry is a vital part of sustainable agriculture and climate-smart agriculture, which aims to provide defended residing zones for food systems in anticipation of climate change. Around the world, agroforestry has been a viable choice for ecological agriculture for hundreds of years. It is made up of trees assigned together over space, light, water and time complemented by cyclical and perennial plants, thereby are more frequently combined with livestock or pasture and offer an arrangement of agroecological perks (Verma et al., 2023; Kumar et al., 2020; Kumar et al., 2021e).

Agroforestry is a reasonable sustainable option which has been used for a long time spanning the world. It encompasses a spatial or temporal association of trees with other periodic and recurring vegetation, which are more typically incorporated with farm animals/pasture, which provides multiple sustainability rewards and in addition, boosts agricultural yield in an ecologically friendly manner while retaining or extending wooded land, fuel wood, fibre, fertilizer, flosses, flower, fruit etc. (Tewari et al., 2022; Kumar et al., 2018; Kumar et al., 2021c&d).

Innovative technology to achieve sustainability in agroforestry involves combining diverse technical tools and methodologies to optimize resource usage, boost effectiveness and strengthen the conservation of the environment.

Agroforestry-related innovations include

12.1.1 Remote sensing and Geographic information system

Remote sensing is the art and science of obtaining information about an object, area, or phenomenon through the analysis of data acquired by a device without being in physical contact. Remote sensing technology involves acquiring information about the earth's surface and atmosphere, using satellite-based sensors, interpretation, of spectral measurements and characterization of land resources and environmental phenomena. A GIS is a computer-based information system for input, storage, analysis, management, manipulation, retrieval and output of geographic data and their attributes. All information in GIS must be linked to a spatial reference (Latitude/longitude) or other spectral coordinates. Thus the GIS uses geo-references as the primary means of storing and accessing information (Joshi, 2013).

Remote sensing technology, such as satellite imaging and drones, joined with Geographic Information Systems (GIS), allows farmers to monitor agroforestry systems at a landscape scale. The deployment of optical remote detection in agroforestry requires geographic extent mapping, production of tree species spectral signature, carbon sequestration assessment and suitability mapping. Concurrently, the recent increase in the use of synthetic aperture radar in conjunction with algorithms based on vegetation photosynthesis and optical data offers a more accurate estimation of gross primary productivity at different scales. However, unmanned aerial vehicles furnished with sensors, such as multispectral, LiDAR, hyperspectral, and thermal, offer a substantially greater aptitude and reliability than satellite-based statistics. Into the foreseeable future, the wellness supervision of agroforestry operations can be a key concern that may be addressed by utilizing hyperspectral and thermal datasets to study vegetation life sciences, fluorescence from chlorophyll, and drought stress. This technique can be used to assess tree health, monitor vegetation cover and identify sites for prospective agroforestry development (Kumar et al., 2022; Pandey et al., 2018; Bisht et al., 2022).

12.1.2 Land Use/Land Cover Mapping

Land Use/Land Cover mapping of agroforestry systems and related interventions is an important approach for developmental planning at local, regional and national levels, given that the demand for up-to-date geospatial information for informed decision-making is increasing exponentially. The ecological conditions, altitudes, geological structure and slope, in addition to the technological, socioeconomic, and institutional setup, all have a major role in determining land use/cover. These factors can also be used to categorize different types of land cover within agroforestry landscapes. The LULC has extremely changed as a result of urbanization, different construction work, industrialization and population expansion. Although variations in land use and cover may not always indicate deterioration of the land, LULC change is one of the main causes of global climate change and has an impact on many aspects of the geoenvironmental and natural ecosystems, including radiation heat wave, drought, biodiversity and water availability. Variations in the state and makeup of the land cover have an impact on the climate, natural resources, biogeochemical cycles, energy flows and human livelihoods (Tewari et al., 2021; Kumar et al., 2021 Kumar et al., 2024). The spatial distribution of trees, crops, and other components of the agroforestry system is better understood by researchers and practitioners thanks to the integration of this data into precise maps of land use and cover created by GIS platforms.

12.1.3 Precision Agriculture

Precision agriculture (PA) is an emerging field of enhancing crop yields and assisting management decisions utilizing high-technology sensor and analysis technologies. PA is a novel concept embraced throughout the world to boost production, decrease work time, and assure the appropriate control of fertilizers and irrigation procedures agriculture leverages data from numerous sources to improve yields in agriculture and improve the effectiveness and costs of crop management tactics involving nutrients ingredients, irrigation oversight, and pesticide application. Although definitions of precision agriculture are relatively varied, the word is often used for techniques and tactics that are useful locally at subfield scales (i.e., < 5 m). Remote sensing has been presented as an essential piece of knowledge that is accessible to promote precision agriculture to feed decades, although implementation is on hold for a variety of reasons. In recent years, technological advances, rather the ubiquity of high-definition satellite imagery, along with the of late, the swift advancement of unmanned aerial vehicle technology (UAV), argue that the usage of remote-sensing data sources in precise agriculture is bound to grow exponentially within the foreseeable future. Satellite- and UAV-based utilization of remote sensing in precision agriculture often uses multispectral observations to figure out high-spatial resolution information relating to soil parameters, plant health, and crop yields. Reflectance spectra from soils provide information relating to a range of soil parameters including soil moisture and organic matter concentration. In addition to the spectral resolution of the device, specific elements such as clay minerals, calcium magnesium carbonate, and oxides of iron that impact the soil's fertility and moisture-holding capacity can also be detected. Each of these soil constituents has unique spectral regions where reflectance (or absorption) is highest and narrowband or hyperspectral images, in combination with approaches such as spectral unmixing algorithms (Tewari et al., 2021; Kumar et al., 2021 Kumar et al., 2024). Either narrowband or hyperspectral imaging, in combination with approaches such as spectral unmixing algorithms or offshoot spectra, is typically necessary to detect these soil elements.

12.2 Detection of Agroforestry Services

Carbon sequestration

Forest sequesters a huge amount of carbon in above-ground biomass, below-ground biomass and soil carbon, which plays an important role in regulating the carbon cycle. The forest sector is both a source, as well as a sink for natural carbon, and various forest policies and programs, have implications on the standing carbon stock in the forests. Forest deforestation, degradation and destruction caused almost 17 % of emissions of CO_2 and other GHGs. According to India's State of Forest Report (ISFR), data indicate that total forest carbon stocks increased during the period from 2011 to 2021, from 6663 Mt in 2011 to 7204 Mt in 2021. Total carbon stocks increased by 541Mt in 10 years, which means an average annual increment of 54.1 Mt of carbon stock. The total carbon sequestration annually by the forest in India is 54.1 x 3.666 = 198.33 Mt of CO_2. From the present study it may be concluded that forests have a very high potential for carbon sequestration on a long-term basis, it will be a milestone for achieving the net zero carbon emissions goal of India, controlling or reducing climate change, global warming, greenhouse gases, environment protection and other benefits.

Carbon credit

Agroforestry plays a significant role in carbon credit markets by providing a sustainable method for carbon sequestration while offering economic benefits to farmers. Here's a detailed overview of how agroforestry is linked to carbon credits, the mechanisms involved, and recent developments in this area.

Carbon Credits and Agroforestry

Carbon credits represent a unit of measurement for greenhouse gas emissions reductions, specifically one metric ton of carbon dioxide (CO_2) or its equivalent in other greenhouse gases. Agroforestry systems, which integrate trees and shrubs into agricultural landscapes, are recognized for their potential to sequester carbon, thus generating carbon credits.

Mechanisms of Carbon Sequestration

1. Tree Planting: Trees absorb CO_2 from the atmosphere during photosynthesis, storing carbon in their biomass (trunks, branches, leaves, and roots). Agroforestry practices enhance this process by increasing tree density and diversity, leading to greater carbon storage.

2. Soil Carbon Storage: Agroforestry systems improve soil health and structure, promoting the accumulation of organic matter and enhancing soil carbon stocks. This is particularly important as healthy soils can store significant amounts of carbon.

3. Biodiversity and Ecosystem Services: The integration of trees in farming systems not only sequesters carbon but also provides additional ecosystem services, such as improved water quality, reduced soil erosion, and enhanced biodiversity, which can further support carbon storage.

12.3 Recent Developments

Global Carbon Market Trends

The global carbon credit market was valued at approximately \$760 billion in 2021 and is projected to reach \$2.68 trillion by 2028. India has emerged as a significant player in this market, issuing 278 million carbon credits between 2010 and 2022, accounting for 17% of global supply.

In Uttar Pradesh, for instance, an agroforestry project is expected to generate 4.5 million carbon credits, translating to an estimated income of Rs 230 crore (approximately \$27.6 million) from carbon credit sales. This initiative aims to enhance rural income while promoting environmental sustainability.

Agroforestry encourages jobs and green growth

Agroforestry is increasingly recognized as a catalyst for job creation and green growth, particularly in rural areas. This multifaceted approach to land use integrates trees with crops and livestock, providing numerous environmental and economic benefits.

Job Creation through Agroforestry

1. Green Jobs in Forestry: The transition towards a green economy is generating new job opportunities in the forestry sector. These jobs span various areas, including recreation, urban forestry, and pest control, moving beyond traditional roles in timber harvesting. Agroforestry practices can create substantial employment, especially in rural communities, where job opportunities are often limited (FAO, 2018).

2. Economic Revitalization: Investment in agroforestry can revitalize rural economies. For instance, increasing funds allocated to forestry initiatives under programs like India's National Rural Employment Guarantee Act could create millions of jobs. Estimates suggest that investments in agroforestry could generate around 8 to 10 million jobs in India alone, addressing chronic unemployment while enhancing environmental sustainability (Matta, 2009).

3. Diverse Income Sources: Agroforestry diversifies income for farmers by providing multiple products such as fruits, nuts, timber, and other tree-based resources. This diversification not only stabilizes farmer incomes but also promotes resilience against market fluctuations in traditional agricultural sectors (Muthri et al., 2023).

Economics of agroforestry system

Agroforestry systems represent a significant shift in agricultural practices, combining trees with crops and livestock to create sustainable and economically viable land-use strategies. The economics of agroforestry encompasses various factors, including profitability, risk reduction, and contributions to ecosystem services.

Economic Benefits of Agroforestry

1. Multiple Outputs: Agroforestry systems produce a variety of products, such as timber, fruits, nuts, and fodder, which can diversify income sources for farmers. This diversification reduces financial risks associated with relying on a single crop and can lead to increased overall income (Akhter et al., 2023)

2. Cost Savings: By integrating trees into agricultural systems, farmers can reduce the need for chemical fertilizers and pesticides, as trees contribute to soil fertility and pest management. This not only lowers input costs but also enhances environmental sustainability (Muhie, 2022).

3. Enhanced Resilience: Agroforestry practices improve the resilience of farming systems to climate variability and market fluctuations. The presence of trees can mitigate soil erosion, improve water retention, and provide shade, which can enhance crop yields and livestock health (Fahad et al., 2023).

4. Long-Term Investment: Although agroforestry may require initial investment and longer timeframes for returns (especially for tree crops), the long-term benefits often outweigh these costs. Studies indicate that agroforestry can yield significant profits over time, particularly in smallholder systems where trees provide essential products for both subsistence and cash income (Perks et al., 2018).

Economic Challenges and Considerations

1. Market Access: The economic viability of agroforestry systems can be hindered by limited access to markets for tree products. Developing infrastructure and market linkages is crucial for ensuring that farmers can sell their diverse outputs profitably (Cialdella et al., 2024; Gold et al., 2014).

2. Complex Management: Implementing agroforestry requires knowledge of managing complex systems that combine multiple species and production methods. Farmers may face challenges in learning and adapting to these systems, which can affect their willingness to adopt agroforestry practices (Do et al., 2024).

3. Policy Support: Government policies and incentives play a critical role in promoting agroforestry. Supportive measures can include financial incentives, technical assistance, and the establishment of markets for ecosystem services, such as carbon credits (Tedesco et al., 2024).

14.4 Agroforestry landscape

Agroforestry landscapes are complex systems that integrate trees with crops and livestock, providing a range of ecological, economic, and social benefits. These landscapes are characterized by their multifunctionality, contributing to sustainable land management and resilience against climate change.

12.4.1 Characteristics of Agroforestry Landscapes

1. Diversity of Species: Agroforestry landscapes typically include a mix of trees, shrubs, and agricultural crops. This biodiversity enhances ecosystem stability and productivity, allowing for various outputs such as timber, fruits, and fodder.

2. Ecosystem Services: These landscapes provide critical ecosystem services, including soil conservation, water management, and carbon sequestration. By improving soil health and reducing erosion, agroforestry contributes to sustainable agricultural practices.

3. Cultural and Social Dimensions: Agroforestry landscapes often reflect local cultural practices and knowledge. They can enhance community resilience by providing food security, income diversification, and opportunities for local employment.

Urban agroforestry for food and nutritional security

Urban agroforestry is a transformative approach that integrates trees and shrubs into urban environments, significantly enhancing food and nutritional security. By cultivating fruits, vegetables, and other edible plants within city landscapes, urban agroforestry directly addresses food availability, particularly in areas known as food deserts where access to fresh produce is limited. This practice not only increases the supply of nutrient-dense foods but also provides economic opportunities for urban residents through the sale of surplus produce, thereby improving household incomes and overall nutritional status. Furthermore, urban agroforestry contributes to ecosystem services, such as improved air quality and biodiversity, which are essential for creating healthier urban environments. By fostering community engagement and utilizing local resources, urban agroforestry promotes resilience against food insecurity and enhances the sustainability of urban living, making it a vital strategy for addressing contemporary urban challenges related to nutrition and food access (Jamnadass et al., 2013; Dawson et al., 2013; Clark and Nicholas, 2013).

12.5 Agroforestry: pests, diseases and weeds

Agroforestry systems, which integrate trees and shrubs with crops and livestock, face various challenges from pests, diseases, and weeds.

Understanding these challenges is crucial for effective management and sustainability.

Pests in Agroforestry

Pests in agroforestry can be categorized into several groups, including insects, nematodes, and vertebrates. The diversity of plant species in agroforestry systems can lead to increased pest populations, but it can also enhance biological control through natural predators.

Common Pests

1. Insects: Various insect pests, such as aphids and beetles, can affect both trees and crops. For example, the cypress aphid has become a significant pest in Kenya, impacting cypress trees, while other pests like the Leucaena psyllid affect *Leucaena leucocephala*, a common agroforestry species.

2. Nematodes: Root-knot nematodes can infest tree roots, leading to reduced growth and vigour. While there are concerns about nematode populations under certain trees like *Sesbania*, evidence of significant damage remains limited.

3. Birds: Some bird species can be beneficial for controlling insect populations, while others, like the red-billed quelea, can damage crops by feeding on grains.

Diseases in Agroforestry

Diseases can significantly impact the health of trees and crops in agroforestry systems. Common diseases include fungal infections, bacterial blights, and viral diseases, which can spread rapidly due to the proximity of different plant species.

Notable Diseases

1. Fungal Diseases: Diseases such as anthracnose and various rusts can affect both trees and crops, leading to decreased yields.

2. Bacterial Diseases: Conditions like bacterial blight and cankers can severely affect tree health, particularly in species like poplars and willows.

3. Viral Diseases: Viruses can spread among closely planted species, necessitating careful monitoring and management to prevent outbreaks.

Weeds in Agroforestry

Weeds can compete with crops and trees for resources, impacting overall productivity. Some agroforestry species can become invasive if not managed properly.

Invasive Species

1. Leucaena: While beneficial in many contexts, *Leucaena leucocephala* can become invasive in certain regions, outcompeting native species and crops if allowed to seed freely.

2. **Other Invasive Species:** Species like *Acacia mearnsii* and *Prosopis* have shown a potential to become weeds in specific environments, complicating management practices in agroforestry systems.

Management Strategies

Effective management of pests, diseases, and weeds in agroforestry involves a combination of cultural practices, biological control, and integrated pest management (IPM) strategies.

1. Cultural Practices: Crop rotation, intercropping, and maintaining biodiversity can help mitigate pest and disease pressures.

2. Biological Control: Encouraging natural predators and parasitoids can help control pest populations without the need for chemical pesticides.

3. Monitoring and Assessment: Regular monitoring of plant health and pest populations is essential for early detection and management of outbreaks.

Roots dynamics in agroforestry

Root dynamics in agroforestry systems are crucial for understanding the interactions between tree and crop roots, their growth patterns, and their impact on soil health and productivity.

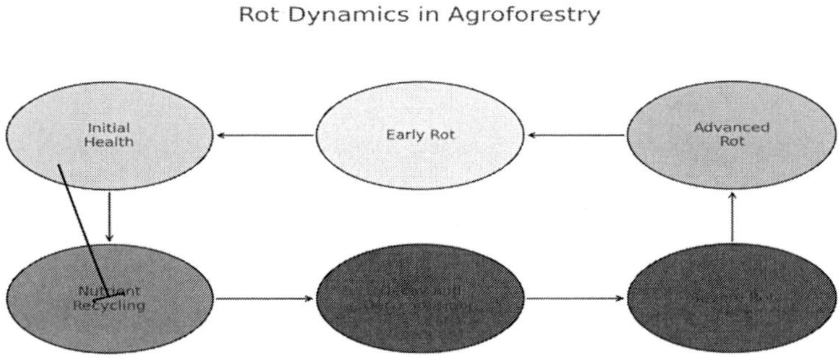

Fig 1: Root Dynamics in Agroforestry

Root Growth and Turnover

Root dynamics encompass the growth, decay, and turnover of roots, which are vital for nutrient uptake and carbon sequestration. In agroforestry, tree roots can enhance soil fertility through carbon inputs and nutrient uptake, yet they may also compete with crops for these resources. The balance between these positive and negative effects is essential for optimizing agroforestry systems (Rolo et al., 2023,).

Key Factors Influencing Root Dynamics

1. Root Length Density (RLD): Studies indicate that higher RLD can improve crop root growth and overall soil health. For instance, in an alley cropping experiment with *Gliricidia sepium*, the presence of tree roots positively influenced the root development of associated crops, particularly when tree roots were less dense, minimizing competition for resources.

2. Seasonal Variability: Root dynamics are subject to seasonal fluctuations. Research has shown pronounced variations in live root length density, which directly affects the nutrient and water uptake capacity of vegetation throughout the growing season. This variability is crucial for understanding the timing of root growth and its relationship with soil conditions.

3. Root Turnover Rates: The turnover of fine roots is significant for carbon cycling in agroecosystems. Fine roots contribute to soil organic matter and provide a food source for soil biota. The dynamics of root turnover can vary widely depending on environmental conditions and management practices, highlighting the need for tailored approaches in different agroforestry systems.

Implications for Agroforestry Management

Understanding root dynamics is essential for effective agroforestry management. Key considerations include:

Resource Allocation: Farmers can optimize root growth by managing tree density and pruning practices, which can enhance nutrient cycling while minimizing competition with crops.

Soil Health: Maintaining a balance between tree and crop roots can improve soil structure and fertility, contributing to sustainable agricultural practices. This is particularly important in regions prone to nutrient leaching, where deeper root systems can help retain soil nutrients.

Modeling and Simulation: There is a growing need for comprehensive models that accurately represent root dynamics in agroforestry systems. Such models can help predict the impact of different management strategies on root growth, turnover, and overall ecosystem health.

12.6 Agroforestry modelling

Agroforestry modelling is an essential tool for understanding the complex interactions within agroforestry systems and optimizing their management for sustainability and productivity. This overview explores the significance of agroforestry modelling, the key models used, and recent developments in the field.

Importance of Agroforestry Modeling

Agroforestry models serve several critical purposes:

Integration of Knowledge: They synthesize experimental and conceptual information about tree-crop interactions, helping researchers and practitioners understand how different components of agroforestry systems interact under various conditions.

Hypothesis Testing: Models provide a framework for formulating and testing hypotheses regarding the effects of management practices, climate change, and other factors on agroforestry systems. This is crucial for developing effective strategies for both productivity and environmental sustainability.

Resource Optimization By simulating different scenarios, agroforestry models can help identify optimal combinations of tree and crop species, management practices, and resource inputs (such as water and nutrients) to maximize yields and minimize environmental impacts.

Key Agroforestry Models

Several models have been developed to address various aspects of agroforestry systems:

1. WaNuLCAS (Water, Nutrient, and Light Capture in Agroforestry Systems): This model focuses on the interactions between water, nutrients, and light in agroforestry systems, helping to predict yields and optimize resource use.

2. APSIM (Agricultural Production Systems Simulator): This versatile model can simulate tree-crop interactions and is widely used to assess the impacts of different management practices on agricultural productivity and sustainability.

3. SCUAF (Soil-Crop-Tree-User Agroforestry Framework): This model estimates soil changes under various land use scenarios, including agroforestry, providing insights into soil health and fertility dynamics.

4. UB Forest Model: This model seeks to understand the complex mechanisms connecting abiotic factors (like soil and climate) to plant growth and interactions within agroforestry systems. It emphasizes the importance of these interactions for predicting yields and ecosystem services.

Table 1: Agroforestry Models

Model Type	Description	Key features	References
Allometric Models	Simple empirical models that relate tree biomass to measurable parameters like	Focus on tree components only; useful for estimating biomass without	Khaine, I., & Woo, K. (2018). Allometric relationships for estimating tree

	height and diameter.	complex growth dynamics.	biomass.
Non-Growth Models	Models that do not simulate growth but focus on static relationships or environmental impacts.	Often used for assessing soil properties and ecosystem services without dynamic growth simulation.	Young, A., et al. (1998). SCUAF: A model to estimate soil changes under agroforestry.
Plot-Scale Models	Models that simulate tree-crop interactions at a specific plot scale.	Can incorporate spatial arrangements and competition dynamics; which are useful for field-level assessments.	Luedeling, E. et al. (2016). Field-scale modelling of tree-crop interactions.
Architectural Models	Models that simulate the 3D structure and growth of trees and crops.	Account for light interception and canopy dynamics; useful for optimizing spatial arrangements.	Grisafi, F. et al. (2022). Insights into fruit tree models for agroforestry.
Farm Decision Models	Models that incorporate farmer decision-making processes into simulations.	Reflect farmers' priorities and risk preferences; useful for assessing management strategies.	Moore, A.D. et al. (1997). GrazPlan: A decision support system for grazing management.
Landscape Agroforestry Models	Models that simulate agroforestry systems at a landscape level.	Integrate multiple land uses and ecosystem services; useful for regional planning and policy-making.	Huth, N.I. et al. (2002). A framework for simulating agroforestry options in low rainfall areas.

Conclusion

In conclusion, the innovations in agroforestry represent a transformative approach to sustainable land management, effectively harnessing technology to enhance productivity, resilience, and environmental health. By integrating advanced practices such as precision agriculture, remote sensing, and genetic improvements, agroforestry systems can optimize resource use, increase biodiversity, and bolster food security in the face of climate change. These technological advancements not only support the economic viability of agroforestry but also contribute significantly to carbon sequestration and ecosystem services, positioning agroforestry as a key strategy for achieving global sustainability goals. As we move forward, fostering collaboration among researchers, policymakers, and practitioners will be essential to fully realize the potential of agroforestry innovations in creating a more sustainable and resilient agricultural landscape.

References

Akter, R., Hasan, M. K., Kabir, K. H., Darr, D., & Roshni, N. A. (2022). Agroforestry systems and their impact on livelihood improvement of tribal farmers in a tropical moist deciduous forest in Bangladesh. *Trees, Forests and People*, *9*, 100315.

Bisht, S., Kumar, A., Kumar, N., Singh, H. and Kumar, P. 2022. Biofuel Production by Using Biomass and Its Application. In book: Renewable Energy and Green Technology in CRC Press.

Cialdella, N., Jacobson, M., & Penot, E. (2023). Economics of agroforestry: links between nature and society. *Agroforestry Systems*, *97*(3), 273-277.

Clark, K. H., & Nicholas, K. A. (2013). Introducing urban food forestry: A multifunctional approach to increase food security and provide ecosystem services. *Landscape Ecology*, *28*, 1649-1669.

Dawson, I. K., Place, F., Torquebiau, E., Malézieux, E., Iiyama, M., Sileshi, G. W., & Jamnadass, R. (2013). Agroforestry, food and nutritional security.

Do, H., Whitney, C., La, N., Storm, H., & Luedeling, E. (2024). Adapting agroforestry to upland farming systems: narratives from smallholder farmers in Northwest Vietnam. *Agronomy for Sustainable Development*, *44*(2), 17.

Dollinger J, Jose S (2018) Agroforestry for soil health. Agrofor Syst 92:213–219. https://doi.org/10.1007/s10457-018- 0223-9

Fahad, S., Chavan, S. B., Chichaghare, A. R., Uthappa, A. R., Kumar, M., Kakade, V., & Poczai, P. (2022). Agroforestry systems for soil health improvement and maintenance. *Sustainability*, *14*(22), 14877.

Gold, M. A., Godsey, L. D., & Josiah, S. J. (2004). Markets and marketing strategies for agroforestry speciality products in North America. In *New Vistas in Agroforestry: A Compendium for 1st World Congress of Agroforestry, 2004* (pp. 371-382). Springer Netherlands.

Jamnadass, R., Place, F., Torquebiau, E., Malézieux, E., Liyama, M., Sileshi, G., & Dawson, I. (2013). Agroforestry, food and nutritional security.

Joshi, D. C. 2013. Remote sensing and GIS-based soil resources appraisal. *Curr. Agric.*, 37(12): 1-116.

Kaushal R, Tewari S., Thapliyal SD., Kumar, A., Roy, T., Islam, S., Lepcha, STS., Durai, J. 2021. Build-up of Labile, and non-labile carbon fractions under fourteen-year-old bamboo plantations in the Himalayan foothills. Heliyon 7(3):e07850

Kumar, A., Dwivedi, G. K., Tewari, S., Sah, V. K., Singh, H., Kumar, P., Kumar, N., Kaushal, R. 2020a. Soil organic carbon pools under *Terminalia chebula* Retz. based agroforestry system in the Himalayan foothills, India. Current Science (00113891), 118(7).

Kumar, A., Dwivedi, G.K., Tewari, S., Paul, J., Anand, R., Kumar, N., Kumar, P., Singh, H. and Kaushal, R. 2020b. Carbon Mineralization and Inorganic Nitrogen Pools under *Terminalia chebula* Retz.-Based Agroforestry System in Himalayan Foothills, India. Forest Science. 66(5), 634-643

Kumar, A., Kumar, P., Singh, H. and Kumar, N., 2021a. Impact of plant functional traits on near-saturated hydraulic conductivity of soil under different forests of Kempty watershed in Garhwal Himalaya. Indian Journal of Soil Conservation, 49(1), pp.38-44.

Kumar, A., Kumar, P., Singh, H., Kumar, N., 2021b. Adaptation and mitigation potential of roadside trees with bio-extraction of heavy metals under vehicle emissions and their impact on physiological traits during seasonal regimes. Urban Forestry & Urban Greening, 58, p.126900.

Kumar, A., Kumar, P., Singh, H., Kumar, N., 2021c. Modulation of Plant Functional Traits under Essential Plant Nutrients During Seasonal Regime in Natural Forests of Garhwal Himalayas. Plant and Soil 465(1):1-16

Kumar, A., Kumar, P., Singh, H., Kumar, N., 2021d. Relationship of physiological plant functional traits with soil carbon stock in the temperate forest of Garhwal Himalaya. Current Science 120(7): 1064-1068.

Kumar, A., Singh, H., Kumari, G., Bisht, S., Malik, A., Kumar, N., Singh, M., Raturi, A., Barthwal, S., Thakur, A. and Kaushal, R., 2022. Adaptive resilience of roadside trees to vehicular emissions via leaf enzymatic, physiological, and anatomical trait modulations. Environmental Pollution, p.120191.

Kumar, A., Tewari, S., Singh, H., Kumar, P., Bisht, S., 2021e. Biomass accumulation and carbon stock in different agroforestry systems prevalent in the Himalayan foothills. Current Science 120(6), 1083-1088

Kumar, A., Tewari, S., Singh, Indra., Panday, R., Kumar, D. and Anand, R. 2018. Effect of nutrient sources on growth, yield and quality of turmeric under harad (*Terminalia chebula*) based agroforestry system Indian J. of Agroforestry, 20(2). 1-6

Kumar, N., Jeena, N., Kumar, A., Khwairakpam, R., Singh, H. 2021. Comparative response of rice cultivars to elevated air temperature in Bhabar region of Indian Himalaya: Status on Yield attributes. Heliyon 7(6):e07474

Kumar, N., Kumar, A., Jeena, N., Singh, R., and Singh, H. 2020. Factors Influencing Soil Ecosystem and Agricultural Productivity at Higher Altitudes. Microbiological Advancements for Higher Altitude Agro-Ecosystems & Sustainability. Springer. p. 55- 71.

Kumar, N., Singh, H., Giri, K., Kumar, A., Joshi, A., Yadav, S., Singh, R., Bisht, S., Kumari, R., Jeena, N., Khairakpam, R., and Mishra, G. 2024. Physiological and molecular insights into the allelopathic effects on agroecosystems under changing environmental conditions. Physiology and Molecular Biology of Plants. 248. 119255.

Kumar, N., Verma, S., Kumar, A., Dashila, H., Suyal, D.C., Kumari, G., Rawat, S., Jeena, N., Singh, M., Kumar, A. 2022. Molecular insights into stress-responsive genes in the mitigation of environmental stresses. Unravelling Plant-Microbe Synergy, Edition: 1, Publisher: Elsevier

Matta, J. R. (2009). Rebuilding rural India: potential for further investments in forestry and green jobs. *Unasylva*, *233*(60), 36-41.

Mercer, D. E., Frey, G. E., & Cubbage, F. W. (2014). Economics of agroforestry. In *Handbook of forest resource economics* (pp. 204-225). Routledge.

Muhie, S. H. (2022). Novel approaches and practices to sustainable agriculture. *Journal of Agriculture and Food Research*, *10*, 100446.

Muthuri, C. W., Kuyah, S., Njenga, M., Kuria, A., Öborn, I., & van Noordwijk, M. (2023). Agroforestry's contribution to livelihoods and carbon sequestration in East Africa: A systematic review. *Trees, Forests and People*, 100432.

Pandey R., Dubey S., Tewari L., Kumar A., Kumar D., Sharma N. and Tewari S.K. 2018. Performance of soybean under poplar-based agroforestry system in Terai region. Indian J. of Agroforestry, 20(1), 31-34.

Pantera A, Mosquera-Losada MR, Herzog F, den Herder M (2021) Agroforestry and the environment. Agrofor Syst 95:767–774. https://doi.org/10.1007/s10457-021-00640-8

Perks, M., Khomik, M., Bathgate, S., Chapman, S., Slee, B., Yeluripati, J., & Morison, J. (2018). Agroforestry in Scotland–potential benefits in a changing climate. *CXC Report, July*.

Rolo, V., Rivest, D., Maillard, É., & Moreno, G. (2023). Agroforestry potential for adaptation to climate change: a soil-based perspective. *Soil Use and Management*, *39*(3), 1006-1032.

Schroth, G., & Zech, W. (1995). Root length dynamics in agroforestry with *Gliricidia sepium* as compared to sole cropping in the semi-deciduous rainforest zone of West Africa. *Plant and Soil*, *170*, 297-306.

Schroth, G., Oliver, R., Balle, P., Gnahoua, G. M., Kanchanakanti, N., Leduc, B., & Zech, W. (1995). Alley cropping with *Gliricidia sepium* on a high base status soil following forest clearing: effects on soil conditions, plant nutrition and crop yields. *Agroforestry Systems*, *32*, 261-276.

Tedesco, A. M., Brancalion, P. H., Hepburn, M. L. H., Walji, K., Wilson, K. A., Possingham, H. P., & Rhodes, J. R. (2023). The role of incentive mechanisms in promoting forest restoration. *Philosophical Transactions of the Royal Society B*, *378*(1867), 20210088.

Tewari, S., Tewari, L., Dubey, A., Kumar, A., Kumar, N., Kaushal, R. 2022. Use of the RAPD marker to determine the genetic diversity of various *Dalbergia sissoo* Roxb. (Shisham) genotypes and their evolutionary relationship in Indian subcontinents. Vegetos,

Tewari, S., Tewari, L., Kumar, A., Bisht, S, Kaushal, R. and Kumar, A. 2024. Germination Behaviour of Teak Genotypes and Effect of Different Treatments on Seed Germination Characteristics in Himalayan Foothills, India. Indian Forestry, 150 (3): 244-254

Tewari, S., Tewari, L., Kumar, A., Kaushal, R. 2021. Performance of *Glycine max* L. varieties under *Tectona grandis* L. and *Dalbergia sissoo* Roxb. based agroforestry system Indian J. of Agroforestry, 20(2) 1-6

Verma, T., Bhardwaj, D.R., Sharma, U., Sharma, P., Kumar, D., Kumar, A. and Kumar, A. 2023. Agroforestry systems in the mid-hills of the north-western Himalaya: A sustainable pathway to improved soil health and climate resilience. Journal of Environmental Management. 348, 119264.

Zomer RJ, Neufeldt H, Xu J, Ahrends A, Bossio D, Trabucco A, van Noordwijk M, Wang M (2016) Global tree cover and biomass carbon on agricultural land: the contribution of agroforestry to global and national carbon budgets. Sci Rep 6:29987. https://doi.org/10.1038/srep29987

CHAPTER THIRTEEN

SOIL SUSTAINABILITY UNDER AGROFORESTRY TO IMPROVE ECOSYSTEM PRODUCTIVITY

NEHA[1], AMIT KUMAR[2] AND NIDHI[3]

[1] DEPARTMENT OF HORTICULTURE & FOOD PROCESSING, GOVERNMENT OF UTTARAKHAND

[2]CSIR-INSTITUTE OF HIMALAYAN BIORESOURCE TECHNOLOGY, PALAMPUR-176061

[3] DEPARTMENT OF ANIMAL HUSBANDRY, FISHERIES & DAIRY DEVELOPMENT, GOVERNMENT OF UTTARAKHAND

Abstract

Global climate change, environmental crises, food crises and security, land degradation, along ever-increasing demand for various products due to the burgeoning population cause serious pressure on the various natural resources of the earth. The pressure to fulfill the increasing food demand, animal feed, fiber as well and fuel, is increasing day by day which leads to the adoption of intensified practices such as intensive tillage, monoculture, excessive and improper use of inorganic fertilizer, use of agro-chemicals, deforestation, over-grazing, among others and not concerning above the food quality. These modern practices of agriculture are disrupting the ecology and natural balance and thus may be considered to be the major unsustainable practices leading to land degradation and deterioration in soil health. Therefore, sustainable land use, production systems, and technologies are needed for the present and future well-being of humans. From this perspective, agroforestry seems to be a promising sustainable and climate-smart farming practice, that fulfills various aspects for farmers and the environment. It not only efficiently utilizes all the available resources but also conserves and improves the ecology and environment of the farming systems. This integration of trees with agriculture production systems provides various soil-related ecological services such as soil fertility enhancements and improvements in soil physical, biological, and chemical properties, carbon dioxide sequestration, along with food, wood, and fodder. Besides their ecological functions, agroforestry helps preserve biodiversity, livelihood improvement, and the economic status of the farmers. It plays a key role in resource utilization, ecosystem productivity, and environment resilience, the basic pillar of agroecosystem sustainability and sustainable development. This chapter reviews the role of agroforestry systems in improving soil biota, soil nutrient enrichment, soil conservation, carbon sequestration, climate change adaptation, biodiversity conservation, and ecosystem services.

Keywords: Agroforestry, Soil fertility, Nutrient cycling, Carbon sequestration, Soil health, Soil sustainability, Ecological services.

Introduction

Soil serves as the most important resource in agriculture to meet the diverse food requirements of the burgeoning population globally, which is estimated to reach about 11 billion by 2050 (Adam, 2021). The history of mankind also reveals that the civilizations that have flourished have always intensively used soil as a resource for food production. Thereby, it focuses on the duty of the entire globe to achieve and maintain food and nutrition security for present and upcoming generations by increasing agricultural land production while significantly reducing the impact on the environment (Turan *et al.,* 2019). Soil and climate are the main determinants of agricultural productivity, profitability, and human well-being (Babu *et al.,* 2022). Soil acts as a source of stability for the plants and raw materials such as water, nutrients, and oxygen which are considered essential for plant growth. On the other hand, climate adversities exhibit a negative impact on crop productivity, natural resources, and eventually food security, and environmental health (Das *et al.,* 2022).

The soil is vital for maintaining global biogeochemical cycling, ecosystem health as well as sustainable food production. It is an indispensable section of ecosystems that plays a crucial role in the delivery of primary ecosystem services. The ability of soils to augment and sustain agricultural output, absorb carbon, reduce nutrient leaching, and conserve biodiversity adds to soil health, which also incorporates the functioning of ecosystem services provided by soils (Fahad *et al.,* 2022). According to the

USDA, soil health can be defined as "the continued capacity of the soil to function as a vital living ecosystem that sustains plants, animals, and humans" (USDA-NRCS 2018). Healthy soils are considered one of the most critical resources for the health and sustainability of ecosystems, including the agroecosystems so that they can sustain food production as well as provision of ecosystem services. Therefore, one of the main objectives in the development of sustainable farming techniques for higher crop production is to improve soil health (Eddy and Yang, 2022). This raises a need to protect and improve soil health, as soil is the most important resource for future generations (Dollinger and Jose, 2018). In this perspective, organic farming, climate-smart agriculture, the use of bio-climatology, conservation farming, and agroforestry are such sustainable practices which ensure the eco-environmental development as well as the prosperity of human beings (Jahariya *et al.,* 2019). Considering this realization in agricultural land use management along with the better income, the role of agroforestry as a sustainable land management strategy came into existence that can not only improve several common indicators that define soil health but also boost farmers' income.

Agroforestry is considered one such multifunctional approach for our food system, which is the intentional combination or compatible association of trees with shrubs, seasonal crops, or livestock. It is basically an agricultural technique that focuses on diversifying various agroecosystem production components such as woody perennials, crops, forages, or animals (Rosatti *et al.,* 2020). The main theme of agroforestry encompasses sustainability in terms of economics: productivity and profitability; ecology: resource conservation and environment; social issues: food safety and security which makes a favorable unparalleled land use system (Pandey, 2007). Apart from meeting rural wood-based demands, agroforestry has various positive effects on agricultural production, soil health, and environmental quality (Babu *et al.,* 2020). Agroforestry has the potential to ameliorate soil quality which has been widely recognized as a major benefit since its inception as a scientifically recognized discipline and practice (Nair, 2011). By implementing systems that mimic nature's functions, agroforestry perceived benefits to remain productive while supporting a range of ecosystem services (Jose, 2009). However, many of the environmental benefits and ecosystem services expected from agroforestry would not be materialized unless these practices improved the capacity of soils to be productive and healthy over the long term. Therefore, agroforestry if developed more, has the potential to improve socio-economically more sustainable and make the landscape better (Kittur and Bargali, 2013).

Agroforestry as a transformative solution

Agroforestry is an age-old land use system whose origin dates long back and has been practiced for thousands of years by farmers all over the world. The most appropriate definition, perhaps, is stated by Nair (1984): "Agroforestry is a land use that involves deliberate retention, introduction, or mixture of trees or other woody perennials in crop/animal production fields to benefit from the resultant ecological and economic interactions." From time immemorial, it has gained wide traditional importance due to its diversified practices of location-specific models in particularly India. Several other definitions are also available. The World Agroforestry Centre defines it as "a dynamic, ecologically based, natural resources management system that, through the integration of trees on farms and in the agricultural landscape, diversifies and sustains production for increased social, economic, and environmental benefits for land users at all levels." Nair (1985) further indicated agroforestry as a system involving trees and agricultural and/or animal crops, simultaneously or sequentially, with the objective of sustainably increasing the total productivity of plants and animals per unit of area.

The agroforestry system is often mentioned as a holistic approach toward enhancing agriculture production and sustainable management of agroecosystems. This system has numerous benefits in economic, social, and environmental contexts. The main objective of agroforestry is to take advantage of the complimentary relationships between trees, crops, and livestock in such a way that the productivity, stability, and sustainability of the total system exceed an individual component. Generally, agroforestry exists around the world in several forms such as silver-arable systems, silvopastoral systems, agrosilvopastoral systems, multipurpose trees, riparian buffers, and improved fallow (Mosquera-Losada *et al., 2009).*

Today, agroforestry is recognized as an integrated applied science that has the potential to address many of the land management and environmental problems found in both developing and industrialized nations. Also, one of the objectives of the 2030 Agenda, in relation to the Millennium Challenges established by the United Nations (UN), is to "Protect, recover and promote the sustainable use of terrestrial ecosystems sustainably manage forests, combat desertification, stop and reverse land degradation, and halt the loss of biodiversity." To achieve this goal by 2030 it is necessary "to take action against desertification and restoration of degraded land and soil, including land affected by desertification, droughts, and floods, and to seek for a world neutral in terms of soil degradation" (UNIC Rio, 2016). The beneficial outcomes of agroforestry include reductions in nutrient cycling, positive influence on soil biota, nutrient, and pesticide runoff, carbon sequestration, erosion control, improved soil quality and soil health, improved wildlife habitat, reduced fossil fuel use, and increasing resilience in the face of an uncertain agricultural future (Dixon *et al.,* 1994, Caudill *et al.,* 2015). It also provides multiple ecosystem services such as enhancing the overall (biomass) productivity, enhancing soil fertility, soil conservation, nutrient cycling, micro-climate improvement, carbon sequestration, and bio drainage (Fanish and Priya, 2013). Moreover, the role and scope of agroforestry are also studied in way of preserving biodiversity, the yield of goods and services to society, augmentation of the carbon storages in agro-ecosystems, climate change adaptations and resource-conserving farming approaches that are socially relevant and relate well to livelihood and ecosystem functions (Pandey, 2007; Kuyah *et al.,* 2016). The fact that it can simultaneously provide economic, ecological, and cultural benefits gives agroforestry a good potential as a land use strategy in both the developing and developed world (Jose *et al.,* 2012).

Role of Agroforestry in soil health management and improving ecosystem productivity

Substantial studies have been conducted and reported on the positive effect and role of agroforestry on soil health management such as in land productivity (Kassie, 2016), soil quality (Dollinger & Jose, 2018), and erosion control (Young, 1987). Moreover, agroforestry has demonstrated the greatest potential for climate change mitigation and enhancing ecosystem productivity (Hart *et al.,* 2017). It can enhance the carbon stock by sequestering greenhouse gases (GHGs) in woody biomass in the soil of cultivated fields (Kim *et al.,* 2016; Lal, 2015), increase soil organic matter, and improve water availability (Alletto *et al.*, 2022). Therefore, in this section, we will delve into the role of agroforestry in soil health management and explore its importance in the context of improving ecosystem productivity.

Agroforestry in maintaining soil physical properties

Important soil physical properties for determining the level of soil fertility and improving soil health consist of solum, texture, structure, consistency, porosity, permeability, infiltration, bulk density, water content, soil moisture, temperature, and soil color. Thereby their role in supporting plant growth is interrelated (Tripathi *et al.,* 2020). Thus, it raises the importance of soil's physical properties to secure food, energy, water, and ecosystem services (Shang *et al.,* 2018). Out of all the soil's physical properties, the maintenance of soil structure is of utmost importance for better health and quality of the respective soil. In this context, agroforestry exhibits more effective resource usage than monocropping and the reason is the structural and functional diversity of the components acquired in a mixed cropping canopy (Hailu, 2015). The soil protection provided by the tree roots and their influence on soil aggregate stability not only helps in maintaining the soil structure but also reduces the rates of soil erosion. The carbon inputs from decomposing litter and decaying tree roots also stabilize the soil structure (Salako *et al.,* 2001). Tree roots can also improve soil structure and porosity where dead tree roots produce pore holes, thereby increasing infiltration and reducing the runoff (Tsufac *et al.,* 2021). Earthworms and mycorrhizal hyphae also contribute to processes of soil aggregate formation and stabilization and simultaneously reduce the potential for soil erosion and increase the carbon sequestration in the soil through the physical protection of soil organic matter (Fonte *et al.,* 2010).

Several studies conducted on the inclusion of trees on farms reported the enhancement in the field capacity (FC), organic matter (OM) (Chatterjee *et al.,* 2018), soil carbon stocks, and lower bulk density (BD) (Hailu, 2015), which by increasing the water holding capacity retain more water and releases the water to plants gradually like a sponge. Furthermore, it was proved that agroforestry practices affect various soil physical properties, such as changes in soil retention, circulation, and available water capacity. It was also observed that the physical properties of soil are related to the dynamics of water and air and similarly related to aspects of soil conservation and land management (Alletto *et al.*, 2022).

Agroforestry in maintaining soil fertility

The increasing pressure to augment food production has resulted in a heavy burden on soil resources (Mosier *et al.,* 2021). A useful path, complementary to chemical fertilizers, to enhance soil fertility offers several alternative land use systems such as agroforestry, agro-horticultural, agro-pastoral, and agro-Silvi-pastoral which are considered effective towards soil organic matter restoration and thus soil fertility (Manna *et al.,* 2003). Soil chemical properties (pH, EC, soil organic matter/carbon, available nitrogen, phosphorus, potassium, and cation exchange capacity, etc) are some of the main parameters of soil fertility which have an important role in supporting plant growth and development (Khaleel *et al.,* 2020). Apart from meeting rural wood-based demands, agroforestry provides a significant impact on the improvement of soil fertility as a long-term benefit, through nitrogen fixation, greater organic matter production, and recycling of nutrients. The accumulation of litter through the shedding of leaves and twigs returned to the floor followed by its decomposition releases to the soil and acts as the main source of nutrients and organic carbon in agroforestry systems (Junaedi *et al.,* 2020). Even in the dry regions, the mean annual litter fall by neem trees can be estimated at 6059 kg ha^{-1} at the density of 400 trees ha^{-1} with the potential return of 98, 2.25, 32, and 131 kg ha^{-1} of available nitrogen, phosphorous, potassium and Calcium (Pandey *et al.,* 2010). The process of litter decomposition and mineralization supplies abundant nutrient stock in agroforestry systems and consequently improves crop yield. The active deep root system of the trees further also increases absorption and high organic matter which results in enhancing the nutrient use efficiency and results in easier availability of nutrients to the plants. Such nutrients become an input upon being transferred to the topsoil via tree litter decomposition and other mechanisms (Sonmez *et al.,* 2016).

Agroforestry as a land management practice also significantly improves the soil cation exchange capacity significantly through the accumulation of crop residues. The cation exchange capacity which can be defined as the capacity of clay to adsorb and exchange cations is influenced by clay content, type of clay mineral, and organic matter content (Dori *et al.,* 2022). Many researchers have also highlighted the adoption of agroforestry as a promising land-use system as it plays a crucial role in conserving soil and water, reducing soil losses, and sustaining higher crop production, productivity, and profitability (Kuyah *et al.,* 2019). These land use practices on upland and sloping lands are not only related to soil fertility but also protect the soil from erosion and degradation (Wawire *et al.,* 2021). A study conducted by Tsuface *et al.* (2021) examined farmers' perceptions toward agroforestry in the southwest region of Cameroon and observed that about 53% of farmers considered the agro-silvopastoral system to be the most successful in increasing soil fertility, with 57% of farmers believing that the agro-silvopastoral system provided the highest degree of soil fertility.

Agroforestry for nutrient cycling and water uptake

Trees play an important role in the recycling of nutrients by recapturing and pumping back leached nutrients via deep roots, which act as a 'safety net' against nutrient losses from the nutrient cycle. It is another way in which agroforestry trees contribute to soil health. Basically, the trees have deep and spreading roots which make them capable of deep capture of subsoil nutrients and water from deeper soil layers usually where crop roots cannot reach. The process of taking up nutrients from deeper soil layers and eventually depositing them on the surface layers through litter-fall and other mechanisms is termed 'nutrient pumping' by trees (Buresh *et al.,* 2004). However, the process mainly depends on the characteristics of tree species and other soil, climatic, and topographic factors. In addition, deep-rooted species could reduce competition for nutrients and moisture with crops by pumping from deeper soil profiles (Das and Chaturvedi, 2008). Tree roots can also recycle the fertilizer applied by farmers that have leached to subsurface soil levels, hence improving the nutrient-use efficiency as well as the economic returns to fertilizer use. Additionally, a great nutrient acquisition capacity is exhibited by mycorrhizal fungi in association with agroforestry trees (Dierks *et al.,* 2021). This association between the fungi and roots of higher plants displays an extensive network of fungal hyphae that extend into the soil and give trees access to pools of recalcitrant phosphorus in the soil. It shows the nutrient-capture function by taking up microbially solubilized phosphorus and exploring it into tree roots from where the plant's vascular system circulates the phosphorus within the tree. The influx and uptake of phosphorus in roots colonized by mycorrhizal fungi is estimated to be three to five times greater than in non-mycorrhizal roots (Smith and Read, 2008). The use of agroforestry practices can be further helpful in maintaining the requisite forest cover to ensure the hydrological forest functions (Van *et al.,* 2020). In this case, the litter maintains the quality of water by filtering soil particles from runoff (Suprayogo *et al.,* 2020). This tree-based system can also be referred to as an "infiltration-friendly" land use system where acting through litter fall, it can maintain high infiltration rates by funneling through stem flow, can ameliorate the hydrologic functions and maximizes the groundwater recharge in the seasonally dry tropics (Ilstedt *et al.,* 2018). In dry areas, deep-rooting trees can play a vital role in enhancing the water availability to crops, where drawing it up from the deep soil profile is known by a mechanism '*hydraulic lift*' (Bayala *et al.,* 2008). Some agroforestry tree species, e.g., *Vitellaria paradoxa*, have been found to take up as much as 50% of the water from below the rooting zone of grown crops, which clearly depicts that they are not competing so much with crops (Bayala and Prieto, 2020).

Agroforestry in maintaining soil biota

The inclusion of trees with agricultural crops not only influences the soil's physical and chemical properties but also influences the microorganisms present in the soil. Soil biota, the soil microorganisms are a crucial part of the soil ecosystem and are considered critical to soil health and ecosystem sustainability. The soil biota plays a significant role in enhancing fertility and productivity as it promotes plant growth indirectly through the decomposition of soil OM, nutrient cycling, and thereby influencing soil chemical and physical properties, which ultimately determines soil fertility and long-term sustainability (Chamorro-Martinez *et al.,* 2022). These rhizospheric bacteria are essential for plant growth and nutrition (Ortiz and Sansinenea, 2022), and contribute significantly to the functioning of stable soil ecosystems (Puskaric *et al.,* 2021). The mycorrhizal fungi as discussed above in brief also influence the nutrient availability (especially phosphorus) and its uptake, increase the photosynthetic rate, and increase the tolerance against varied environmental stress, as well as have a positive impact on both the primary and secondary metabolites and thereby favors the increased growth, yield, and productivity of crops (Khan *et al.,* 2022). The higher beneficial bacterial and/or fungal population can act as a substitute for chemical pesticides and mineral fertilizers and can be helpful in agriculture because of its organic and eco-friendly nature (Ortiz and Sansinenea, 2022). The tree also releases certain root exudates that also play a significant role in the enrichment of soil fertility due to its capability to act as an organic carbon and energy source for microbiology and determine nutrient P solubilization (Pantigoso *et al.,* 2020).

Agroforestry and Organic matter (OM)

Soil organic matter acts as the energy source of soil organisms and influences soil biodiversity, associated soil biological functions, and soil physical and chemical properties. Incorporation of trees in agroforestry enhances the soil organic matter by adding litter to both above and belowground tree biomass to the soil as well as by rhizodeposition mediated by the activity of decomposer organisms (Barrios, 2023). The speed of decomposition of organic materials into simpler molecules and the related opportunities for managing organic matter depend on the chemical composition and quality of that matter (Palm *et al.,* 2001). The replenishment of soil organic matter is the cornerstone in regenerating soil health and increasing soil OM is the most prior goal of adopting agroforestry as a land use system. The addition of various organic inputs into the soil by trees integrated into agroforestry sustains the functional capacity of soil to accrue the soil health benefits of soil OM by its positive impact on carbon transformations, nutrient cycling, and maintenance of soil structure, and indirectly, in the biological regulation of pests and diseases through habitat creation (Barrios *et al.,* 2012a). The nitrogen content, lignin content, and polyphenol contents of soil OM are considered important indicators of its quality (Vanlauwe *et al.,* 2006). The soil OM also plays a key role in soil aggregation and in reducing the soil bulk density. Thereby the reduced soil bulk density influences the air circulation, and water distribution in the rhizosphere, and in improving the groundwater recharge and nutrient quality of the soil in arid and semi-arid areas, respectively (Surki *et al.,* 2020). Additionally, the increased microbial diversity due to the addition of OM (Lacombe et al., 2009) probably provides mycorrhizae, releasing P and making it accessible to the crops (Lorenz and Lal, 2014).

Agroforestry and biological nitrogen fixation (BNF)

Indiscriminate use of nitrogenous fertilizers in intensive agriculture is constrained by issues like high costs, micro-nutrient deficiencies, and environmental pollution, and thus cannot be relied upon for a long-time purpose. Biological nitrogen fixation by trees plays a key role in providing the natural source of nitrogen inputs in agroforestry systems in minimizing the higher synthetic inputs as nitrogenous fertilizers and combating the soil health problem (Giller, 2001). Symbiotic nitrogen fixation is mainly facilitated by nitrogen-fixing bacteria in which atmospheric nitrogen is transformed into ammonia or ureides within root nodules, specialized root structures from which fixed nitrogen is taken up by the host plant and assimilated by trees in agroforestry systems (Sprent, 2009). The nitrogen-fixing trees particularly the leguminous ones, comprising of *Acacia spp., Dalbergia sissoo,* etc. can contribute to adding nitrogen to the soil naturally and have been found to range from 56 to 675 kg N ha^{-1} yr^{-1} depending on climate, tree species, and management system (Nygren *et al.,* 2012). Being phosphorus deficient in Indian soils, the lower availability of soil phosphorus often limits the amount of BNF achieved by agroforestry trees, which may be reduced by enhancing the supply of phosphorus through three-way symbioses that include enhancing the phosphorus solubilization and mobilization by mycorrhizal fungi (Gianninazi *et al.,* 2010).

Agroforestry and soil erosion

One of the most important environmental concerns in today's time is soil erosion and land degradation, which results in weathering of nutrients and loss in fertility and altering the soil structure and texture, impacting the nutrient uptake of plants and their growth and development and adversely inducing the lower productivity and sustainability of natural and agricultural ecosystems (Cao *et al.,* 2020). Agroforestry has been recognized as a viable system for decreasing runoff and preventing soil erosion to some extent, thereby assisting in the restoration of the environment and providing economic benefits (Fahad *et al.,* 2022). Trees have been observed to reduce soil loss by the surface litter brought on by perennial plants may aid in reducing splashing and limiting the soil detachment, while the inclusion of tree trunks in an agroforestry system can induce low runoff rates which diminishes the sediment-carrying capacity (Du *et al.,* 2022). It could also be helpful by forming barriers that delay runoff and capture sediments (Angima *et al.,* 2002), protecting soil aggregates by capturing raindrops (Lal 1989; Omoro and Nair, 1993), improving soil structure (Lal 1989), increasing transpiration and water retention in the soil, and help retain the soil and soil nutrients (Parwada *et al.,* 2022). Without soil cover, direct raindrops on bare soils increase the detachment of soil particles, which lowers infiltration and can stimulate runoff and soil loss. Further deposition of biomass through root turnover and aboveground litter fall acts as a food source for soil organisms enhances the biological activity (Kamau *et al.,* 2017), and influences the soil's structural stability and macroporosity, which are essential in decreasing soil erosion rates under agroforestry indirectly (Muchane *et al.,* 2020). Purwaningsih *et al.* (2020) further reported that trees and crops in agroforestry systems should necessarily be arranged according to the morphological units formed by the previous landslides, residual zones, and erosional and sedimental zones, as different crops and trees have varied ecological functions on different units.

Agroforestry and carbon sequestration

Carbon dioxide contributes as a major greenhouse gas and the continued increase in its concentration in the atmosphere is mainly accelerated by human activities such as burning of fossil fuels and deforestation (IPCC,

2007). One of the approaches to reduce the concentration of CO_2 in the atmosphere is by a process commonly referred to as 'carbon sequestration', which according to UNFCC is the process of removing carbon from the atmosphere and depositing it in a reservoir. The largest international agreement to stabilize greenhouse gas concentrations by UNFCC also allowed the use of carbon sequestration through afforestation and reforestation as a form of greenhouse gas-offset activities. In that sequence, agroforestry systems attracted attention as a carbon sequestration strategy from both industrialized and developing countries (Makundi and Sathaye, 2004). It is not only an important indicator of soil health but also a mechanism for climate change mitigation. In an agroforestry system, carbon can be seen sequestered in several ways, such as carbon sequestered in tree biomass, soil organic carbon enhancement, and carbon stored in block and boundary plantations (Murthy *et al.,* 2013). Agroforestry systems are believed to have a higher potential to sequester carbon better than agricultural annual crops due to the higher belowground C inputs and less soil disturbance (Pretty, 2018). Priano *et al.* (2018) conducted a study and compared soil organic carbon storage and methane fluxes among prairies, woodlots, and silvopastoral systems in Argentina. They further observed that soil organic carbon was found greater under systems containing trees than under pasture. This land use system is mostly practiced by subsistence farmers in developing countries, which offers an opportunity for those farmers to benefit economically from agroforestry by selling the sequestered carbon to developed countries.

Agroforestry and climate change

Agroforestry has great potential as a mitigation strategy for the changing climate because it involves several activities that have been shown to increase carbon absorption and hence lower greenhouse gas emissions (Bai *et al.,* 2019). More specifically, agroforestry can mitigate the atmospheric greenhouse gas concentration as well as support local farmers to adapt to climate change (Yadav *et al.,* 2021b), as these systems can store comparatively more carbon than arable systems by enhancing the soil organic carbon pool (Lorenz and Lal, 2014; Babu *et al.,* 2020). Carlsson *et al.* (2017) also reported the tolerance of perennial fruit trees against environmental stresses with their ability to withstand weather aberrations and quick biomass gain under resource-limited situations over annual crops. Several studies have also proven that there is a significant association between agroforestry systems and environmental services, *viz* carbon sequestration, biodiversity, soil erosion, and the improvement of water quality (Alavalapati *et al.,* 2004). Research has also illustrated that agroforestry being a low-carbon farming system combines the provision of food security in a changing climate with the sequestration of carbon in soil and vegetation with the help of management of natural resources such as light, land, water, and nutrients (Yadav *et al.,* 2017).

Moreover, the important elements of agroforestry systems provide a variety of eco-friendly methods that are being adopted to mitigate climate change including changes in the microclimate, protection through the provision of permanent cover, opportunities for diversification of the agricultural systems, improving the efficient use of soil, water and climatic resources, contribution to soil fertility enrichment, reducing carbon emissions and increasing sequestration, and promoting gender equity. Furthermore, the system can also support biodiversity by the inclusion of several plant/crop species that could serve as home for a varied source of wildlife (Santos *et al.*, 2019).

Agroforestry for abatement of soil pollution

Soil health can also be impacted by pollutants and toxic elements. Improper use of agrochemicals, its management, and the mass application of agrochemicals including pesticides and fertilizers in agriculture are significant sources of non-point source groundwater and surface water pollution globally and may result in soil deterioration (Pavlidis *et al.*, 2020). Agroforestry is one such practice that reduces the use of agrochemicals, results in accelerated mineralization, and remediates soils and shallow groundwater (Dawson *et al.*, 2014). It was further reported by a study that the roots of trees in agroforestry systems were able to reduce the residues of nitrogen and phosphorus in the soil by 20–100%, with up to 100% pesticide transport (Pavlidis and Tsihrintzis, 2017). Examples of trees that have the potential to absorb contaminants, including pesticides are *Platanus* spp., *Populus* spp., and willow tree species etc which help in the breakdown of contaminants, immobilizing them in the woody parts of the trees or leaves (Licht and Isebrands, 2005); this potential also varies with the physicochemical properties of the used agrochemicals. Agroforestry, in certain circumstances and with certain plant species, can also be used for phytoremediation of contaminated sites. The prerequisite for phytoremediation is primarily the tolerance towards pollutants and heavy metals. Some species can survive, tolerate, and accumulate pollutants and/or heavy metals in their roots, stems, and leaves; hence, perennial species are preferred for phytoremediation (Hussain *et al.*,

2020). These pollutants were basically retained or stabilized in the rhizosphere through a process of phytostabilization. Through phytodegradation, the pollutant is somewhere degraded or converted to a less harmful form by the plants, and whole contaminants are harvested through storage in the leaves or stems of the plant during phytoremediation (Greipsson, 2011). The other mechanisms adopted by the roots in agroforestry systems are decontamination, hyperaccumulation, and hydraulic lift that removes pollutants and heavy metals from contaminated soil (Atangana *et al.*, 2014).

Agroforestry for socio-economic benefits

To be effective and promote sustainability, agroforestry needs improved integration: not only the integration of agriculture with trees but also the integration of trees with people, considering the suitability of technical, economic, and social conditions (Khatri *et al.,* 2023). Several studies have also proven that agroforestry; a natural resources management system that integrates trees on farms in the agricultural landscape has the capacity to provide multifarious benefits and therefore now has become a viable option for overcoming these above-discussed challenges. Agroforestry is a relatively undemanding form of development with no serious supply constraints and is also a highly practicable management option at the farm level. It does not require any substantial capital or machinery, and limited formal education is required while adopting this practice. In the context of socioeconomic advantages of agroforestry for rural populations in addition to its beneficial effects on the environment, it enables farmers to have the opportunity to work and earn from various plant species throughout the year on a daily, monthly, or yearly basis for their families, from fruits, cash crops, food, wood-based and other by-products (Octavia *et al.,* 2022). However, the farmer's understanding of the procedure to be adopted, mainly regarding how to choose the best plants or trees for their system, may be considered a major factor in how this outcome turns out. Reynolds *et al.* (2007) also indicated the benefits of growing some trees alongside crops that are complementary to them. Contrarily, the incorrect choice of tree or crop components can impact the competition for nutrient uptake which consequently reduces yield and ultimately farmers' profit. It can also help in mainstreaming gender in rural areas, as women may also benefit from more job opportunities since they can participate directly in production activities (Kiptot *et al.,* 2014). Therefore agroforestry practices can improve livelihood by enhancing food and nutrition security for communities living near forests while also generating revenue.

Agroforestry and ecosystem services

Agroforestry includes several ecological practices that have the potential to improve ecosystem services. The fundamental advantage of these tree-based multifunctional land use systems over monocropping is the resource complementarity between the trees and crops (Augustine *et al.,* 2007). Plant diversity increases the ecosystem productivity and functioning in natural ecosystems mainly by two mechanisms: (i) the phenomenon of niche complementarity, in which positive associations (non-overlapping resource requirements) between the species lead to “stable multispecies coexistence or”; (ii) species selection, wherein the possibility that a species will provide useful ecological activities or services rises with diversification (Eddy and Yang, 2022).

The integration of trees, agricultural crops, and animals into an agroforestry system has the potential to improve ecosystem services by enhancing soil physical and chemical fertility, improving soil biota and microclimate, addition of organic matter, reducing erosion, improving water quality, enhancing biodiversity and sequester carbon in rainfed areas (Octavia, 2022). Furthermore, soil organisms are considered the major drivers of key ecosystem processes that underpin crop productivity and the provision of ecosystem services (Balvanera *et al.,* 2016). The studies however clearly depicted the importance of trees in boosting the functional resilience of agroecosystems, providing resources and habitat that conserve soil biota, particularly during environmental stress periods and recolonization and recovery of soil organisms again in the favorable period for desirable agroecosystem functioning (Kamau *et al.,* 2017).

In this context, the agroforestry system is an overall win-a-win strategy that ameliorates soil health through continuous additions and decompositions of dead plants, leaf litter, pruned leaves, twigs, and branches that ultimately enhance the organic matter content in the soils. It leads to soil health and soil quality sustenance by decomposition of organic residues which releases the essential nutrients, uptake by plants which raises the crop productivity, produces quality food, and enhances farmer's profitability along with maintaining quality food, good soil health and climate security for better ecosystem through carbon sequestration. Thus, agroforestry has the greatest potential and is gaining wider recognition due to its continuous supply of wood and quality food in sustainable ways along with resource conservation and uncountable ecosystem services.

Conclusion

Healthy soil is undoubtedly one of the most critical resources for healthy ecosystems, including the agroecosystems, which provide goods and services. Globally, the health of the soil has been threatened for the past several decades due to reasons such as indiscriminate use of fertilizers and pesticides, land degradation, erosion, etc. Agroforestry, as a natural resource management practice, has shown solid evidence of its great potential in improving the soil and soil health for current and future generations. This sustainable practice can restore soil-based ecosystem services in degraded soil and enhance the ecosystem productivity by providing a viable pathway for intensification to make agriculture more sustainable because of its nutrient pumping and cycling, soil fertility enrichment, enhancement in organic matter as well as the biological, chemical, and physical properties of the soil, through its ability for microclimate modification. Moreover, their role in above-ground carbon sequestration and agroforestry systems also shows beneficial effects in climate change, carbon sequestration, water conservation, and abatement of soil pollution. The ideal form of integrating trees in an agroecosystem can be achieved by the right selection of plant species and planting pattern configurations to help address benefits related to productivity, sustainability, and adaptability. The multi-benefit of smart agroforestry has been widely proven, but agroforestry should also be developed by making the best use of local resources and use of both modern and indigenous knowledge for effective and efficient agricultural practices in order to gain environmental benefits and high economic profits. Thus, its implementation on a larger scale still requires awareness among farmers and collaboration of multistakeholders involved in realizing sustainable forest and land management. Hence, further research should focus on addressing the optimal ways to identify and diversify multiple agroforestry components, their management strategies, assessing the ecosystem services exhibited by various agroforestry systems, and the contributions of urban agroforestry to ecosystem preservation and climate change mitigation. Furthermore, it is imperative that the broader agriculture community and policymakers pay increased attention to agroforestry as a viable strategy to restore and sustain soil health.

References

Adam, D. (2021). How far will the global population rise? Researchers can't agree. *Nature*, 597, 462–465.

Alavalapati, J.R., Shrestha, R.K., Stainback, G.A. and Matta, J.R., (2004). Agroforestry development: An environmental economic perspective. In New Vista in Agroforestry: A Compendium for 1st World Congress of Agroforestry, 2004 (pp. 299-310). Springer Netherlands.

Alletto, L., Cueff, S., Brechemier, J., Lachaussee, M., Derrouch, D., Page, A., Gleizes, B., Perrin, P., and Bustillo, V., (2022). Physical properties of soils under conservation agriculture: A multi-site experiment on five soil types in south-western France. *Geoderma*, 428, 116228.

Angima, S.D., Stott, D.E. and O'Neill, M.K., (2002) Use of *Calliandra*-Napier grass contour hedges to control erosion in central Kenya. *Agric Ecosyst Environ*, 91:15–23.

Atangana, A., Khasa, D.; Chang, S. and Degrande, A., (2014). Phytoremediation in Tropical Agroforestry. *In*: Tropical Agroforestry; Springer: Dordrecht, The Netherlands, pp. 343–351.

Augustine, C.M.J., Vogt, K.A., Harrison, R.B. and Hunsaker, H.M. (2007). Nitrogen-fixing trees in small-scale agriculture of mountainous southeast Guatemala: Effects on soil quality and erosion control. *J. Sustain. For*. 23, 61–80.

Babu, S., Mohapatra, K.P., Yadav, G.S., Lal, R., Singh, R., Avasthe, R.K., Das, A., Chandra, P., Gudade, B.A. and Kumar, A., (2020). Soil carbon dynamics in diverse organic land use systems in the North Eastern Himalayan ecosystem of India. *Catena,* 194, 104785.

Babu, S., Singh, R., Yadav, D., Rathore, S.S., Raj, R., Avasthe, R., Yadav, S.K., Das, A., Yadav, V., Yadav, B., Shekhawat, K., Upadhyay, P.K., Yadav, D.K. and Singh, V.K., (2022). Nanofertilizers for agricultural and environmental sustainability. *Chemosphere*.

Balvanera, P., Quijas, S., and Martin-Lopez, B., (2016). The links between biodiversity and ecosystem services, *In* Routledge Handbook of Ecosystem Services, eds. Potschin, M. *et al.,* 45–49, Routledge, London.

Bai, X., Yang, J., and Matocha, C., (2019). Responses of soil carbon sequestration to climate-smart agriculture practices: A meta-analysis. Global change biology, 25(8), pp.2591-2606.

Barrios, E., Sileshi, G.W., Shepherd, K., and Sinclair, F., (2012). Agroforestry and soil health: Linking trees, soil biota, and ecosystem services, in Soil Ecology and Ecosystem Services, eds. Wall, D.H. *et al.,* 315–330, Oxford University Press, Oxford, UK.

Bayala, J., and Prieto, I., (2020). Water acquisition, sharing and redistribution by roots: Applications to agroforestry systems, *Plant Soil*, 453, 17–28.

Bayala, J., Heng, L.K., Van Noordwijk, M., and Ouedraogo, S.J., (2008). Hydraulic lift study in two native tree species of agroforestry parklands of West African dry savanna, *Acta Oecol.,* 34, 370–378.

Buresh, R.J., Rowe, E.C., Livesley, S.J., Cadisch, G., and Mafongoya, P., (2004). Opportunities for the capture of deep soil nutrients, *In* BelowGround Interactions in Tropical Agroecosystems, eds. Van Noordwijk, M., Cadisch, G., and Ong, C., 109–123, CAB International, Wallingford, UK.

Cao, L., Wang, S., Peng, T., Cheng, Q., Zhang, L., Zhang, Z. Yue, F. and Fryer, A.E. (2020). Monitoring of suspended sediment load and transport in an agroforestry watershed on a karst plateau, Southwest China. Agric *Ecosyst. Environ*. 299, 106976.

Carlsson, G., Martensson, L.M., Prade, T., Svensson, S.E. and Jensen, E.S., (2017). Perennial species mixtures for multifunctional production of biomass on marginal land. *Bioenergy* 9, 191–201.

Caudill, S.A., DeClerck, F.J.A., Husband, T.P., (2015). Connecting sustainable agriculture and wildlife conservation: Does shade coffee provide habitat for mammals? *Agric. Ecosyst. Environ*., 199, 85–93.

Chamorro-Martınez, Y., Torregroza-Espinosa, A. C., Pallares, M. I. M., Osorio, D. P., Paternina, A. C., & EcheverrıaGonzalez, A. (2022). Soil macrofauna, mesofauna and microfauna and their relationship with soil quality in agricultural areas in northern Colombia: Ecological implications. Revista Brasileira de Ciencia Do Solo, 46.

Chatterjee, N., Nair, P.R., Chakraborty, S. and Nair, V.D., (2018). Changes in soil carbon stocks across the Forest-Agroforest-Agriculture/Pasture continuum in various agroecological regions: A meta-analysis. Agric. Ecosyst. Environ. 266, 55–67.

Das, D. and Chaturvedi, O. (2008). Root biomass and distribution of five agroforestry tree species, *Agrofor. Syst*. 74(3):223-230.

Das, A., Babu, S., Datta, M., Sanjeev, K., Singh, R., Avasthe, R.K., Rathore, S.S., Yadav, S. K. and Singh, V.K., (2022). Restoring soil carbon in marginal land of Indian Himalayas: Impact of crop intensification and conservation tillage. *J. Environ*. Manage.

Dawson, I.K., Leakey, R., Clement, C.R., Weber., J.C., Cornelius, J.P. and Roshetko, J.M., (2014). The management of tree genetic resources and the livelihoods of rural communities in the tropics: Non-timber forest products, smallholder agroforestry practices, and tree commodity crops. *For. Ecol. Manag.*, 333, 9–21.

Dierks, D., Blaser-Hart, W.J., and Gamper, H.A., (2021). Trees enhance the abundance of arbuscular mycorrhizal fungi, soil structure, and nutrient retention in low-input maize cropping systems. *Agric. Ecosyst. Environ.,* 318, 107487.

Dixon, R.K.; Winjum, J.K.; Andrasko, K.J.; Lee, J.J. and Schroeder, P.E., (1994). Integrated land-use systems: Assessment of promising agroforest and alternative land-use practices to enhance carbon conservation and sequestration. *Clim. Change,* 27, 71–92.

Dollinger, J. and Jose, S., (2018). Agroforestry for soil health. *Agrofor. Syst.* 92, 213–219.

Dori, T., Asefaw, Z., and Kippie, T. (2022). Soil characteristics under dominant agroforestry systems along toposequence of Gedeo, Southeastern Ethiopia. *Environmental and Sustainability Indicators, 15(June),* 100191.

Du, X., Jian, J., Du, C. and Stewart, R.D., (2022). Conservation management decreases surface runoff and soil erosion. *Int. Soil Water Conserv. Res.* 10, 188–196.

Eddy, W.C. and Yang, W.H., (2022). Improvements in soil health and soil carbon sequestration by an agroforestry for food production system. *Agric. Ecosyst. Environ.* 333, 107945.

Barrios, E., Coe, R., Place, F., Gudeta, W., Sileshi and Sinclair, F. (2023). Nurturing Soil Life through Agroforestry The Roles of Trees in the Ecological Intensification of Agriculture. *In:* Biological Approaches to Regenerative Soil Systems. 265-277.

Fahad, S., Chavan, S.B., Chichaghare, A.R., Uthappa, A.R., Kumar, M., Kakade, V., Pradhan, A., Jinger, D., Rawale, G., Yadav, D.K., Kumar, V., Farooq, T.H., Ali, B., Sawant, A., Saud, S., Chen, S. and Poczai, P. (2022). Agroforestry Systems for Soil Health Improvement and Maintenance, *Sustainability*, 14, 14877.

Fanish, S.A. and Priya, R.S., (2013). Reviewon Benefits of Agrio Forestry System. *Int. J. Educ. Res.,* 1(1):1-12.

Fonte, S.J., Barrios, E., and Six, J., (2010). Earthworms, soil fertility and aggregate-associated soil organic matter dynamics in the Quesungual agroforestry system, *Geoderma*, 155, 320–328.

Giller, K.E., (2001). Nitrogen Fixation in Tropical Cropping Systems, 2nd ed., CAB International, Wallingford UK.

Gianninazi, S., Gollotte, A., and Binet, M.N., (2010). Agroecology: The key role of arbuscular mycorrhizas in ecosystem services, *Mycorrhiza*, 20, 519–530.

Greipsson, S. (2011). Phytoremediation. *Nat. Educ. Knowl.* 3, 7.

Hailu, G. (2015). A review on the comparative advantage of intercropping systems. *J. Biol. Agricu. Healthc.* 5, 2224–2320.

Hart, K., Allen, B., Keenleyside, C., Nanni, S., Maréchal, A., Paquel, K., Nesbit, M., and Ziemann, J., (2017). Research for agri committee-the consequences of climate change for EU agriculture—follow-Up to the COP21-UN Paris Climate Change Conference.

Hussain, Z., Rasheed, F., Tanvir, M.A., Zafar, Z., Rafay, M., Mohsin, M., Pulkkinen, P. and Ruffner, C., (2020). Increased antioxidative enzyme activity mediates the phytoaccumulation potential of Pb in four agroforestry tree species: A case study under municipal and industrial wastewater irrigation. *Int. J. Phytoremed.* 23, 704–714.

Ilstedt, U., Bargues Tobella, A., and Bazie, H.R., (2018). Intermediate tree cover can maximize groundwater recharge in the seasonally dry tropics. *Sci. Rep.,* 6, 21930.

IPCC (2007): Climate Change 2007: Mitigation of Climate Change. Working Group III contribution to the Intergovernmental Panel on Climate Change, Fourth Assessment Report. Bangkok, Thailand.

Jhariya, M.K., Yadav, D.K. and Banerjee, A. (2019). Agroforestry and climate change: issues and challenges Eds.

Jose, S., (2009). Agroforestry for ecosystem services and environmental benefits: an overview. *Agrofor. Syst.* 76:1–10

Jose, S., Gold, M.A. and Garrett, H.E., (2012). The Future of Temperate Agroforestry in the United States. In Agroforestry—The Future of Global Land Use; Nair, P.K.R., Garrity, D., Eds.; Springer: Dordrecht, The Netherlands; pp. 217–245.

Junaedi, A., Hardiwinoto, S., Supriyo, H., and Mindawati, N., (2020). Litter productivity and leaf litter nutrient return of three native tree species in drained tropical peatland, RiauIndonesia. IOP Conference Series: Earth and Environmental Science, 533(1), 012007.

Kassic, G.W. (2016). Agroforestry and land productivity: Evidence from rural Ethiopia. Cogent Food & Agriculture, 2, 1.

Kamau, S., Barrios, E., Karanja, N., Ayuke, F. and Lehmann, J. (2017). Soil macrofauna under dominant tree species increases along a soil degradation gradient. *Soil Biol. Biochem.* 112, 35–46.

Khaleel, A.A., Sauer, T.J., and Tyndall, J.C. (2020). Changes in deep soil organic carbon and soil properties beneath tree windbreak plantings in the US Great Plains. *Agrofor. Syst.* 94, 565–581.

Khan, Y., Shah, S., and Tian, H., (2022). The roles of arbuscular mycorrhizal fungi in influencing plant nutrients, photosynthesis, and metabolites of cereal crops—A review. *Agronomy,* 12(9), 2191.

Khatri, N.D., Paudel, D., Bhusal, P., Ghimere S., and Bhandari, B., (2023). Determinants of farmers' decisions to adopt agroforestry practices:

insights from the Mid-hills of Western Nepal. *Agrofor. Syst.* 97, 833–845.

Kim, D.G., Kirschbaum, M.U.F. and Beedy, T.L., (2016). Carbon sequestration and net emissions of CH_4 and N_2O under agroforestry: Synthesizing available data and suggestions for future studies. *Agric. Ecosyst. Environ.* 226, 65–78.

Kiptot, E., Franzel, S. and Degrande, A., (2014). Gender, agroforestry, and food security in Africa. *Curr. Opin. Environ. Sust.* 6, pp.104-109.

Kittur, B.H. and Bargali, S.S. (2013). Perspectives of agroforestry: Present and future facts. *J. Progress. Agric.* 4(2):91-94.

Kuyah, S., Oborn, I., Jonsson, M., (2016) Trees in agricultural landscapes enhance provision of ecosystem services in Sub-Saharan Africa. *Ecosyst. People*, 12: 255–273.

Kuyah, Cory, W., Whitney, Jonsson, M., Gudeta, W., Sileshi, Oborn, I., Muthuri, C.W. and Luedeling, E., (2019). Agroforestry delivers a win-win solution for ecosystem services in sub-Saharan Africa. A meta-analysis In *Agron. Sustain. Dev.* 39: 47.

Lacombe, S., Bradley, R.L., Hamel, C. and Beaulieu, C. (2009). Do tree-based intercropping systems increase the diversity and stability of soil microbial communities? *Agric. Ecosyst. Environ.* 131, 25–31.

Lal, R. (1989). Agroforestry systems and soil surface management of a tropical alfisol, water runoff, soil erosion, and nutrient loss. *Agrofor. Syst.* 8:97–111.

Lal, R. (2015). Restoring soil quality to mitigate soil degradation. *Sustainability*, 7, 5875–589.

Licht, L. and Isebrands, J.G., (2005). Proceedings of the joint IEA bioenergy task 30 and task 31, Workshop on sustainable bioenergy production systems: Environmental, operational and social implications linking phytoremediated pollutant removal to biomass economic opportunities. *Biomass Bioenerg.* 28, 203–218.

Lorenz, K. and Lal, R., (2014). Soil organic carbon sequestration in agroforestry systems- A review. *Agron. Sustain. Dev.* 34, 443–454.

Lorenz, K. and Lal, R., (2014). Biochar application to soil for climate change mitigation by soil organic carbon sequestration. *Plant Nutri. Soil Sci.* 177 (5), 651–670

Makundi, W.R. and Sathaye, J.A. (2004). GHG mitigation potential and cost in tropical forestry – relative role for agroforestry. *Environ. Dev. Sustain.* 6, 235–260.

Manna, M.C., Ghosh, P.K. and Acharya, C.L., (2003). Sustainable crop production through management of soil organic carbon in semiarid and tropical India, *J. Sustain. Agric.* 21:87-116.

Mosier, S., Cordova, S.C., and Robertson, G.P. (2021). Restoring soil fertility on degraded lands to meet food, fuel, and climate security needs *via* perennialization. *Front. Sustain. Food Syst.* 5, 706142.

Mosquera-Losada, M.R., McAdam, J.H., Romero-Franco, R., Santiago-Freijanes, J.J. and Rigueiro-Rodróguez, A., (2009). Definitions and components of agroforestry practices in Europe. *In*: Agroforestry in Europe: Current Status and Future Prospects. 3-19.

Muchane, M.N., Sileshi, G.W., Gripenberg, S., Jonsson, M., Pumarino, L. and Barrios, E., (2020). Agroforestry boosts soil health in the humid and sub-humid tropics: A meta-analysis. *Agric. Ecosyst. Environ.* 295,106899.

Murthy, I.K., Gupta, M., Tomar, S., Munsi, M., Tiwari, R., Hegde, G.T. and Ravindranath, N.H., (2013). Carbon sequestration potential of agroforestry systems in India. *J. Earth Sci. Climate Change.* 4(1): 1-7.

Nair, P.K., (1985). Classification of Agroforestry Systems. *Agrofor. Syst.* 3, 97–128.

Nair, P.K.R. (Eds.) (1984). Agroforestry system descriptions 1-26. *Agrofor. Syst.* Various issues, 1984-88.

Nair, P.K.R., (2011). Agroforestry systems and environmental quality: introduction. *J. Environ. Qual*. 40:784–790.

Nygren, S.E.P., Fernandez, M., Harmand, J.M., and Leblanc, H.A., (2012). Symbiotic dinitrogen fixation by trees: An underestimated resource in agroforestry systems? *Nutr. Cycling Agroecosyst.,* 94, 123–160.

Octavia, D., Suharti, S., Murniati, M., Dharmawan, I.W.S., Nugroho, H.Y.S.H., Supriyanto, B., Rohadi, D., Njurumana, G.N., Yeny, I., Hani, A., and Mindawati, N., (2022). Mainstreaming Smart Agroforestry for Social Forestry Implementation to Support Sustainable Development Goals in Indonesia: A Review. *Sustainability*, 14(15), 9313.

Ortiz, A., and Sansinenea, E., (2022). The role of beneficial microorganisms in soil quality and plant health. *Sustainability*, 14(9), 5358.

Omoro, L.M.A. and Nair, P.K.R., (1993). Effects of mulching with multipurposetree prunings on soil and water run-off under semi-arid conditions in Kenya. *Agrofor. Syst.* 22:225–239.

Palm, C.A., Gachengo, C.N., Delve, R.J., Cadisch, G., and Giller, K.E., (2021). Organic inputs for soil fertility management in tropical agroecosystems: Application of an organic resource database, *Agric. Ecosyst. Environ.* 83, 27–42.

Pandey, D.N. (2007). Multifunctional agroforestry systems in India. *Curr. Sci.* 455-463.

Pandey, A.K., Gupta, V.K. and Solanki, K.R., (2010). Productivity of neem-based agroforestry system in semi-arid region of India, *Range Manag. Agrofor.* 31(2):144-149.

Pantigoso, H.A., Yuan, J., He, Y., Guo, Q., Vollmer, C., and Vivanco, J.M., (2020). Role of root exudates on assimilation of phosphorus in young and old Arabidopsis thaliana plants. PLoS One, 15(6), e0234216.

Parwada, C., Chipomho, J. and Mapope, N., (2022). Role of agroforestry on farmland productivity in semi-arid farming regions of Zimbabwe. *Res. World Agric. Econ.* 3, 515.

Pavlidis, G., Karasali, H., Tsihrintzis, V.A., (2020). Pesticide and Fertilizer Pollution Reduction in Two Alley Cropping Agroforestry Cultivating Systems. *Water Air Soil Pollut.* 231, 241.

Pavlidis, G. and Tsihrintzis, V. (2017). Pollution control by agroforestry systems: A short review. *Eur. Water*, 59, 297–301.

Pretty, J.N. (2018.) Intensification for redesigned and sustainable agricultural systems. *Science*, 362.

Priano, M.E., Fuse, V.S., Mestelan, S., Berkovic, A.M., Guzman, S.A., Gratton, R. and Juliarena, M.P., (2018). Afforested sites in a temperate grassland region: influence on soil properties and methane uptake. *Agrofor. Syst.*

Purwaningsih, R. Sartohadi, J., Setiawan, M.A., (2020). Trees and Crops Arrangement in the Agroforestry System Based on Slope Units to Control Landslide Reactivation on Volcanic Foot Slopes in Java, Indonesia. *Land*, 9, 327.

Puskaric, J., Jovic, J., Ivezic, V., Popovic, B., Paponja, I., & Brmez, M. 2021. The Communities of the Nematodes, Bacteria, and Fungi and the Soil's Organic Matter in an Agroforestry Ecosystem in Croatia. Poljoprivreda, 27(1), 66– 74.

Reynolds, P.E., Thevathasan, N.V. and Gordon, A.M., (2007). Effects of tree competition on corn and soybean photosynthesis, growth, and yield in a temperate tree-based agro forestry intercropping system in southern Ontario, Canada. *Ecological Engin.* 29(4), pp.362-371.

Rosati, A., Borek, R. and Canali, S., (2020). Agroforestry and organic agriculture. *Agrofor. Syst.* 95, 805–821.

Salako, F.K., Hauser, S., Babalola, O. and Tian, G., (2001). Improvement of the physical fertility of a degraded Alfisol with planted and natural fallows under humid tropical conditions. *Soil Use Manag*. 17:41–47.

Santos, P.Z.F., Crouzeilles, R. and Sansevero, J.B.B., (2019). Canagro forestry systems enhance biodiversity and ecosystem service provision in agricultural landscapes? A meta-analysis of the Brazilian Atlantic Forest. *For. Ecol. Manag.* 433, pp.140-145.

Shang, J., Zhu, Q. and Zhang, W., (2018). Advancing soil physics for securing food, water, soil and ecosystem services. *Vadose Zone J.* 17(1), 1–7.

Smith, S.E., and Read, D., (2008). *Mycorrhizal* Symbioisis, 3rd ed., Academic Press, Cambridge MA.

Sprent, J.I. (2009). Legume Nodulation: A Global Perspective, WileyBlackwell, Hoboken, NJ.

Sonmez, O., Turan, V. and Kaya, C., (2016). The effects of sulfur, cattle, and poultry manure addition on soil phosphorus. *Turk. J. Agric. For.* 40, 536–541.

Suprayogo, D., Van Noordwijk, M., Hairiah, K., Meilasari, N., Rabbani, A.L., Ishaq, R.M. and Widianto, W., (2020). Infiltration-friendly agroforestry land uses on volcanic slopes in the Rejoso Watershed, East Java, Indonesia. *Land*. 9, 240.

Surki, A.A., Nazari, M., Fallah, S., Iranipour, R., and Mousavi, A. (2020). The competitive effect of almond trees on light and nutrients absorption, crop growth rate, and the yield in almond–cereal agroforestry systems in semi-arid regions. *Agrofor. Syst.* 94, 1111–1112.

Tripathi, D., Mani, S., and Singh, M., (2020). Fundamental of Soil Science. Agrimoon.com.

Tsufac, A.R., Awazi, N.P. and Yerima, B.P.K., (2021). Characterization of agroforestry systems and their effectiveness in soil fertility enhancement in the south-west region of Cameroon. *Curr. Res. Envirn. Sustain.* 3, 100024.

Turan, V., Schroder, P., Bilen, S., Insam, H. and Juarez, M.F., (2019). Co-inoculation effect of *Rhizobium* and *Achillea millefolium* L. oil extracts on growth of common bean (*Phaseolus vulgaris* L.) and soil microbial-chemical properties. *Sci. Rep.* 9, 15178.

Vanlauwe, B., Tittonell, P., and Mukalama, J., (2006). Within-farm soil fertility gradients affect response of maize to fertiliser application in Western Kenya, *Nutr. Cycl. Agroecosyst.* 76, 171–182.

Van'Noordwijk, M., Ekadinata, A., Leimona, B., Catacutan, D., Martini, E., Tata, H.L., Oborn, I., Hairiah, K., Wangpakapattanawong, P. and Mulia, R., (2020). Agroforestry options for degraded landscapes in Southeast Asia. In Agroforestry for Degraded Landscapes: Recent Advances and Emerging Challenges; Dagar, J.C., Gupta, S.R., Teketay, D., Eds.; Springer: Singapore.

Wawire, A.W., Csorba, A., Toth, J.A., Micheli, E., Szalai, M., Mutuma, E., and Kovacs, E., (2021). Soil fertility management among smallholder farmers in Mount Kenya East region. Heliyon, 7(3), e06488.

UNIC Rio (2016). Transformando Nosso Mundo: A Agenda 2030 para o Desenvolvimento Sustentável. AmbientalMENTEsustentable 25, 171–190.

USDA NRCS. 2018. Healthy soil for life. https://www.nrcs.usda.gov/wps/portal/nrcs/main/soils/health/. Accessed 17 March 2018.

Yadav, G.S., Debbaram, C. and Datta, M., (2017). Effects of godawariphosgold and single supper phosphate on groundnut (Arachis hypogaea) productivity, phosphorus uptake, and phosphorus use efficiency, and economics. *Indian J. Agric. Sci.* 87(9), pp.1165-1169.

Yadav, G.S., Kandpal, B.K., Das, A., Babu, S., Mohapatra, K.P., Devi, A.G., Devi, H.L., Chandra, P., Singh, R. and Barman, K.K., (2021). Impact of 28-year-old agroforestry systems on soil carbon dynamics in Eastern Himalayas. *J. Environ. Manage.* 283.

Young, A. (1987). Agroforestry for soil conservation, CAB International with cooperation ICRAF Nairobi, Kenya.

CHAPTER FOURTEEN

AGROFORESTRY AND INDIGENOUS KNOWLEDGE: BRIDGING TRADITIONAL WISDOM WITH MODERN PRACTICES

GARGI VISHNU NAVPUTE[1], KARUNA GUSAIN[2], HUKUM SINGH[1], AND SANTAN BARTHWAL[1]

[1]GENETICS AND TREE IMPROVEMENT DIVISION, ICFRE-FOREST RESEARCH INSTITUTE, DEHRADUN, UTTARAKHAND, 248006

[2]G.B. PANT NATIONAL INSTITUTE FOR HIMALAYAN ENVIRONMENT, KOSI-KATARMAL, ALMORA-263643

Abstract

This chapter delves into the pivotal role of indigenous knowledge within agroforestry, emphasizing the imperative of respecting traditional wisdom while integrating contemporary methodologies to foster sustainable development. With roots entrenched in centuries-old ecological insights and cultural heritage, indigenous agroforestry systems offer comprehensive solutions to present-day challenges, encompassing biodiversity depletion, climate variability, and food insecurity. Through collaborative research and community partnerships, this chapter scrutinizes the ecological, societal, and economic facets of indigenous agroforestry, spotlighting its capacity to enrich ecosystem services, elevate livelihoods, and fortify cultural resilience. By acknowledging historical inequities and embracing fair resource distribution, we can bridge the gap between ancestral knowledge and modern methodologies, paving avenues towards inclusive, regenerative, and culturally attuned agroforestry domains. This chapter advocates for interdisciplinary cooperation, participatory governance, and policy reinforcement to unlock the complete potential of indigenous wisdom within agroforestry, thus heralding a more sustainable and equitable trajectory for the future.

14.1 Introduction

Agroforestry is a type of land use management that combines forestry and agricultural techniques to create a variety of sustainable production systems by incorporating trees and shrubs into agricultural landscapes. This strategy promotes environmental sustainability and resilience by utilising the beneficial interactions between trees, crops, and livestock to increase production, biodiversity, and ecosystem services (Nair, 2012; Dinesha et al., 2023). Agroforestry offers numerous environmental benefits, including soil conservation, water management, and carbon sequestration. By integrating trees into agricultural landscapes, agroforestry helps prevent soil erosion, improve soil fertility, and reduce nutrient runoff, thereby enhancing long-term soil health and productivity (Nair, 2012). Additionally, agroforestry systems can contribute to climate change mitigation by sequestering carbon in trees and soil, mitigating greenhouse gas emissions, and enhancing ecosystem resilience to climate variability (Amonum et al., 2009; Kumar et al., 2020; Kaushal et al., 2021).

Agroforestry also improves biodiversity by giving a variety of plant and animal species a place to live and food sources. In comparison to traditional monoculture agriculture, the structural complexity of agroforestry systems generates a wider range of microhabitats, supporting a higher diversity of species. Improved crop yields and pest management are made possible by this biodiversity, which also contributes to ecosystem services including pollination, pest control, and natural pest regulation (Sollen-Norrlin et al., 2020). Through crop resilience and income source diversification, agroforestry can improve livelihoods, food security, and rural development (Wilson & Lovell, 2016; Dinesha et al., 2023). Additionally, agroforestry encourages the sharing of traditional knowledge and the preservation of culture, strengthening communal bonds and increasing resistance to outside forces (Albuquerque, 2020). Overall, agroforestry represents a promising approach to sustainable land management that reconciles agricultural production with environmental conservation and socio-economic development. By harnessing the synergies between trees, crops, and livestock, agroforestry offers a pathway towards more resilient and sustainable food systems, contributing to global efforts to address pressing challenges such as climate change, biodiversity loss, and rural poverty.

14.1.1 Importance of indigenous knowledge in Agroforestry

Indigenous knowledge must be included into agroforestry techniques to increase the sustainability and resilience of agricultural systems, particularly in diverse and resource-constrained countries. Indigenous knowledge in agroforestry offers crucial insights into traditional land management techniques, ecosystem dynamics, and plant species selection—all of which are commonly overlooked in conventional agricultural systems (Méndez et al., 2010). Indigenous communities possess extensive knowledge of the local plant species, their ecological roles, and their interactions within agroecosystems, which enables them to identify resilient and multipurpose tree species that can enhance soil fertility, provide shade and fodder, and support biodiversity conservation (Cruz-Garcia & Price, 2014). Furthermore, indigenous knowledge systems offer comprehensive perspectives on landscape management that support socio-ecological resilience and cultural continuity by fusing agriculture with water conservation, forest care, and spiritual values (Altieri, 2004). By recognising and valuing traditional knowledge, practitioners in agroforestry can foster participatory decision-making processes, harness the adaptive capacity of local communities, and together develop innovative solutions to contemporary agricultural challenges (Nerlich et al., 2012). By doing this, the incorporation of indigenous knowledge promotes social justice and environmental stewardship in a world that is changing quickly. It also increases the ecological sustainability and productivity of agroforestry systems while strengthening the cultural resilience and empowerment of indigenous peoples (Tewari et al., 2024; Kumar et al., 2021a&b; Kumar et al., 2022).

14.2 Traditional agroforestry practices

The historical significance of agroforestry systems goes beyond their ability to increase agricultural output, despite its old roots in indigenous methods. Indigenous agroforestry methods have shaped landscapes and fostered a strong sense of connection between communities and their ecosystems by being deeply entwined with cultural and spiritual beliefs (Davidson-Hunt et al 1999). For instance, for thousands of years, native tribes in the Amazon basin have been managing intricate agroforestry systems known as "forest gardens" or "chacras," which integrate food crops, medicinal plants, and sacred locations inside forested environments (Posey, 1985). These systems, which represented the close bond between humans and environment, not only supplied food but also functioned as hubs for social cohesiveness and cultural identity (Coomes et al., 2015). Indigenous agroforestry techniques were uprooted and marginalised as colonial authorities viewed them as archaic or outdated, as modernization and colonialism swept the world (Zent, 2001). Traditional land management practices were upset by land expropriation, forced relocation, and the adoption of Western agricultural models. This resulted in biodiversity loss, environmental degradation, and the loss of cultural knowledge (Richards, 1985). Many indigenous groups have persevered their agroforestry traditions in spite of these obstacles, fighting against cultural absorption and claiming their rights to resources and land (Mistry & Berardi, 2016).

Scientists, decision-makers, and development professionals have come to appreciate the importance of indigenous knowledge in agroforestry during the past few decades (Cruz-Garcia & Price, 2011). According to Berkes et al. (2000), indigenous agroforestry systems are becoming more and more recognised as models of sustainable land management, repositories of traditional knowledge, and reservoirs of biodiversity. The goal of improving agricultural sustainability, protecting biodiversity, and empowering indigenous communities has led to an increase in the documentation, validation, and promotion of indigenous agroforestry methods (Wiersum, 1997).

Despite these challenges, there is growing recognition of the need for respectful engagement with indigenous knowledge holders, recognizing their rights, values, and priorities (Menzies, 2006). Participatory research methodologies, community-based co-management approaches, and Indigenous-led initiatives are emerging as promising pathways for bridging traditional wisdom with modern agroforestry practices (Berkes, 2009). By fostering mutual learning, reciprocal exchange, and collaborative decision-making, these approaches seek to harness the complementary strengths of indigenous and scientific knowledge systems, fostering more equitable and sustainable agroecological transitions (Lele et al., 2013). Nonetheless, there are several obstacles to overcome in order to incorporate indigenous knowledge into conventional agroforestry research and development. Reyes-García et al. (2013) suggested that effective collaboration between scientists and indigenous populations might be impeded by language difficulties, cultural disparities, and power imbalances, which may result in misinterpretation or exploitation of indigenous knowledge. Furthermore, traditional knowledge is frequently implicit and context-specific, which makes it difficult to apply standard scientific methods and necessitates creative approaches to validation and documentation (Gadgil et al., 1993). Nevertheless, participatory and Indigenous-led approaches are emerging as promising pathways for bridging traditional wisdom with modern agroforestry practices, fostering more equitable and sustainable agroecological transitions (Tewari et al., 2022; Kumar et al., 2018; Kumar et al., 2021c&d).

In summary, the historical background of indigenous agroforestry highlights the enduring relationships that exist between people, environments, and biodiversity. Over thousands of years, indigenous agroforestry systems have developed, incorporating complex knowledge structures, flexible tactics, and cultural beliefs. Many indigenous groups have managed to preserve and revive their agroforestry traditions in spite of centuries of colonisation and modernity. They have also managed to oppose cultural absorption and stand up for their rights to land and resources. A growing number of people are realising now how important indigenous knowledge is to agroforestry, motivated by the goals of improving agricultural sustainability, protecting biodiversity, and strengthening the position of indigenous communities. However, there are several obstacles to overcome in order to incorporate indigenous knowledge into mainstream agroforestry research and development, such as power disparities, language hurdles, and cultural differences (Verma et al., 2023; Kumar et al., 2020; Kumar et al., 2021e).

14.2.1 Traditional agroforestry systems around the world

Indigenous tribes all across the world have been using a variety of traditional agroforestry methods for millennia, proving how adaptable and versatile it is to incorporate trees into agricultural landscapes. These systems, which represent a richness of traditional knowledge and sustainable land management techniques, have developed in response to regional ecological circumstances, cultural traditions, and resource demands (Young, 1988). Traditional agroforestry techniques have been crucial to maintaining agricultural output throughout a range of agro-ecological zones in Africa and many other countries. Traditional agroforestry methods have flourished in a variety of habitats. Indigenous cultures' inventiveness, tenacity, and cultural diversity are reflected in traditional agroforestry systems across the globe. These systems combine trees with crops, cattle, and other landscape features to create multipurpose landscapes that support ecological sustainability and cultural continuity while producing food, fuel, fibre, and medicinal resources. Policymakers, practitioners, and researchers can address modern issues like climate change, biodiversity loss, and rural poverty by valuing and acknowledging traditional agroforestry practices

(Doddabasawa et al., 2018). This will promote more equitable and sustainable agricultural development. Some of the important traditional agroforestry systems practiced around the worlds are discussed below:

1. Taungya Agroforestry: Taungya is a traditional agroforestry practice that originated in Southeast Asia, specifically in Myanmar and Thailand. Farmers grow young trees alongside annual crops in wooded areas while cultivating annual crops. This method guarantees the long-term availability of timber resources while also enabling the production of food right away. Food crops including rice, maize, and vegetables are frequently cultivated alongside trees like acacia, mahogany, and teak. According to Tewari et al. (2008), this strategy encourages biodiversity, sustainable land use, and soil conservation.

2. Parkland Agroforestry: In the Sahelian region of Africa, parkland agroforestry is common. Farmers there intercrop valuable tree species like *Faidherbia albida* with cereal crops. In addition to fixing nitrogen in the soil, this approach improves the soil and gives crops shade. In order to provide the soil with organic matter and nitrogen during the rainy season, *Faidherbia albida* drops its leaves, allowing crops to enjoy sunshine during the dry season. By improving soil fertility and drought resistance, this method supports food security and livelihoods in dry regions (Sileshi et al., 2007).

3. Forest Gardens (Chacras): The Amazon basin's indigenous tribes have created complex agroforestry systems called "forest gardens," or "chacras." Within forested settings, these systems incorporate food crops, medicinal plants, and spiritual locations. Many vegetation layers, such as ground cover plants, shrubs in the understory, and canopy trees, produce a variety of microhabitats that are home to a large number of species. A holistic approach to land management is embodied by forest gardens, which combine ecological sustainability with spiritual and cultural benefits. They support the preservation of biodiversity and the resilience of ecosystems while giving indigenous tribes access to resources, food, and medical care (Posey, 1985).

4. Alley Cropping: Planting alleys of pasture or annual crops in between rows of trees or bushes is known as alley cropping. Both tropical and temperate regions frequently employ this technique, which has several advantages including improved biodiversity, nitrogen cycling, and soil conservation (Nair, 1993).

5. Silvopasture: Through the integration of trees and livestock grazing, silvopasture enables the simultaneous production of cattle, feed, and timber. This method gives cattle shade and protection while improving soil fertility, animal welfare, and carbon sequestration (Nair et al., 2009).

6. Windbreaks: According to Garrity et al. (2010), windbreaks, often referred to as shelterbelts or hedgerows, are rows of trees or shrubs planted to protect crops and livestock, reduce wind speed and erosion, and provide microclimatic conditions that are favourable for agricultural productivity.

7. Riparian Buffer Strips: Planting trees and bushes along rivers, streams, and lakes creates riparian buffer strips that preserve the quality of the water, stabilise the banks, and improve the habitat of both aquatic and terrestrial animals (Lehmann et al., 2005).

14.2.2 Agroforestry Practices in India

1. Traditional Agroforestry: Agroforestry techniques, such as home gardens, taungya, and agro-silvopastoral systems, have a long history in India. Home gardens are tiny agroforestry plots with a variety of fruit trees, herbs, and vegetables that are incorporated into residential neighbourhoods (Saha et al., 2018).

2. Agro-Silvopastoral Systems: Agro-silvopastoral systems blend crop production, livestock grazing, and tree cultivation. Farmers engage in agroforestry in areas such as the Eastern Himalayas and the Western Ghats, planting teak, eucalyptus, and bamboo alongside food crops and grazing fields (Negi et al., 2019).

3. Community Forest Management: Community forest management projects in multiple Indian states use afforestation, agroforestry, and participatory decision-making to engage local communities in the sustainable management of forest resources (Gadgil et al., 1993).

4. Bamboo Agroforestry: In bamboo agroforestry, various crops are grown alongside bamboo for a variety of purposes, including building, handicrafts, and erosion control. States like Assam, Mizoram, and Tripura, where bamboo is important to the economy and culture, are known for this technique (Singh et al., 2014).

14.3 Indigenous Knowledge Systems

Indigenous knowledge systems (IKS) are the body of information, practices, beliefs, and technology that indigenous cultures have amassed over many generations. The close interaction that exists between indigenous peoples and their environs is reflected in these systems, which have their roots in local settings, cultural traditions, and social structures (Davidson-Hunt et al 1999). IKS are known for their holistic approach to knowledge, which incorporates social, spiritual, and ecological aspects. For instance, Dreamtime stories serve as cultural narratives that communicate moral values and spiritual beliefs as well as ecological information about ecosystems, species interactions, and resource management among Indigenous Australian communities (Altman, 2012).

IKS are also dynamic and adaptable, which enables ongoing learning, experimenting, and innovation in response to shifting environmental circumstances (Fikret et al., 2018). Indigenous societies use adaptive management practices that have developed over centuries of observation and trial, such as agroforestry, rotational farming, and water harvesting techniques. Quechua farmers in the Andes employ "ayllu" systems, which provide food security and increase resilience to climate variability through cooperative land management and crop diversity (Herrera et al., 2008). These adaptable tactics are a reflection of the profound ecological knowledge ingrained in indigenous civilizations, and they offer insightful solutions to today's sustainability problems (Tewari et al., 2021; Kumar et al., 2021 Kumar et al., 2024).

In addition, IKS are distinguished by their integration into regional customary laws and social structures that control resource utilisation, rights, and conservation measures (Berkes et al., 2000). Traditional leadership systems, elder councils, and tribal councils are examples of indigenous governance frameworks that are essential to sustaining social cohesion, managing natural resources, and settling disputes. For instance, traditional grazing rights and land tenure systems among the Maasai people of East Africa control cattle migration and rangeland management, guaranteeing sustainable land use practices (Goldman et al., 2017). The efficacy and legitimacy of conservation and development projects in indigenous lands can be improved by acknowledging the authority of local knowledge holders and incorporating customary rules with institutional governance systems.

14.3.1 Understanding Indigenous Knowledge

Native American cultures have acquired a wide range of insights, behaviours, and beliefs through direct observation, experimentation, and cultural transmission. These have been compiled into what is known as indigenous knowledge (IK) (Davidson-Hunt et al 1999). Recognising IK's contextual and placed nature—which is shaped by particular ecological, social, and historical contexts—is essential to comprehending it (Agrawal, 1995). Indigenous knowledge reflects the complex understanding of ecological processes, biodiversity, and climate variability because it is closely linked to the local ecosystems. For instance, knowledge about weather patterns, animal behaviour, and sea ice dynamics is used by Inuit communities in the Arctic to guide their subsistence activities, which include fishing, hunting, and navigating (Berkes, 2008). Furthermore, in order to comprehend IK, one must recognise that its worldview and epistemological underpinnings diverge from conventional Western scientific paradigms (Warren et al., 1995). Indigenous knowledge systems, in contrast to the reductionist and Cartesian methods of modern science, frequently emphasise relationality, interconnection, and reciprocity between humans and nature. Indigenous Amazonian cosmologies, for example, emphasise ecological integrity and cultural continuity in stewardship practices because they view forests as living things with spiritual agency (Posey, 1985). Engaging with indigenous ontologies and cosmologies can help one understand the intricate relationships that exist between environmental ethics, cultural beliefs, and sustainable resource management.

Moreover, realising IK's dynamic character and the continuous processes of transmission, adaptation, and modification among indigenous groups are essential to comprehending it (Sillitoe, 2007). IK is dynamic and constantly changing, incorporating new technology, information sources, and cultural influences in response to environmental, social, and economic shifts. For instance, in the Pacific Islands, old navigating methods based on astronomical cues and ocean currents have survived alongside contemporary navigational aids, representing a fusion of traditional knowledge with cutting-edge discoveries (Lewis, 1972). Researchers and practitioners can cultivate cooperative collaborations with indigenous populations, co-generating knowledge and co-designing solutions to urgent environmental concerns, by recognising the dynamic nature of Indigenous knowledge (IK).

14.3.2 Bridging the traditional wisdom with modern practices

Combining traditional agricultural knowledge with contemporary methods is a dynamic and widely accepted way to honour indigenous knowledge systems and meet today's ecological concerns. Traditional knowledge, which is frequently ingrained in indigenous cultures, includes cultural values that have upheld communities for many generations as well as a thorough awareness of regional ecosystems and adaptive resource management techniques (Davidson-Hunt et al 1999). In rural environments, it is possible to improve ecological resilience, advance social fairness, and maintain cultural continuity by fusing ancient wisdom with contemporary farming practices.

An essential component of integrating ancient knowledge with contemporary methods is recognising the complementary advantages of both scientific understanding and indigenous wisdom. Indigenous knowledge systems provide important insights into agroecological concepts, climate adaption tactics, and sustainable land management (Berkes et al., 2000). Alley cropping, home gardens, and rotational cropping are a few examples of traditional agroforestry techniques that have been demonstrated to improve soil fertility, biodiversity, and water retention, all of which support resilient agricultural systems (Nair, 2012). The integration of indigenous knowledge with contemporary scientific research and innovation can facilitate the development of evidence-based, context-specific solutions that respect cultural variety and community autonomy while addressing local concerns.

Furthermore, it takes collaborations and cooperative decision-making processes involving indigenous people, scholars, politicians, and practitioners to bridge ancient wisdom with contemporary practices (Berkes, 2009). To ensure that local stakeholders are meaningfully engaged and that benefits are shared fairly, participatory approaches, community-based co-management projects, and conservation activities led by Indigenous peoples are crucial (Reyes-García et al., 2013). For example, programmes like indigenous seed banks, community-based forest management, and documentation projects for traditional ecological knowledge enable indigenous peoples to take back authority over their lands, resources, and cultural legacy (Mistry & Berardi, 2016). It is possible to close the knowledge gap between conventional wisdom and contemporary methods and create more resilient, equitable, and sustainable agricultural systems by encouraging courteous and inclusive discourse.

14.4 Challenges in integrating indigenous knowledge with modern agroforestry

The problem of ownership, benefit-sharing, and intellectual property rights presents another difficulty in combining traditional knowledge with contemporary agroforestry techniques. According to Posey (2002), indigenous knowledge is frequently susceptible to misuse, exploitation, and biopiracy by outside parties looking to profit from traditional resources or innovations without permission or fair pay. To preserve indigenous knowledge systems and guarantee that benefits reach indigenous peoples, mechanisms for defending indigenous intellectual property rights—such as customary laws, community protocols, and international agreements like the Nagoya Protocol—are crucial (Kohler & Scott, 2007). Implementing these processes, however, necessitates resolving past injustices and power disparities in addition to navigating difficult legal, ethical, and cultural issues.

Furthermore, issues with generational transitions, social upheaval within indigenous groups, and cultural change provide obstacles to the adaptation and integration of indigenous knowledge into contemporary agroforestry techniques. Traditional knowledge systems, languages, and cultural practices are being undermined by the rapid processes of globalisation, urbanisation, and cultural homogenization, especially in the younger generations (Gavin et al., 2015). Reviving cultural practices, bolstering community institutions, and establishing avenues for youth engagement and empowerment are all necessary to sustain the intergenerational transmission of indigenous knowledge (Berkes, 2009). Furthermore, fostering gender-equitable and socially inclusive methods to agroforestry development requires addressing gender inequities and making sure marginalised groups, such Indigenous women and youth, are included (Reyes-García et al., 2013).

In summary, combining indigenous knowledge with contemporary agroforestry techniques requires negotiating a challenging landscape of institutional, sociocultural, and ethical issues. In order to overcome these obstacles, it is necessary to implement rights-based, culturally aware, participatory strategies that respect indigenous peoples' rights and agency, value their knowledge, and encourage inclusive decision-making processes. Through the promotion of cooperation, reciprocal education, and mutual respect amongst researchers, policymakers, practitioners, and indigenous communities, it becomes feasible to use the combined knowledge base's collective expertise and attain more equitable and sustainable agroforestry results.

14.5 Case studies of successful integration

14.5.1 World context

1. Maori Agroforestry Practices in New Zealand: Sustainable land management initiatives and effective ecological restoration have resulted from the integration of Maori traditional ecological knowledge (TEK) into contemporary agroforestry practices in New Zealand. In order to improve soil fertility, water retention, and biodiversity in agroecosystems, Maori communities have revived traditional practices including "māra kai" (food gardens) and "rimurapa" (stone walls) (Tipa & Nelson, 2008). Projects such as the Waikereru Ecosanctuary have restored native forests, wetlands, and cultural landscapes by combining Maori TEK with scientific research and community-led conservation initiatives, thereby promoting ecological resilience and cultural revival (Robinson et al., 2016).

2. Mayan Agroforestry Systems in Guatemala: Mayan communities in Guatemala have effectively combined traditional agroforestry methods with cutting-edge conservation techniques to improve livelihoods, food security, and ecosystem services. Initiatives like the Maya Nut Institute encourage the growth of native tree species in agroforestry systems, such as *Brosimum alicastrum* (ramón) and Inga spp. (inguarán), giving rural people access to nutrient-dense food, fodder, and revenue (García-Frapolli et al., 2009). These programmes have enabled Mayan farmers to manage their landscapes sustainably and adjust to environmental change by fusing traditional knowledge of forest management, soil conservation, and seed preservation with market-based incentives and participatory research techniques (Peters et al., 2014).

3. Indigenous Agroforestry in Brazil's Atlantic Forest: In the Atlantic Forest region of Brazil, indigenous groups have developed agroforestry systems that fuse traditional knowledge with contemporary ecological concepts. Communities support biodiversity conservation, improved soil fertility, and sustainable livelihoods by integrating native tree species like Brazil nuts (*Bertholletia excelsa*) and açaí palms (*Euterpe oleracea*) into agroecosystems (Peters et al., 2017). In order to increase well-being and environmental stewardship, organisations like as the Socio-Environmental Institute (ISA) collaborate with indigenous people

to support agroforestry initiatives, advance land rights, and fortify local government systems (Isendahl et al., 2013).

4. Aboriginal Agroforestry Practices in Australia: For millennia, Australian Aboriginal people have perpetuated traditional agroforestry techniques, such as "bush tucker gardens" and "firestick farming," exemplifying sustainable land management approaches tailored to a variety of environments (Jones et al., 2019). To improve soil fertility, biodiversity, and cultural ties to the land, these methods combine fire, crops, and water management. Indigenous knowledge has the potential to mitigate climate change and promote ecosystem resilience. Collaborative research projects, like the Indigenous Fire and Carbon Initiative, support Aboriginal-led land management initiatives, carbon sequestration projects, and cultural revitalization efforts (Russell-Smith et al., 2013).

5. Indigenous Agroforestry in Cameroon: To improve food security and environmental sustainability, indigenous people in Cameroon have created agroforestry systems that combine traditional knowledge with contemporary agricultural practices. For example, the Baka people use agroforestry in conjunction with shifting farming methods, combining a variety of tree species such cocoa, plantain, and oil palm (Wiersum et al., 2005). Through the integration of scientific knowledge on crop diversification, water management, and soil conservation with customary methods, these communities sustain robust agricultural systems that sustain their way of life while protecting biodiversity.

6. Indigenous Agroforestry in the Philippines: Indigenous communities in the Philippines practice agroforestry systems known as "swidden agriculture," or kaingin, which involve the rotational clearing of forests for cultivation. These systems integrate traditional knowledge of forest management, crop selection, and fire management with modern agroecological principles (Raintree, 1987). Agroforestry initiatives supported by organizations like the Philippine Agroforestry Education and Research Network (PAFERN) promote sustainable land use practices, improve farmers' incomes, and conserve natural resources across diverse landscapes.

7. Indigenous Agroforestry in Ethiopia: Agroforestry systems, such "home gardens" and "church forests," are used by indigenous communities in Ethiopia to incorporate food crops, fruit trees, and medicinal plants into shared landscapes (Lemenih et al., 2005). These systems represent

conventional understandings of water conservation strategies, soil fertility management, and plant diversification. Organisations that work with local communities to document traditional agroforestry techniques, advance sustainable land management, and strengthen climate resilience in the face of environmental change include the Ethiopian Environment and Development Research Institute (EEDRI).

8. Maori Agroforestry Practices in New Zealand: Sustainable land management initiatives and effective ecological restoration have resulted from the integration of Maori traditional ecological knowledge (TEK) into contemporary agroforestry practices in New Zealand. In order to improve soil fertility, water retention, and biodiversity in agroecosystems, Maori communities have revived traditional practices including "mara kai" (food gardens) and "rimurapa" (stone walls) (Tipa & Nelson, 2008). Projects such as the Waikereru Ecosanctuary have restored native forests, wetlands, and cultural landscapes by combining Maori TEK with scientific research and community-led conservation initiatives, thereby promoting ecological resilience and cultural revival (Robinson et al., 2016).

Table 1 Indigenous Agroforestry Practices and Collaborations Worldwide

Sr. No.	Regions	Indigenous group	Agroforestry Practices	Key Initiatives and Collaborations
1	New Zealand	Maori	Revival of "māra kai" (food gardens) and "rimurapa" (stone walls)	Waikereru Ecosanctuary: Integrating Maori TEK with scientific research and community-led conservation initiatives
2	Guatemala	Mayan	Integration of native tree species in agroforestry systems	Maya Nut Institute: Promoting growth of native trees for food security and ecosystem services
3	Brazil (Atlantic Forest)	Indigenous Communities	Integration of native tree species into agroecosystems	Socio-Environmental Institute (ISA): Collaboration to support agroforestry initiatives and advance land rights
4	Australia	Aboriginal	Traditional techniques like "bush tucker	Indigenous Fire and Carbon Initiative: Supporting Aboriginal-

			gardens" and "firestick farming"	led land management and cultural revitalization efforts
5	Cameroon	Indigenous Peoples	Agroforestry combined with shifting farming methods	Incorporation of scientific knowledge on crop diversification, water management, and soil conservation
6	Philippines	Indigenous Communities	"Swidden agriculture" integrating traditional forest management	Philippine Agroforestry Education and Research Network (PAFERN): Supporting sustainable land use and resource conservation
7	Ethiopia	Indigenous Communities	Use of "home gardens" and "church forests"	Ethiopian Environment and Development Research Institute (EEDRI): Advancing sustainable land management and climate resilience

14.5.2 Indian context

1. Adivasi Agroforestry Practices in Central India: For many years, agroforestry systems, which integrate traditional knowledge with contemporary agricultural techniques to sustainably manage wooded environments, have been utilised by Adivasi (indigenous) populations in central India. For instance, the Gond and Baiga tribes engage in shifting cultivation known as "dongar kheti," which entails the rotational cultivation of tree species like mahua (Madhuca longifolia), tendu (Diospyros melanoxylon), and amla (Phyllanthus emblica) alongside crops like rice, millet, and pulses (Dwivedi & Pandey, 2019). These agroforestry systems preserve biodiversity and ecological services while promoting food security, revenue generating, and cultural preservation.

2. Tribal Agroforestry Systems in Northeast India: Tribal communities in Northeast India, such as the Khasi, Jaintia, and Garo tribes in Meghalaya, practice agroforestry systems that integrate traditional knowledge with modern conservation practices. These systems, known as "jhum" or "shifting cultivation," involve the rotation of crops and fallow periods in forested areas, allowing for natural regeneration of vegetation and

maintenance of soil fertility (Lal, 2005). Initiatives like the Meghalaya Basin Development Authority (MBDA) work with indigenous communities to promote sustainable land use practices, improve livelihoods, and conserve biodiversity in the region.

3. Traditional Agroforestry Practices in Kerala: The Kani tribes and Malayalis are two groups that have been engaging in traditional agroforestry techniques like "homegardens" and "thottam" in Kerala, South India, for millennia. In smallholder farming systems, these home gardens incorporate a variety of tree species, fruit trees, medicinal plants, and vegetable crops (Nair, 1985). The Kani tribe of Kerala, for example, preserves biodiversity and improves soil fertility by cultivating home gardens that resemble natural forest ecosystems and offer food, medicine, and cultural values.

Table 2 Indigenous Agroforestry Practices and Collaborations in India

Sr. No.	**Region**	**Indigenous Group**	**Agroforestry Practices**	**Key Initiatives and Collaborations**
1	Central India	Adivasi	"Dongar kheti" involves rotational cultivation of trees and crops	Preservation of biodiversity, food security, and cultural heritage
2	Northeast India	Tribal Communities	"Jhum" or shifting cultivation with crop rotation and fallow periods	Meghalaya Basin Development Authority (MBDA): Promoting sustainable land use practices and biodiversity conservation
3	Kerala, South India	Kani tribes, Malayalis	"Homegardens" and "thottam" with diverse tree and crop species	Preservation of biodiversity, soil fertility, and cultural values

14.6 Socio economic importance

Integrating indigenous knowledge with agroforestry techniques enhances soil health, ecosystem resilience, and biodiversity conservation—all of which are major contributors to environmental sustainability. Indigenous groups have created complex knowledge systems that highlight the relationship between humans and the natural world. This has led to the development of sustainable land management techniques that preserve ecological equilibrium (Berkes et al., 2000). Agricultural landscapes can be made more resilient to natural pressures like droughts, floods, and pests by implementing native agroforestry practices including intercropping, mixed-species planting, and forest gardening (Nair, 2012). Furthermore, according to García-Frapolli et al. (2009), native agroforestry systems frequently resemble natural forest ecosystems, increasing habitat diversity and assisting in the preservation of native flora and animals.

There are several socio-economic advantages to incorporating indigenous knowledge into agroforestry methods, such as increased food security, better livelihoods, and increased community resilience. According to studies, native agroforestry systems encourage a variety of cropping strategies and revenue sources, which lessens the susceptibility of households to changes in the market and environmental hazards (Dev & Shankar, 2018; Gheyssens & Speelman, 2019). These systems preserve biodiversity and natural resources while increasing production through the use of conventional wisdom, such as intercropping and mixed-species planting (Brodt & Merrick, 2017). Furthermore, communal land tenure and group decision-making procedures are frequently given priority in indigenous agroforestry projects, which promote social cohesiveness and cultural preservation (Rasul, 2014). Additionally, by offering chances for skill development, leadership training, and participatory decision-making, community-led agroforestry programmes empower marginalised populations, including women and youth (Wezel et al., 2009).

14.7 Cultural preservation and empowerment of indigenous communities

Since they preserve traditional values, fortify communal identities, and encourage self-determination among indigenous peoples, cultural preservation and empowerment are essential components of incorporating indigenous knowledge into agroforestry methods. Communities sustain intergenerational knowledge transmission and cultural heritage preservation by integrating indigenous methods into agroforestry programmes. For instance, "kaingin," or shifting cultivation, is a method used by indigenous people in the Philippines. It entails alternating between crop rotation and periods of fallow land in forested areas. In addition to maintaining agricultural output, this custom strengthens cultural linkages to the land

(Fernandez & Alvez, 2018). Similar to this, tribal groups like the Khasi, Jaintia, and Garo tribes in Northeastern India engage in "jhum" farming, upholding sustainable land management while conserving cultural traditions (Lal, 2005).

In addition, the incorporation of indigenous knowledge into agroforestry projects strengthens communities by encouraging group decision-making, improving social harmony, and advancing fair resource allocation. To promote agroforestry projects, uphold land rights, and empower local populations, for example, the Socio-Environmental Institute (ISA) in Brazil's Atlantic Forest region works with indigenous groups (Isendahl et al., 2013). Comparably, Aboriginal people in Australia engage in "bush tucker gardens" and "firestick farming," which support traditional ties to the land while also managing landscapes sustainably (Jones et al., 2019). In addition to empowering indigenous people and promoting cultural preservation, policymakers and practitioners can achieve more inclusive and sustainable agroforestry outcomes by acknowledging the significance of indigenous knowledge systems and collaborating in partnerships.

14.8 Policy and government consideration

The establishment of enabling environments that promote meaningful engagement of indigenous groups in decision-making processes, safeguard indigenous knowledge, and enable collaboration are contingent upon the implementation of effective policies and governance structures (Tsing et al., 2017). For instance, the Indigenous and Tribal Peoples Convention (ILO Convention 169) and the Recognition of Traditional Knowledge (RTK) Law in Brazil and Ecuador, respectively, offer legal recognition and protection for indigenous rights, such as traditional knowledge, land tenure, and cultural heritage (Brosius et al., 2018). These laws provide indigenous tribes the ability to defend their rights.

Moreover, trust-building, social inclusion, and resolving power imbalances between indigenous peoples and outside actors depend on governance mechanisms that support multi-stakeholder collaboration, participatory decision-making, and free, prior, and informed consent (FPIC) (Corbera et al., 2020). Meaningful engagement and partnership between indigenous communities, government agencies, research institutions, and civil society organisations are facilitated by initiatives like co-management agreements, community-based resource management, and indigenous land rights recognition processes (Baleé, 2006). Policymakers can guarantee that agroforestry initiatives are culturally appropriate, environmentally sustainable, and socially inclusive by incorporating indigenous perspectives into the processes of policy formulation, implementation, and monitoring. This will help to create more resilient and equitable socio-ecological systems (Borrini-Feyerabend et al., 2013).

In summary, the integration of indigenous knowledge into agroforestry methods, the advancement of indigenous rights, and the promotion of sustainable development outcomes all depend on policy and governance concerns. Policymakers may foster an atmosphere that values indigenous knowledge, honours cultural diversity, and helps indigenous populations take charge of their own destiny by embracing inclusive and rights-based strategies.

14.8.1 Policy frameworks supporting the integration of indigenous knowledge in India

Governments and organisations have a critical role to play in India's sustainable development, biodiversity conservation, and traditional knowledge systems preservation by supporting indigenous agroforestry methods. The Indian government has launched a number of initiatives to promote and acknowledge traditional agroforestry methods through institutional support, policy interventions, and capacity-building efforts. The historic Forest Rights Act (FRA), also known as the Scheduled Tribes and Other Traditional Forest Dwellers (Recognition of Forest Rights) Act, 2006, seeks to grant recognition to forest-dwelling Scheduled Tribes and other traditional forest dwellers who have been living in such forests for generations, as well as the right to occupy and exercise their forest rights. This act gives communities who live near forests legal recognition and ownership rights over their ancestral lands, acknowledging the past injustices they have experienced and enabling them to take part in decision-making procedures regarding. Furthermore, the National Agroforestry Policy-2014, which acknowledges the significance of agroforestry in raising agricultural productivity, encouraging environmental sustainability, and boosting rural livelihoods. The policy supports farmers and people by offering financial incentives, technical assistance, and capacity-building programmes. This stimulates the adoption of agroforestry practices, including traditional indigenous agroforestry systems.

In addition, the National Biodiversity Authority (NBA) and the Biological Diversity Act, 2002 of India support the preservation and sustainable use of biological resources, including traditional knowledge related to agroforestry techniques and biodiversity protection. By facilitating the recording, preservation, and just distribution of advantages resulting from the use of traditional knowledge, the NBA guarantees that indigenous people are fairly compensated for their contributions to sustainable development and biodiversity preservation. On the other hand, several state governments have created their own policies and initiatives that are customised to the unique ecological and sociocultural circumstances in various parts of the nation. These policies support inclusive development in rural regions, environmental sustainability, and cultural preservation by recognising the rights and contributions of indigenous groups. Additionally, through research, extension services, and capacity-building programmes, institutions like the Indian Council of Agricultural Research (ICAR), National Institute of Agroforestry (NIA), and non-governmental organisations (NGOs) are instrumental in advancing indigenous agroforestry practices. To create and share best practices, hold training sessions, and support knowledge exchange platforms for exchanging indigenous knowledge and experiences, these organisations work in conjunction with governments, academic institutions, and local communities.

Agroforestry is also encouraged as a way to improve land productivity, preserve natural resources, and create jobs in rural areas through government-sponsored programmes like the National Rural Employment Guarantee Act (NREGA) and the Mahatma Gandhi National Rural Employment Guarantee Scheme (MGNREGS). These initiatives serve smallholder farmers and indigenous people nationwide by providing funding for land development, plantation operations, and sustainable land management techniques (Ministry of Rural Development, 2005).

To sum up, the promotion of native agroforestry techniques in India by governments and organisations is essential to accomplishing sustainable development objectives, protecting biodiversity, and strengthening the position of native communities. India can leverage the potential of agroforestry to address modern issues while preserving traditional wisdom and cultural legacy by acknowledging the contributions of indigenous knowledge systems and offering legislative support, institutional mechanisms, and financial resources.

1. Incorporating native agroforestry techniques has great potential for promoting climate resilience, biodiversity preservation, and sustainable

development. A number of critical factors will probably influence how indigenous agroforestry projects develop in the future.

2. Policy Support and Institutional Strengthening: Mainstreaming indigenous agroforestry practices will require sustained efforts to fortify institutional frameworks and policy frameworks. According to Lamb et al. (2019), governments must make sure that new policies are created to address opportunities and difficulties that arise as well as that current policies are executed successfully.

3. Research and Innovation: To improve our knowledge of indigenous agroforestry systems, find best practices, and create new technologies and methods, research and innovation expenditures are crucial (FAO, 2019). According to Reed et al. (2020), research partnerships that foster collaboration among indigenous people, academics, and institutions have the potential to stimulate innovation and facilitate the adaptation of traditional knowledge to evolving environmental conditions.

4. Capacity Building and Knowledge Exchange: According to Larsen et al. (2014), information exchange platforms and capacity development initiatives are essential for empowering indigenous communities, bolstering regional institutions, and encouraging peer-to-peer learning. The transmission of indigenous knowledge and practices across geographies and generations can be aided by training sessions, field demonstrations, and community-led projects (Rist et al., 2020).

5. Climate Change Adaptation and Mitigation: By storing carbon, boosting ecosystem resilience, and lessening susceptibility to extreme weather events, indigenous agroforestry systems have the potential to slow down global warming (Chazdon et al., 2016). Upgrading climate-smart agroforestry techniques, utilising indigenous knowledge to create adaptation plans, and securing funding for climate-resilient agroforestry initiatives should be the main priorities of future efforts (IPCC, 2019).

6. Market Access and Value Chains: Indigenous people can be financially motivated to manage their lands sustainably by expanding market access and fortifying the value chains for agroforestry products (Miller et al., 2018). Market connections, fair trade policies, and certification programmes are a few examples of initiatives that might help indigenous farmers access premium markets and enhance their standard of living (Zomer et al., 2017).

7. Social Inclusion and Equity: Promoting the rights and well-being of women, marginalised groups, and indigenous peoples requires ensuring social inclusion and equity in agroforestry efforts (Dawson et al., 2020). In agroforestry landscapes, gender-sensitive initiatives, participatory techniques, and community-led decision-making processes can assist resolve power disparities and advance social justice (Bennett et al., 2018).

8. International Cooperation and Partnerships: International cooperation between governments, organisations, and indigenous groups is essential for information exchange, resource mobilisation, and the resolution of transboundary problems (Zurayk et al., 2020). Global discourse and collaboration on indigenous agroforestry projects can be facilitated by platforms like the United Nations Forum on Forests (UNFF), the International Union for Conservation of Nature (IUCN), and the Global Environment Facility (GEF) (UNEP, 2020).

In summary, embracing indigenous knowledge and fostering collaboration among diverse stakeholders can unlock the full potential of agroforestry to achieve sustainable development goals and build resilient landscapes for future generations.

14.9 Opportunities for future research and collaboration

In the field of indigenous agroforestry methods, there are many opportunities for additional study and collaboration. These opportunities present means of promoting innovation, expanding knowledge, and tackling urgent issues. Agroforestry landscapes can undergo revolutionary change as a result of cooperative efforts involving scholars, indigenous groups, governments, and organisations. The following are some significant prospects for additional study and cooperation in the Indian setting:

1. Traditional Knowledge Systems: It is crucial to do research aimed at preserving and comprehending the complex traditional knowledge systems ingrained in native agroforestry techniques in India. The ecological, cultural, and social aspects of traditional agroforestry systems can be better understood through cooperative research that treats indigenous populations as equal participants (Gadgil & Berkes, 1991).

2. Ecosystem Services and Benefits: It is necessary to measure and evaluate the ecosystem services and advantages that India's native

agroforestry systems offer. To assess the multifunctional roles of agroforestry landscapes, such as carbon sequestration, biodiversity conservation, improved soil fertility, and water regulation, collaborative research projects can utilise interdisciplinary methodologies (Kumar et al., 2018).

3. Climate Change Adaptation: It is imperative to conduct research on how resilient indigenous agroforestry systems are in the Indian setting to variability and climate change. Collaborative research partnerships have the capacity to investigate customary adaptation strategies, evaluate their efficacy, and pinpoint prospects for merging native wisdom with climate-smart agroforestry methodologies customised for India's varied agro-climatic zones (Rai et al., 2021).

4. Socio-economic Impacts: It is crucial to comprehend the socioeconomic effects of native agroforestry techniques on regional livelihoods, food security, and the reduction of poverty in India. The socio-economic dynamics of agroforestry value chains, market integration, and prospects for income generation for indigenous communities in various parts of India can all be studied through collaborative studies (Rao et al., 2020).

5. Policy and Governance: Evidence-based decision-making and institutional reforms pertaining to policy and governance aspects of indigenous agroforestry in India can be influenced by research. In addition to identifying strategies to improve governance processes that protect indigenous land tenure and rights, collaborative activities can assess the favourable conditions for incorporating indigenous knowledge into national, state, and local policy frameworks (Kala, 2005).

6. Capacity Building and Knowledge Exchange: In order to empower indigenous groups and establish research partnerships in India, it is imperative that capacity-building efforts and knowledge-sharing platforms be strengthened. According to Baloguni and Raina (2012), collaborative programmes can support the co-production of knowledge, the transfer of skills, and participatory learning processes that improve communities' and institutions' ability to adapt.

In a nutshell, there are many chances for additional study and cooperation in native agroforestry methods in India, providing a route forward for sustainable development, biodiversity preservation, and cultural resilience. Researchers, practitioners, and policymakers can fully utilise indigenous agroforestry to address complex socio-environmental challenges and create resilient landscapes for future generations by promoting interdisciplinary collaboration, community engagement, and knowledge co-production.

Conclusion

Indigenous knowledge systems are the result of generational transmission and represent centuries of creativity, wisdom, and environmental adaptation. One of the key strengths of indigenous knowledge lies in its holistic approach to land stewardship, which emphasizes the interconnectedness of ecological, social, and spiritual dimensions. Indigenous agroforestry practices are often rooted in deep ecological understanding, incorporating diverse tree species, crop varieties, and livestock breeds to create resilient and biodiverse landscapes. Moreover, indigenous knowledge systems emphasize principles of reciprocity, community cooperation, and respect for nature, fostering harmonious relationships between humans and the environment. We can improve ecosystem services like carbon sequestration, soil fertility, and water conservation while advancing food security, diversifying livelihoods, and reducing poverty by utilising indigenous knowledge in agroforestry. It has been demonstrated that indigenous agroforestry systems, especially on marginal and degraded soils, increase agricultural productivity, improve soil health, and lessen the effects of climate change. Incorporating native agroforestry techniques into mainstream development plans can also support social inclusion, cultural renewal, and the acknowledgement of indigenous rights.

The historical injustices, marginalisation, and displacement that indigenous populations have experienced must be acknowledged, though, as their traditional knowledge systems have frequently been ignored or disregarded. We must form genuine partnerships with indigenous peoples while respecting their rights, autonomy, and self-determination if we are to properly honour traditional knowledge in agroforestry. Co-designing and executing agroforestry projects that represent the goals and objectives of indigenous people requires cooperative research, democratic decision-making, and fair resource sharing.

References

Agrawal, A. 1995. Dismantling the divide between indigenous and scientific knowledge. Development and Change, 26(3), 413-439.

Altieri, M. A. 2004. Linking ecologists and traditional farmers in the search for sustainable agriculture. *Frontiers in Ecology and the Environment*, *2*(1), 35-42.

Altieri, M.A., Toledo, V.M., Funes-Monzote, F.R., & Lana, M.A. 2003. The agroecological revolution in Latin America: rescuing nature, ensuring food sovereignty and empowering peasants. Journal of Peasant Studies, 40(4), 587-612.

Altman, J.C. 2012. The significance of 'caring for country' in Australian Aboriginal and Torres Strait Islander social policy. Journal of Indigenous Policy, 13(1), 1-15.

Amonum, J. I., Babalola, F. D., & Agera, S. I. N. 2009. Agroforestry systems in Nigeria: Review of concepts and practices. *Journal of Research in Forestry, Wildlife and Environment*, *1*(1), 18-30.

Asare, R., Franco, J., Gyau, A., & Guisse, A. 2010. Cocoa agroforestry is less resilient to sub-canopy climate variability than cocoa in full sun. Agriculture, Ecosystems & Environment, 138(3-4), 218-224.

Baleé, W. 2006. The research programme of historical ecology. Annual Review of Anthropology, 35, 75-98.

Balooni, K., & Raina, R. 2012. Developing rural landscapes for meeting sustainable development goals: A case of participatory agroforestry initiative in Uttaranchal Himalaya. Forest Policy and Economics, 21, 60-67.

Bennett, N.J., et al. 2018. Mainstreaming the social sciences in conservation. Conservation Biology, 32(1), 56-66.

Berkes, F. 2008. Sacred ecology: Traditional ecological knowledge and resource management. Routledge.

Berkes, F. 2009. Indigenous ways of knowing and the study of environmental change. Journal of the Royal Society of New Zealand, 39(4), 151-156.

Berkes, F., Colding, J., & Folke, C. 2000. Rediscovery of traditional ecological knowledge as adaptive management. Ecological Applications, 10(5), 1251-1262.

Bisht, S., Kumar, A., Kumar, N., Singh, H. and Kumar, P. 2022. Biofuel Production by Using Biomass and Its Application. In book: Renewable Energy and Green Technology in CRC Press.

Borrini-Feyerabend, G., Kothari, A., & Oviedo, G. 2013. Indigenous peoples and conservation: From rights to resource management. Conservation International.

Brodt, S., & Merrick, L. 2017. Agroforestry solutions to address food security and climate change challenges in Africa. Current Opinion in Environmental Sustainability, 26-27, 1-7.

Brosius, J.P., Tsing, A.L., & Zerner, C. 2018. Communities and Conservation: Histories and Politics of Community-Based Natural Resource Management. Rowman Altamira.

Chazdon, R.L., et al. 2016. Carbon sequestration potential of second-growth forest regeneration in the Latin American tropics. Science Advances, 2(5), e1501639.

Coomes, O. T., Takasaki, Y., & Rhemtulla, J. M. 2011. Land-use poverty traps identified in shifting cultivation systems shape long-term tropical forest cover. *Proceedings of the National Academy of Sciences*, *108*(34), 13925-13930.

Coomes, O.T., Takasaki, Y., & Rhemtulla, J.M. 2015. Land-use poverty traps identified in shifting cultivation systems shape long-term tropical forest cover. Proceedings of the National Academy of Sciences, 112(43), 13245-13250.

Corbera, E., Martin, A., & Brown, K. 2020. Governing REDD+ through 'policies of exception': Lessons from Cambodia. World Development, 125, 104681.

Cruz-Garcia, G. S., & Price, L. L. 2011. Ethnobotanical investigation of'wild'food plants used by rice farmers in Kalasin, Northeast Thailand. *Journal of ethnobiology and ethnomedicine*, *7*, 1-21.

Davidson-Hunt, I. J., & Berkes, F. 1999. Changing Resource Management Paradigms, Traditional Ecological Knowledge, and Non-timber Forest Products. *Forest Communities in the Third Millennium: Linking Research, Business, and Policy Toward a Sustainable Non-Timber Forest Product Sector*, *1*, 78.

Dawson, N., et al. 2020. Transformational learning for sustainable development: An empirical assessment of its drivers. Journal of Cleaner Production, 261, 121023.

Dev, A., & Shankar, U. 2018. Agroforestry and rural livelihood security: A case study of farmers' perceptions from Chhattisgarh, India. Agroforestry Systems, 92(1), 227-237.

Dinesha, S., Hosur, S. R., Toushif, P. K., Bodiga, D., Deepthi Dechamma, N. L., Ashwath, M. N., & Pradhan, D. 2023. Sustaino-resilient agroforestry for climate resilience, food security and land degradation neutrality. *Land and Environmental Management through Forestry*, 217-245.

Doddabasawa, Chittapur, B. M., & Murthy, M. M. 2019. Structural analysis and mapping of agroforestry systems under irrigated ecosystem in north-eastern part of Karnataka, India. *Agroforestry systems*, *93*, 1701-1716.

Dwivedi, S., & Pandey, D. N. 2019. Indigenous practices for biodiversity conservation in India. In Indigenous Knowledge for Sustainable Development (pp. 29-55). Springer.

Fajardo Cavalcanti de Albuquerque, M. 2020. Linking agrobiodiversity and culture through the adoption of agroforestry practices: The Agroforestry Indigenous Agents. In *Parks Stewardship Forum* (Vol. 36, No. 1).

Fernandez, E.C., & Alvez, J.P. 2018. The Indigenous People and their Knowledge in Philippine Agriculture. In Agricultural Research Updates (pp. 85-106). Nova Science Publishers.

Fikret, B., Ozdemir, S., & Cakmak, I. 2018. Traditional agricultural systems of rural communities in Turkey: The case of Bayramiç District. *Journal of Ethnobiology and Ethnomedicine*, 14(1), 16.

Food and Agriculture Organization (FAO). 2019. Agroforestry for landscape restoration: Towards sustainable land use. FAO.

Gadgil, M., & Berkes, F. 1991. Traditional resource management systems. *Resource management and Optimization*, *8*(3-4), 127-141.

Gadgil, M., Berkes, F., & Folke, C. 1993. Indigenous knowledge for biodiversity conservation. Ambio, 22, 151-156.

García-Frapolli, E., Toledo, V. M., Martinez-Alier, J., & J. M. 2009. The pathways of rare and little-known products: non-timber forest resources and agroecosystem management among the Maya Chontal of Tabasco, Mexico. Journal of Latin American Geography, 8(1), 29-51.

Garrity, D.P., Akinnifesi, F.K., Ajayi, O.C., Weldesemayat, S.G., & Mowo, J.G. 2010. Evergreen Agriculture: A Solution to Climate Change, Poverty and Hunger. World Agroforestry Centre (ICRAF).

Gavin, M.C., McCarter, J., Mead, A., Berkes, F., Stepp, J.R., Peterson, D., & Tang, R. 2015. Defining biocultural approaches to conservation. Trends in Ecology & Evolution, 30(3), 140-145.

Gheyssens, J., & Speelman, S. 2019. Agroforestry adoption among smallholder farmers in Ghana: A facilitation perspective. Environmental Development, 31, 77-89.

Goldman, M.J., Roque de Pinho, J., Perry, J., & Shove, E. 2017. Caring for conservationists: challenges and opportunities for assessing Indigenous-led land management. Conservation and Society, 15(1), 84-96.

Herrera, E., Ortega, A., & Astier, M. 2008. Traditional knowledge and agroecology: contributions in defence of maize (*Zea mays* L.). Agronomy for Sustainable Development, 28(1), 65-73.

Intergovernmental Panel on Climate Change (IPCC). (2019). Climate change and land: An IPCC special report on climate change, desertification, land degradation, sustainable land management, food security, and greenhouse gas fluxes in terrestrial ecosystems. IPCC.

Isendahl, N., Brewer, S., & Hörnberg, G. 2013. The Eco-cultural Octagon: A perspective on the Middle Holocene in Central Amazonia. Antiquity, 87(337), 940-952.

Jones, R., Marshall, N., & Reser, J. 2019. Indigenous values and water markets: Survey insights from northern Australia. Australian Geographer, 50(2), 219-236.

Kala, C.P. 2005. Indigenous uses, population density, and conservation of threatened medicinal plants in protected areas of the Indian Himalayas. Conservation Biology, 19(2), 368-378.

Kaushal R, Tewari S., Thapliyal SD., Kumar, A., Roy, T., Islam, S., Lepcha, STS., Durai, J. 2021. Build-up of Labile, non-labile carbon fractions under fourteen-year-old bamboo plantations in the Himalayan foothills. Heliyon 7(3):e07850

Kiptot, E., & Franzel, S. 2012. Gender and agroforestry in Africa: A review of women's participation. Agroforestry Systems, 84(1), 35-58.

Kohler, J., & Scott, P. 2007. The application of traditional knowledge: safeguards and pitfalls. Journal of Environmental Law, 19(2), 151-176.

Kumar, A., Dwivedi, G. K., Tewari, S., Sah, V. K., Singh, H., Kumar, P., Kumar, N., Kaushal, R. 2020. Soil organic carbon pools under *Terminalia chebula* Retz. based agroforestry system in Himalayan foothills, India. Current Science (00113891), 118(7).

Kumar, A., et al. 2018. Assessing carbon sequestration potential of agroforestry systems in India: A review. Agroforestry Systems, 92(3), 659-681.

Kumar, A., Kumar, P., Singh, H. and Kumar, N., 2021a. Impact of plant functional traits on near saturated hydraulic conductivity of soil under different forests of Kempty watershed in Garhwal Himalaya. Indian Journal of Soil Conservation, 49(1), pp.38-44.

Kumar, A., Kumar, P., Singh, H., Kumar, N., 2021b. Adaptation and mitigation potential of roadside trees with bio-extraction of heavy metals under vehicle emissions and their impact on physiological traits during seasonal regimes. Urban Forestry & Urban Greening, 58, p.126900.

Kumar, A., Kumar, P., Singh, H., Kumar, N., 2021c. Modulation of Plant Functional Traits under Essential Plant Nutrients during Seasonal Regime in Natural Forests of Garhwal Himalayas. Plant and Soil 465(1):1-16

Kumar, A., Kumar, P., Singh, H., Kumar, N., 2021d. Relationship of physiological plant functional traits with soil carbon stock in the temperate forest of Garhwal Himalaya. Current Science 120(7): 1064-1068.

Kumar, A., Singh, H., Kumari, G., Bisht, S., Malik, A., Kumar, N., Singh, M., Raturi, A., Barthwal, S., Thakur, A. and Kaushal, R., 2022. Adaptive resilience of roadside trees to vehicular emissions via leaf enzymatic, physiological, and anatomical trait modulations. Environmental Pollution, p.120191.

Kumar, A., Tewari, S., Singh, H., Kumar, P., Bisht, S., 2021e. Biomass accumulation and carbon stock in different agroforestry systems prevalent in the Himalayan foothills. Current Science 120(6), 1083-1088

Kumar, A., Tewari, S., Singh, Indra, Panday, R., Kumar, D. and Anand, R. 2018. Effect of nutrient sources on growth, yield and quality of turmeric under harad (*Terminalia chebula*) based agroforestry system Indian J. of Agroforestry, 20(2). 1-6

Kumar, B. M., & Nair, P. K. R. (2004). The enigma of tropical homegardens. Agroforestry Systems, 61(1-3), 135-152.

Kumar, B. M., Singh, A. K., & Dhyani, S. K. 2012. South Asian agroforestry: traditions, transformations, and prospects. *Agroforestry-The future of global land use*, 359-389.

Kumar, N., Jeena, N., Kumar, A., Khwairakpam, R., Singh, H. 2021. Comparative response of rice cultivars to elevated air temperature in Bhabar region of Indian Himalaya: Status on Yield attributes. Heliyon 7(6):e07474

Kumar, N., Kumar, A., Jeena, N., Singh, R., and Singh, H. 2020. Factors Influencing Soil Ecosystem and Agricultural Productivity at Higher Altitudes. Microbiological Advancements for Higher Altitude Agro-Ecosystems & Sustainability. Springer. p. 55- 71.

Kumar, N., Singh, H., Giri, K., Kumar, A., Joshi, A., Yadav, S., Singh, R., Bisht, S., Kumari, R., Jeena, N., Khairakpam, R., and Mishra, G. 2024. Physiological and molecular insights into the allelopathic effects on agroecosystems under changing environmental conditions. Physiology and Molecular Biology of Plants. 248. 119255.

Kumar, N., Verma, S., Kumar, A., Dashila, H., Suyal, D.C., Kumari, G., Rawat, S., Jeena, N., Singh, M., Kumar, A. 2022. Molecular insights into stress-responsive genes in the mitigation of environmental stresses. Unravelling Plant-Microbe Synergy, Edition: 1, Publisher: Elsevier

Kumar, N., Verma, S., Kumar, A., Dashila, H., Suyal, D.C., Kumari, G., Rawat, S., Jeena, N., Singh, M., Kumar, A. 2022. Molecular insights into stress-responsive genes in the mitigation of environmental stresses. Unravelling Plant-Microbe Synergy, Edition: 1, Publisher: Elsevier

Lal, R. 2005. Forest soils and carbon sequestration. Soil Science Society of America Journal, 69(4), 1406-1414.

Lamb, D., et al. 2019. Regreening the planet: A global agenda for ecological restoration. Routledge.

Larsen, H.O., et al. 2014. Participatory environmental governance: A comparison of three cases in India. Environmental Policy and Governance, 24(3), 192-207.

Lehmann, A., Hoffmann, C.C., & Rietdorf, U. 2005. Riparian buffer strips as a multifunctional management tool in agricultural landscapes: Introduction. Journal of Environmental Quality, 34(2), 757-758.

Lele, S., Springate-Baginski, O., Lakerveld, R., Deb, D., Dash, P., & Atela, J. 2013. Ecosystem services: origins, contributions, pitfalls, and alternatives. Conservation and Society, 11(4), 343-358.

Lemenih, M., Karltun, E., & Olsson, M. 2005. Assessing farmers' perceptions on the role of agroforestry in land management practices in Gedeb, south-eastern highlands of Ethiopia. Environmental Management, 36(6), 868-877.

Lewis, D. 1972. We, the navigators: The ancient art of landfinding in the Pacific. University of Hawaii Press.

Méndez, V. E., Bacon, C. M., Olson, M., Morris, K. S., & Shattuck, A. 2010. Agrobiodiversity and shade coffee smallholder livelihoods: a review and synthesis of ten years of research in Central America. *The Professional Geographer*, *62*(3), 357-376. (Méndez et al 2010)

Menzies, C.R. 2006. Our knowledge is not primitive: Decolonizing botanical Anishinaabe teachings. Syracuse University Press.

Miller, D.C., et al. 2018. The role of agroforestry in building livelihood resilience to floods and droughts in semi-arid Kenya. Journal of Rural Studies, 64, 134-143.

Ministry of Agriculture and Farmers' Welfare. 2014. National Agroforestry Policy. Retrieved from https://agricoop.nic.in/sites/default/files/national_agro_forestry_policy_2014.pdf

Ministry of Rural Development. 2005. Mahatma Gandhi National Rural Employment Guarantee Act, 2005. Retrieved from https://nrega.nic.in/netnrega/mgnrega_new/Nrega_home.aspx

Mistry, J., & Berardi, A. 2016. Bridging indigenous and scientific knowledge. Science, 352(6291), 1274-1275.

Montagnini, F., & Nair, P.K.R. 2004. Carbon sequestration: An underexploited environmental benefit of agroforestry systems. Agroforestry Systems, 61-62(1-3), 281-295.

Nair, P. K. R. 1985. Classification of agroforestry systems. Agroforestry Systems, 3(2), 97-128.

Nair, P.K.R. 1993. An Introduction to Agroforestry. Springer Science & Business Media.

Nair, P.K.R. 2012. Agroforestry for ecosystem services and environmental benefits: An overview. Agroforestry Systems, 86(1), 1-8.

Nair, P.K.R., Garrity, D., & Hossain, M. 2009. Agroforestry: The Future of Global Land Use. Springer Science & Business Media. National Institute of Agroforestry (NIA). (n.d.). Retrieved from http://www.nia.gov.in/

Negi, J.S., Singh, R.D., & Chandra, S. 2019. Role of agroforestry in enhancing the livelihood security of mountain communities in Indian Himalayan Region. Indian Journal of Agroforestry, 21(1), 1-8.

Nerlich, K., Graeff-Hönninger, S., & Claupein, W. 2013. Agroforestry in Europe: a review of the disappearance of traditional systems and development of modern agroforestry practices, with emphasis on experiences in Germany.

Pandey R., Dubey S., Tewari L, Kumar A., Kumar D., Sharma N. and Tewari S.K. 2018. Performance of soybean under poplar based agroforestry system in Terai region. Indian J. of Agroforestry, 20(1), 31-34.

Peters, C. M., Balick, M. J., Kahn, F., Anderson, A. B., & Lees, H. 2014. Oryza: From Mesoamerica to West Africa: The story of rice in two Atlantic slave trading worlds. Ethnobiology Letters, 5(1), 45-53.

Peters, C. M., Gentry, A. H., & Mendelsohn, R. O. 1989. Valuation of an Amazonian rainforest. Nature, 339(6227), 655-656.

Posey, D.A. 1985. Indigenous management of tropical forest ecosystems: The case of the Kayapo Indians of the Brazilian Amazon. Agroforestry Systems, 3(2), 139-158

Posey, D.A. 2002. Intellectual property rights and just compensation for indigenous knowledge. Development, 45(4), 45-49.

Pulhin, J.M., Dressler, W., & Pasicolan, P. 2019. Governing forest landscapes: Interplay of tenure, market, and policies. Journal of Environmental Management, 244, 119-128.

Rai, R.K., et al. 2021. Climate-smart agroforestry practices for enhancing resilience in India: A review. Forests, Trees and Livelihoods, 1-20.

Raintree, J.B. 1987. Agroforestry systems in the humid tropics. In Agroforestry: A Decade of Development (pp. 49-60). World Bank.

Rao, P.S., et al. 2020. Socioeconomic impact assessment of agroforestry systems in India: A review. Journal of Agroforestry Systems, 94(3), 801-817.

Rasul, G. 2014. Farmers' indigenous knowledge and climate change adaptation strategies in the Himalayas. Mountain Research and Development, 34(2), 127-140.

Reed, J., et al. 2020. Learning and change in agroforestry: Viewpoints and opportunities from different actors. World Development, 133, 104964.

Reyes-García, V., Gueze, M., Luz, A.C., Paneque-Gálvez, J., Macía, M.J., Orta-Martínez, M., & Huanca, T. 2013. Evidence of traditional knowledge loss among a contemporary indigenous society. Evolution and Human Behavior, 34(4), 249-257.

Richards, P. 1985. *Indigenous agricultural revolution: ecology and food production in West Africa*. Routledge.

Rist, S., et al. 2020. Co-producing sustainability: Reordering the governance of science, policy, and practice. Annual Review of Environment and Resources, 45, 289-314.

Robinson, B., Wehi, P., Nelson, K., McAlpine, K., Tipa, G., & Macpherson, C. 2016. Co-developing cultural ecosystem services and wellbeing indicators in the context of an indigenous ecosanctuary. Ecosystem Services, 22, 141-157.

Robinson, B., Wehi, P., Nelson, K., McAlpine, K., Tipa, G., & Macpherson, C. 2016. Co-developing cultural ecosystem services and wellbeing indicators in the context of an indigenous ecosanctuary. Ecosystem Services, 22, 141-157.

Russell-Smith, J., Whitehead, P., Cooke, P., Sutton, S., Edwards, A., Meier, P., & Gill, A. M. 2013. Managing fire regimes in north Australian savannas: applying Aboriginal approaches to contemporary global problems. Frontiers in Ecology and the Environment, 11(s1), e55-e63.

Saha, D., Kuar, A., & Parthasarathy, N. 2018. Homegardens as a sustainable land-use system in North East India: an evaluation based on botanical parameters. Agroforestry Systems, 92(6), 1659-1672.

Sileshi, G.W., Akinnifesi, F.K., Ajayi, O.C., & Matakala, P.W. 2007. Tackling poverty through tree-based systems in sub-Saharan Africa: reflections on the future. Natural Resources Forum, 31(3), 212-222.

Sillitoe, P. 2007. Local science vs global science: approaches to indigenous knowledge in international development. Routledge.

Singh, P.K., Rana, V.K., & Singh, R.D. 2014. Bamboo Agroforestry: A Panacea for Livelihood Security. Indian Forester, 140(7), 754-760.

Sollen-Norrlin, M., Ghaley, B. B., & Rintoul, N. L. J. 2020. Agroforestry benefits and challenges for adoption in Europe and beyond. *Sustainability*, *12*(17), 7001.

Tewari, D.D., Campbell, B.M., & Tewari, L. 2008. Agroforestry and biodiversity conservation in tropical landscapes. Biodiversity, 9(3-4), 44-47.

Tewari, S., Tewari, L., Dubey, A., Kumar, A., Kumar, N., Kaushal, R. 2022. Use of the RAPD marker to determine the genetic diversity of various *Dalbergia sissoo* Roxb. (Shisham) genotypes and their evolutionary relationship in Indian subcontinents. Vegetos,

Tewari, S., Tewari, L., Kumar, A., Bisht, S, Kaushal, R. and Kumar, A. 2024. Germination Behaviour of Teak Genotypes and Effect of Different Treatments on Seed Germination Characteristics in Himalayan Foothills, India. Indian Forestry, 150 (3): 244-254

Tewari, S., Tewari, L., Kumar, A., Kaushal, R. 2021. Performance of *Glycine max* L. varieties under *Tectona grandis* L. and *Dalbergia sissoo* Roxb. based agroforestry system Indian J. of Agroforestry, 20(2) 1-6

Tipa, G., & Nelson, K. C. 2008. Indigenous knowledge and values in planning for sustainability: A case study on developing a cultural health index in Ngāti Kuri. Ecological Economics, 68(1-2), 301-312.

Tsing, A.L., Swanson, H.A., Gan, E., & Bubandt, N. 2017. Arts of Living on a Damaged Planet: Ghosts and Monsters of the Anthropocene. Univ of Minnesota Press.

United Nations Environment Programme (UNEP). 2020. UN Decade on Ecosystem Restoration. Retrieved from https://www.decadeonrestoration.org/

Verma, T., Bhardwaj, D.R., Sharma, U., Sharma, P., Kumar, D., Kumar, A. and Kumar, A. 2023. Aroforestry systems in the mid-hills of the north-western Himalaya: A sustainable pathway to improved soil health and climate resilience. Journal of Environmental Management. 348, 119264.

Warren, D.M., Slikkerveer, L.J., & Brokensha, D. 1995. The cultural dimension of development: indigenous knowledge systems. Intermediate Technology Publications.

Wezel, A., Bellon, S., & Doré, T. 2009. Agroecology as a science, a movement and a practice. A review. Agronomy for Sustainable Development, 29(4), 503-515.

Wiersum, K. F., Ingram, V. J., & Ros-Tonen, M. A. F. 2005. Hunting, foraging, and fire: landscape utilization and agroforestry by the Baka Pygmies of Cameroon. African Study Monographs, 26(3), 131-150.

Wiersum, K.F. 1997. Indigenous exploitation and management of tropical forest resources: an evolutionary continuum in forest–people interactions. Agriculture, Ecosystems & Environment, 63(1), 1-16.

Wilson, M. H., & Lovell, S. T. 2016. Agroforestry—The next step in sustainable and resilient agriculture. *Sustainability*, *8*(6), 574.

Young, A. 1988. Agroforestry and its potential to contribute to land development in the tropics. *Journal of Biogeography*, 19-30.

Zent, S. 2001. Acculturation and ethnobotanical knowledge loss among the Piaroa of Venezuela: demonstration of a quantitative method for the empirical study of TEK change. *On biocultural diversity: Linking language, knowledge, and the environment*, 190-211.

Zomer, R.J., et al. 2017. Global tree cover and biomass carbon on agricultural land: The contribution of agroforestry to global and national carbon budgets. Scientific Reports, 7(1), 1-12.

Zurayk, R., et al. 2020. Transboundary water management in the Arab region: The state of the art and the way forward. Water International, 45(5), 501-524

CHAPTER FIFTEEN

IMPROVEMENT OF LIVELIHOOD THROUGH THE INTEGRATION OF MEDICINAL PLANTS IN AGROFORESTRY

NARESH SINGH[1], R.S. BALI[1], PRIYANKA THAKUR[1], PANKAJ LAL[1] AND RAM GOPAL[2]

[1]COLLEGE OF FORESTRY, RANICHAURI, TEHRI GARHWAL, UTTARAKHAND

[2]NORTH EASTERN HILL UNIVERSITY, TURA CAMPUS, MEGHALAYA

Abstract

The integration of medicinal plants in agroforestry systems offers a promising pathway for improving livelihoods and enhancing sustainable land-use practices. This book chapter examines the key benefits and strategies for incorporating medicinal plants into agroforestry, drawing insights from case studies and empirical research. The findings highlight the ability of medicinal plant-agroforestry systems to diversify income sources, promote food security, and contribute to biodiversity conservation and climate change mitigation. The chapter also explores the socioeconomic and institutional challenges, as well as the policy and regulatory considerations, that shape the adoption and scaling up of these integrated systems. Ultimately, the chapter underscores the importance of collaborative efforts among policymakers, practitioners, and researchers to support the sustainable cultivation of medicinal plants within agroforestry frameworks, thereby fostering livelihood improvements, environmental sustainability, and the supply of valuable herbal products.

15.1 Introduction

15.1.1. Overview of livelihood challenges in rural areas

The livelihood challenges faced by individuals in rural areas encompass several key aspects. These challenges include a need for entrepreneurial skills, low earnings, limited access to land and capital, inadequate skill training, lack of technical support, lack of knowledge, lack of awareness, natural disasters, disease outbreaks, and poverty (Kumar et al., 2020a&b: Kaushal et al., 2021). Rural livelihoods are predominantly dependent on activities like small-scale farming, fishing, raising livestock, and non-farm activities. Diversification of livelihood activities is crucial for improving social status and standard of living in rural communities. Addressing these challenges requires a focus on sustainable livelihoods, skill development, access to education, and the provision of basic amenities like clean drinking water, sanitation facilities, and healthcare (Kanyagui et al., 2023; Kundra and Prakashan, 2020).

15.1.2. Importance of agroforestry for sustainable livelihoods

Agroforestry plays a crucial role in promoting sustainable livelihoods by offering a multifaceted approach that addresses various aspects of agricultural and environmental sustainability. Through the integration of trees with crops and sometimes livestock, agroforestry systems provide a range of benefits that contribute to the well-being of farmers and rural communities. Some key points highlighting the importance of agroforestry for sustainable livelihoods include:

- **Increased Soil Fertility:** Agroforestry practices help improve soil fertility by enhancing nutrient cycling, preventing erosion, and promoting soil health, leading to higher crop yields and improved agricultural productivity (Fahad et al., 2022).

- **Income Diversification:** By providing farmers with multiple sources of income through tree products, crops, and livestock, agroforestry reduces economic vulnerability and enhances livelihood resilience (Kassie, 2018).

- **Climate Change Resilience:** Agroforestry systems contribute to climate change mitigation by sequestering carbon in the soil and biomass, thereby helping to offset greenhouse gas (GHG) emissions. This resilience to climate-related shocks and stresses is vital for sustainable livelihoods (Yirga, 2019).

- **Biodiversity Promotion:** Agroforestry supports biodiversity by creating diverse habitats for wildlife and beneficial insects, which can aid in natural pest and disease management. The diversity of crops and trees in agroforestry systems enhances ecosystem health and resilience (Yirga, 2019).

- **Food Security:** Agroforestry systems can contribute to food security by providing a variety of food products, including fruits, nuts, and vegetables, alongside traditional crops. This diversified food production helps to ensure a more stable and nutritious diet for farmers and their communities (Duffy et al., 2021).

15.1.3. Introduction to medicinal plants and their potential in agroforestry systems

Medicinal plants play a significant role in agroforestry systems, offering a range of benefits that contribute to the sustainability and productivity of these integrated farming approaches. These plants, often with therapeutic properties, are cultivated alongside trees, crops, and sometimes livestock, enhancing the overall biodiversity and functionality of agroforestry systems. The potential of medicinal plants in agroforestry lies in their ability to provide additional sources of income for farmers, improve soil health through their unique properties, and contribute to the conservation of traditional knowledge related to herbal medicine. Furthermore, the cultivation of medicinal plants in agroforestry systems can support local healthcare practices, promote biodiversity conservation, and offer opportunities for sustainable land use management (Kumar et al., 2021a&b; Kumar et al., 2022; Tewari et al., 2024). Overall, the integration of medicinal plants in agroforestry systems represents a valuable strategy for enhancing the ecological, economic, and social sustainability of rural livelihoods (Rao et al., 2004; Thakur, and Dutt, 2007).

Table 1: Medicinal plants in different agroforestry system

Medicinal Plant	Uses	Agroforestry System	Reference
Aloe vera	Skin healing, digestive aid	Silvipasture	(Jemal et al., 2021)
Neem (*Azadirachta indica*)	Antiseptic, insect repellent, anti-inflammatory	Homegarden Agrisilviculture	(Pushpakumara et al., 2012)
Ginger (*Zingiber officinale*)	Digestive aid, anti-inflammatory	Alley cropping	(Bawa and Seidler, 1998)
Turmeric (*Curcuma longa*)	Anti-inflammatory, antioxidant	Homegarden agroforestry	(Kumar et al., 2019)
Brahmi (*Bacopa monnieri*)	Memory enhancer, stress reliever	Taungya system	(Jewel et al., 2019)
Tulsi (*Ocimum tenuiflorum*)	Immunity booster, respiratory aid	Homegarden agroforestry	(Khakurel et al., 2024)

15.1.4. Purpose and structure of the chapter

The chapter on "Improvement of livelihood through the integration of medicinal plants in agroforestry" aims to explore the synergistic relationship between agroforestry practices and the cultivation of medicinal plants for sustainable livelihoods. It delineates the various dimensions of this integration, including ecological, economic, and socio-cultural aspects. The chapter examines how the strategic incorporation of medicinal plants within agroforestry systems can enhance biodiversity, provide additional income streams, and contribute to community health and well-being. Through case studies, empirical evidence, and theoretical frameworks, it elucidates the potential pathways for leveraging agroforestry as a platform for livelihood improvement and holistic development, thereby fostering resilience and prosperity in rural landscapes (Tewari et al., 2022; Kumar et al., 2018; Kumar et al., 2021c&d).

15.2 Agroforestry Systems and Practices

15.2.1. Definition and types of agroforestry systems

Several definitions have been proposed to agroforestry, of which the most commonly used are those of Lundgreen and Raintree (Agricultural research for development: potentials and challenges in Asia, 1982). Agroforestry is any land-use system, practice or technology, where woody perennials are integrated with agricultural crops and/or animals in the same land management unit, in some form of spatial arrangement or temporal sequence. Agroforestry is also a dynamic and ecologically-based natural resource management system. Agroforestry refers to the deliberate introduction or retention of trees on farms to increase, diversify, and sustain production for increased social, economic, and environmental benefits. Agroforestry system classification can be based on vegetation structure, function of woody perennials in the system, levels of input management, and environmental conditions and ecological suitability of the system. Agroforestry practices rather than systems are also used as the unit of an ecologically-based classification that is rooted in the role of trees in agricultural landscape. Lundgreen and Raintree (1982) classification of agroforestry systems is necessary in order to provide a framework for evaluating systems and developing action plans for their improvement. The AF Systems Inventory (AFSI) being undertaken by ICRAF provides the background information for an approach for classification. The words 'system', 'sub-system' and 'practice' are commonly used in AF literature. An AF system refers to a type of AF land-use system that extends over a locality to the extent of forming a land utilization type of the locality. Sub-system and practice are lower-order terms in the hierarchy with lesser magnitudes of role, content and complexity. In common parlance, however, these terms are used loosely, and almost synonymously. Several criteria can be used to classify and group AF systems. The most commonly used ones are the system's structure (composition and arrangement of components), its function, its socio-economic scale and level of management, and its ecological spread. Structurally, the system can be grouped as agrisilviculture. Silvopastoral, and agrosilvopastoral. Other specialized AF systems such as apiculture with trees, aquaculture in mangrove areas, multipurpose tree lots, and so on, can also be specified. Arrangement of components can be in time (temporal) or space (spatial) and several terms are used to denote the various arrangements. Functional basis refers to the main output and role of components, especially the woody ones. These can be productive functions (production of 'basic needs' such as food, fodder, fuelwood, other products, etc.) and protective roles (soil conservation, soil fertility improvement, protection offered by windbreaks and shelterbelts, and so on). On an ecological basis, systems can be grouped for any defined agro-ecological zone such as lowland humid tropics, arid and semi-arid tropics, tropical highlands, and so on. The socio-economic scale of production and level of management of the system can be used as the criteria to designate systems as commercial, 'intermediate', or subsistence. Each of these criteria has merits and applicability in specific situations, but they have limitations too so that no single classification scheme can be accepted as universally applicable (Verma et al., 2023; Kumar et al., 2020; Kumar et al., 2021e). Classification will depend upon the purpose for which it is intended.

15.2.2. Benefits of agroforestry for livelihood improvement

The key benefits of agroforestry for livelihood improvement include:

- **Improved Food and Nutrition Security:** Agroforestry systems provide a diverse range of food products, including fruits, nuts, vegetables, and livestock products, improving the availability and access to nutritious foods for rural communities (Zerihun, 2021; Akter et al., 2022).

- **Increased Income and Asset Generation:** The integration of trees, crops, and livestock in agroforestry systems offers multiple income streams, reducing economic vulnerability and enhancing livelihood resilience (Zerihun, 2021; Akter et al., 2022).

- **Enhanced Land Management and Soil Health:** Agroforestry practices improve soil fertility, prevent erosion, and promote nutrient cycling, leading to higher agricultural productivity and sustainability (Zerihun, 2021; Akter et al., 2022).

- **Climate Change Resilience:** Agroforestry systems help mitigate climate change by sequestering carbon in the soil and biomass, while also making livelihoods more resilient to climate-related shocks and stresses (Zerihun, 2021; Akter et al., 2022).

- **Biodiversity Conservation:** The diverse nature of agroforestry systems supports a greater variety of plant and animal species, enhancing overall ecosystem health and providing additional livelihood benefits (Zerihun, 2021; Akter et al., 2022.).

- **Improved Access to Resources:** Agroforestry can provide farmers with access to resources like firewood, fodder, and non-timber forest products, improving their overall well-being and livelihood security (Zerihun, 2021; Akter et al., 2022).

15.2.3. Examples of successful agroforestry models around the world

Agroforestry Model	Location	Key Features	References
Parkland Planting	Claveria, Misamis Oriental, Philippines	- Combines fruit-bearing trees with crops - Improves food security and access to food - Increases income levels and builds assets for farmers - Provides additional harvest for marketing and food for farm families	(Kraft et. al., 2021)
Agri-Silviculture Models	Various regions globally	- Poplar trees combined with annual crops like wheat, maize, banana, turmeric, and yam - Suitable for well-drained loam or sandy-loam soil conditions - Provides higher economic returns for farmers - Enhances economic viability and sustainability	(Kraft et. al., 2021)
Agri-Silviculture Models	Various regions globally	- Poplar trees combined with annual crops like wheat, maize, banana, turmeric, and yam - Suitable for well-drained loam or sandy-loam soil conditions - Provides higher economic returns for farmers - Enhances economic viability and sustainability	(Kraft et. al., 2021)
Alley Cropping	Worldwide	- Intercropping of trees with crops in alleys - Provides multiple	(Kraft et. al., 2021)

		sources of income - Improves soil fertility and prevents erosion	
Silvopastoral Systems	Various regions globally	- Integration of trees with pastures or animals - Enhances livestock production and provides shade for animals - Improves soil health and carbon sequestration	(Kraft et. al., 2021)
Homegardens	Southeast Asia, Africa	- Diverse mix of trees, crops, and sometimes livestock near homes - Enhances food security and nutrition - Provides additional income sources for households	(Kraft et. al., 2021)
Agroforestry Parklands	Africa, Latin America	- Trees planted in park-like settings with crops - Enhances biodiversity and ecosystem services - Provides timber, fruits, and other products for local communities	(Kraft et. al., 2021)
Taungya Systems	Southeast Asia, Africa	- Temporary intercropping of trees with food crops - Provides income during tree establishment phase - Enhances forest restoration and agricultural productivity	(Kraft et. al., 2021)
Windbreak Agroforestry Systems	Worldwide	- Planting trees as windbreaks in agricultural fields - Protects crops from wind damage - Improves microclimate and soil conditions for crop growth	(Kraft et. al., 2021)
Agroforestry Buffer Strips	North America, Europe	- Planting trees along field edges or water bodies - Reduces soil erosion and nutrient runoff - Enhances biodiversity and	(Kraft et. al., 2021)

		provides habitat for beneficial insects and wildlife	
Riparian Agroforestry Systems	Worldwide	- Planting trees along water bodies - Improves water quality and reduces erosion - Provides habitat for aquatic species - Enhances flood control and ecosystem resilience	(Kraft et. al., 2021)

15.2.4. Challenges and constraints in agroforestry implementation

Some key challenges and constraints in the implementation of agroforestry include:

- **Interference of trees and competition with crops**: The presence of trees can decrease crop yields due to competition for light, moisture, and nutrients (Majaura et al., 2024). Trees can act as alternate hosts for pests that affect food crops (Majaura et al., 2024).

- **Access to resources and inputs:** Lack of access to credit, improved seeds/varieties, and soft loans can hinder adoption of agroforestry by farmers (Ibrahim et al., 2019). Unavailability of labor can be a constraint, as educated and well-to-do farmers may not be interested in agroforestry (Ibrahim et al., 2019).

- **Technological and knowledge gaps:** Insufficient understanding of the ecophysiological dynamics and internal dynamics of agroforestry systems (Patel-Weynand et al., 2019). Lack of information on how to optimize productivity and ecological benefits from agroforestry practices (Patel-Weynand et al., 2019).

- **Institutional and policy barriers:** Policies that favor large-scale monoculture operations over small-scale traditional agroforestry systems (Ajayi and Place, 2012). Lack of institutional support and extension services to promote agroforestry adoption (Ajayi and Place, 2012).

- **Socioeconomic factors:** Resistance from farmers to replace food crops with trees, especially where land is scarce (Wienhold and Goulao, 2023). Longer gestation period for trees leading to delayed returns, which can discourage farmers (Wienhold and Goulao, 2023). Increased population and industrialization leading to competition for land and resources (Ibrahim et al., 2019).

15.3 Medicinal Plants: Diversity and Importance

15.3.1. Diversity of medicinal plants and their traditional uses

The traditional use of medicinal plants is deeply rooted in various cultures worldwide, serving as a primary source of healthcare for many communities. These plants, rich in bioactive compounds, have been utilized for centuries in traditional medicine practices. Across different regions, including Iran, Kenya, India, and Nepal, indigenous tribes and communities have developed a profound knowledge of medicinal plants, relying on them for treating a wide array of ailments. In Iran's Alamut region, for example, 16 medicinal plants from 12 families are commonly used in traditional medicine. Similarly, in Cherangani Hills, Western Kenya, local communities have a strong reliance on medicinal plants for healthcare needs. India, known for its rich biodiversity, uses around 2500 plant species in indigenous medicine, highlighting the significance of these plants in traditional healing practices (Kumar et al., 2022; Pandey et al., 2018; Bisht et al., 2022). In Nepal, ethnic groups in the Kavrepalanchok District heavily depend on the traditional use of medicinal plants due to limited access to modern medical care, emphasizing the vital role these plants play in primary healthcare. The diverse array of medicinal plants and their traditional uses across these regions underscores the profound cultural and therapeutic significance of these natural resources in maintaining community health and well-being (Ahvazi et al., 2012; Mbuni et al., 2020; Chandra, 2016 and Ambu, 2020).

15.3.2. Economic value of medicinal plants

The economic value of medicinal plants in India is substantial:

- The estimated business of medicinal plants in India is worth Rs. 550 crores and is expected to reach Rs. 1000 crores (Taleuzzaman et al., 2022).

- The global market for medicinal plants processing products is growing rapidly, with an average increase of 3-4 times the national economies in the same regions, and in some cases, reaching 20% (Kumar, 2004).

15.3.3. Importance of conserving and sustainably managing medicinal plant resources

The key points regarding the importance of conserving and sustainably managing medicinal plant resources are:

- Medicinal plants are globally valuable sources of herbal products, but they are disappearing at a high rate due to overharvesting and habitat destruction (Leaman, 2007).
- It is estimated that there are between 50,000 and 80,000 flowering plant species used for medicinal purposes worldwide, and around 15,000 of these species are threatened with extinction (Chen et al., 2016).
- Medicinal plants are critical for healthcare, especially in developing countries where up to 80% of people depend on herbal medicines for primary healthcare (Sofowora et al., 2013).
- Over 25% of prescribed medicines in developed countries are derived from wild plant species (Hamilton, 2004).
- Sustainable use and conservation of medicinal plants is important to prevent the loss of potential new drugs and maintain access to these valuable natural resources (Sofowora et al., 2013).
- Both in situ conservation (e.g. natural reserves, wild nurseries) and ex situ conservation (e.g. botanic gardens, seed banks) are recommended strategies to conserve medicinal plant resources (Kadam and Pawar, 2020).
- Cultivation of medicinal plants can help conserve wild populations and ensure a reliable supply of quality raw materials for the herbal products industry (Asigbaase et al., 2023).

- Involving local communities in the management and conservation of medicinal plants in their native habitats is essential for long-term sustainability (Hamilton, 2004).

15.4 Integration of Medicinal Plants in Agroforestry

15.4.1. Concept and principles of integrating medicinal plants in agroforestry

The key principles and concepts for integrating medicinal plants in agroforestry systems are:

- Utilizing shade-tolerant medicinal and aromatic plant (MAP) species as understory or lower strata components in multi-strata agroforestry systems (Ramanan et al., 2024).
- Cultivating short-cycle MAP species as intercrops within existing tree plantations or new forest plantations (Rao et al., 2004).
- Growing medicinal trees as shade providers, boundary markers, and on soil conservation structures (Kala et al., 2006)
- Intercropping MAPs with food crops to diversify production and income sources for smallholder farmers (Maitra et al., 2021).
- Involving MAPs in social forestry programs to promote their cultivation and conservation (Rao et al., 2004).

Concepts of Integrating Medicinal Plants in Agroforestry

Many MAPs are traditionally grown under forest cover and are shade-tolerant, making agroforestry a suitable strategy for their cultivation (Rao et al., 2004). Integrating MAPs can increase resource use efficiency, generate supplemental income, and reduce pressure on wild plant populations (Vinodhini et al., 2023). The congenial microclimatic conditions in plantations (e.g. partial shade, moist soil, high organic matter) are suitable for cultivating numerous medicinal plant species (Phand et al., 2023). Agroforestry provides an opportunity to combine the conservation of medicinal plant diversity with their sustainable utilization through cultivation (Sistla et al., 2016). Collaborative approaches like joint forest management with farmers and contract farming with drug companies can promote the large-scale cultivation of medicinal plants (Bera et al., 2010).

15.4.2. Selection criteria for suitable medicinal plant species

The selection criteria for suitable medicinal plant species for agroforestry include various factors based on farmers' preferences and the multiple functions of the plants. Some key criteria identified from the sources are:

- **Value for Different Uses:** Medicinal plant species are ranked based on how well they meet farmers' selection criteria for various uses such as fruit, fuelwood, fodder, timber, medicinal use, other income sources, soil fertility, soil protection, carbon sequestration, and more (He et al., 2016).

- **High Yield Potential:** The selected tree species should have high yield potential to maximize output per unit area in agroforestry systems (Acheampong et al., 2014).

- **Compatibility with Crops:** Trees should not compete with main agricultural crops for resources like water and nutrients. Deep-rooted trees that draw water from deep soil layers and trees with a light shade that can be managed to reduce shading are preferred (Chavan et al., 2017).

- **Ease of Propagation:** Easy methods of propagation, such as on-farm methods or cuttings, contribute to lower costs and are preferred for agroforestry tree species (Awotedu et al., 2021).

- **Long Life Span:** Trees with a long-life span reduce the relative cost of propagation and allow for multiple harvests from a single establishment procedure (Krishnan et al., 2014).

- **Low Risk:** Indigenous and well-known species are preferred over exotic or unknown species to minimize risks related to failures or negative impacts on the agroforestry system (Suárez et al., 2012).

- **Farmer Preferences:** Involving farmers in the selection process is crucial, considering their preferences, traditional knowledge, and the perceived value of direct and indirect production from the trees (Sánchez Toledano, 2017).

- **Adaptation to Local Conditions:** Selection should consider the climate, soil type, and biotic factors of the specific region to ensure the chosen species are well-suited to the local environment (Macel et al., 2007).

15.4.3. Design and layout of agroforestry systems incorporating medicinal plants

The design and layout of agroforestry systems incorporating medicinal plants involve strategic planning to optimize the arrangement of trees and crops for sustainable cultivation (Tewari et al., 2021; Kumar et al., 2021 Kumar et al., 2024). Based on the provided sources, the design typically includes the following elements:

- **Multistrata Systems:** Integrating shade-tolerant medicinal and aromatic plants (MAPs) as lower strata species in multistrata systems to maximize land use efficiency and promote cultivation and conservation (Rao et al., 2004).

- **Cultivating Short Cycle MAPs:** Incorporating short-cycle MAPs as intercrops within existing stands of plantation tree-crops and new forest plantations to diversify production and income sources (Rao et al., 2004).

- **Medicinal Trees as Shade Providers**: Growing medicinal trees to provide shade, act as boundary markers, and contribute to soil conservation within the agroforestry system (Rao et al., 2004).

- **Interplanting with Food Crops:** Interplanting MAPs with food crops to enhance biodiversity, increase productivity, and provide additional income opportunities for farmers (Rao et al., 2004).

- **Social Forestry Programs:** Involving MAPs in social forestry programs to promote their cultivation, conservation, and sustainable utilization within the community (Rao et al., 2004).

- **Joint Forest Management and Contract Farming**: Collaborative approaches like joint forest management with farmers and contract farming with drug companies can facilitate the cultivation of medicinal plants in agroforestry systems (Rao et al., 2004).

- **Buyback Arrangements:** Establishing buyback arrangements with drug companies to ensure a market for medicinal plants grown in agroforestry systems, promoting their cultivation and economic viability (Rao et al., 2004).

15.4.4. Management practices for optimizing medicinal plant growth and productivity

The management practices for optimizing medicinal plant growth and productivity in agroforestry systems include several key strategies based on the provided sources:

- **Microclimate Amelioration:** Adjusting the local climate to optimize conditions for plant growth, especially for medicinal plants that require specific conditions like partial shade, moist soils, high organic matter, humidity, and mild temperatures (Von et al., 2013).

- **Fertilization:** Applying fertilizers to enhance nutrient availability and support the growth of medicinal plants in agroforestry systems.

- **Mulching or Manure Application:** Using mulch or organic manure to improve soil structure, fertility, and moisture retention, which are essential for the growth of medicinal plants (Sileshi et al., 2014).

- **Irrigation:** Providing supplemental water to meet the water needs of medicinal plants, especially during dry periods, to ensure optimal growth and productivity (Joint, 2008).

- **Soil Tillage:** Cultivating the soil to improve aeration, nutrient availability, and root development, which are crucial for the growth of medicinal plants in agroforestry systems (Joint, 2008).

- **Adapted Species Selection:** Choosing tree and crop species that are well-adapted to the specific agroforestry system and can support the growth of medicinal plants effectively (Joint, 2008).

- **Supplemental Feeding:** Providing additional nutrients or supplements to enhance the growth and productivity of medicinal plants in agroforestry systems (Joint, 2008).

- **Pruning and Pollarding:** Implementing pruning and pollarding techniques to control growth, shape plants, and optimize the productivity of medicinal plants within the agroforestry system (Joint, 2008).

- **Root Pruning and Trenching:** Trimming or cutting tree roots and creating trenches to manage root expansion and optimize the growth of medicinal plants (Joint, 2008).

- **Herbicide Use:** Using herbicides selectively to control unwanted plant growth and manage vegetation in agroforestry systems supporting medicinal plants (Joint, 2008).

15.5 Livelihood Impacts of Medicinal Plant-Agroforestry Integration

15.5.1 Economic benefits for smallholder farmers and rural communities

The economic benefits for smallholder farmers and rural communities in agroforestry systems include:

- **Income Diversification:** Agroforestry provides multiple products for small farmers, offering a diverse income stream and reducing dependency on a single crop, thereby enhancing economic stability (Mercer et al., 2014).

- **Risk Reduction:** Agroforestry systems help mitigate financial risks by spreading income sources across various products, reducing vulnerability to market fluctuations or crop failures (Mercer et al., 2014).

- **Environmental Payments:** Payments for ecosystem services may be necessary to increase the adoption of agroforestry systems to socially desirable levels, recognizing the environmental benefits provided by these systems (Mercer et al., 2014).

- **Increased Returns:** Agroforestry systems have been found to offer returns greater than monoculture forestry, providing economic benefits that surpass traditional agricultural practices (Mercer et al., 2014).

- **Net Benefits for Small-Scale Farms:** Small-scale subsistence farms in developing countries are likely to benefit from agroforestry due to the availability of multiple livelihood products, on-farm labor, and fertilization benefits from trees and livestock, leading to clear net benefits (Mercer et al., 2014).

- **Marketability:** Agroforestry systems produce outputs like fruits, nuts, and bananas that can be sold in local markets, enhancing marketability and economic viability for smallholder farmers (Mercer et al., 2014).

- **Financial and Biophysical Risk Reduction:** Agroforestry systems reduce financial and biophysical risks, improve crop yields, reduce inputs like fertilizers, enhance livestock health, and increase biodiversity, contributing to economic sustainability and resilience (Mercer et al., 2014).

15.5.2. Diversification of income sources and risk reduction

The diversification of income sources and risk reduction in medicinal agroforestry can be achieved through the integration of medicinal plants with other crops and trees within agroforestry systems. By diversifying income sources, farmers can reduce their dependency on a single product, spread financial risks, and enhance economic stability. Additionally, the cultivation of medicinal plants alongside other agricultural products can provide multiple income streams, improving livelihoods and reducing vulnerability to market fluctuations or crop failures. This integration not only contributes to biodiversity conservation and ecosystem services but also offers sustainable economic opportunities for rural communities, enhancing overall resilience and prosperity in medicinal agroforestry system (Ndah et al., 2023; Kassie, 2018; Staton, 2022)

15.5.3. Enhancement of ecosystem services and environmental sustainability

The enhancement of ecosystem services and environmental sustainability through medicinal agroforestry involves leveraging the interactions between trees, crops, and medicinal plants to provide multiple benefits for the environment and society. By integrating medicinal plants within agroforestry systems, farmers can enhance biodiversity, improve soil health, promote carbon sequestration, and support sustainable land management practices. This approach not only contributes to the conservation of natural resources but also provides economic opportunities for farmers and rural communities. Through the cultivation of medicinal plants in agroforestry systems, ecosystem services such as soil fertility, water conservation, biodiversity conservation, and climate change mitigation can be optimized, leading to enhanced environmental sustainability and overall ecosystem health (Van, 2021; Handa et al., 2020: Sobola et al., 2015)

15.5.4. Case studies or examples showcasing successful livelihood improvements through medicinal plant-agroforestry integration

Case Study/Example	Location	Livelihood Improvement	Reference (s)
1. Jivika Project	Bihar, India	Enhanced income and improved health outcomes through agroforestry with medicinal plants such as Aloe Vera, Brahmi, and Ashwagandha.	(Adhikari et al., 2018)
2. Sustainable Forest Management in Cameroon	Cameroon	Increased income for local communities by integrating medicinal plants like *Prunus africana* and *Aframomum melegueta* with timber and food crops.	(Endamana et al., 2016)
3. Medicinal Plants in Sustainable Agriculture in Costa Rica	Costa Rica	Improved livelihoods and conservation through the cultivation of medicinal plants like Echinacea and Arnica within agroforestry systems.	(Lee and Goulson, 2014)
4. Community-Based Agroforestry in Ghana	Ghana	Poverty reduction and improved food security by combining agroforestry with medicinal plants such as Moringa, Turmeric, and Ginger.	(Asare, 2014)

5. Ayurvedic Medicine Gardens in Kerala	Kerala, India	Economic empowerment and healthcare provision through the cultivation of medicinal plants like Tulsi, Turmeric, and Amla in home gardens and agroforests.	(Kadam and Pawar, 2020).

15.5 Challenges and Opportunities

15.5.1. Challenges

- Unfavourable policy environment for agroforestry adoption (Paudel et al., 2017)
- Lack of scientific knowledge and public awareness about medicinal agroforestry (Sharma *et al.*, 2017).
- Pressure on wild medicinal plant populations due to overharvesting (Sharma et al., 2017).
- Lack of standardized agro-techniques for integrating medicinal plants in different agroforestry systems (Sharma et al., 2017).
- Difficulty in ensuring a reliable supply of quality raw materials for the herbal products industry (Sharma et al., 2017).

15.5.2. Opportunities (Paudel et al., 2017)

- High demand for medicinal plants and potential for livelihood improvements through sustainable cultivation.
- Ability of agroforestry to enhance resource use efficiency and generate supplemental income.
- Congenial microclimatic conditions in plantations suitable for cultivating numerous medicinal plant species.
- Potential for agroforestry to combine conservation of medicinal plant diversity with sustainable utilization.

- Collaborative approaches like joint forest management and contract farming to promote large-scale medicinal plant cultivation.

- Agroforestry's potential to contribute to food security, rural economies, and ecological health of agricultural landscapes.

15.5.3. Socioeconomic and institutional challenges in promoting medicinal plant-agroforestry systems

Based on the provided sources, the key socioeconomic and institutional challenges in promoting medicinal plant-agroforestry systems include:

- **Knowledge Barriers:**
    - Lack of awareness and scientific knowledge among farmers about the benefits and best practices of integrating medicinal plants in agroforestry systems (Mukhlis et al., 2022).

- **Policy and Institutional Support:**
    - Unfavorable policy environment that does not provide adequate support for the adoption of agroforestry practices (Mukhlis et al., 2022).
    - Lack of coordinated efforts and institutional mechanisms to promote the cultivation of medicinal plants in agroforestry (Mukhlis et al., 2022).

- **Access to Resources:**
    - Limited access to quality planting materials, technical assistance, and financial resources for smallholder farmers to establish medicinal plant-agroforestry systems (Mukhlis et al., 2022).

- **Tenure and Land Rights:**
    - Unclear or conflicting land and tree tenure rights that create disincentives for long-term investments in agroforestry (Bosselmann et al., 2023).

- Insecurity of land tenure, especially for smallholder farmers, hindering their willingness to adopt agroforestry practices (Bosselmann et al., 2023).

- **Market and Value Chain Challenges:**

    - Difficulty in ensuring a reliable supply of quality raw materials for the herbal products industry, limiting the commercial viability of medicinal plant cultivation (Alavalapati et al., 2001)

    - Lack of well-developed market linkages and value addition opportunities for medicinal plants produced in agroforestry systems (Bosselmann et al., 2023).

- **Coordination and Collaboration:**

    - Limited coordination and collaboration among stakeholders, such as farmers, researchers, policymakers, and the private sector, to promote the integration of medicinal plants in agroforestry (Mukhlis et al., 2022).

15.5.4. Policy and regulatory considerations

The sources provided highlight the importance of policy and regulatory considerations in promoting agroforestry and medicinal plant cultivation. Specifically, the National Agroforestry Policy in India emphasizes the need for uniform guidelines, simplified regulations, and institutional support to facilitate the cultivation of agroforestry tree species, including medicinal plants (Chavan et al., 2015). Additionally, the policy aims to address issues related to quality planting material, tree insurance, transit restrictions, and marketing of agroforestry produce, underscoring the significance of policy initiatives in supporting agroforestry practices and maximizing their economic potential (Chavan et al., 2015). These examples underscore the critical role of policy frameworks in creating an enabling environment for the successful integration of medicinal plants in agroforestry systems, ensuring sustainable cultivation practices and livelihood improvements (Sing and Pandey, 2011).

15.5.5. Research gaps and areas for further exploration

The key research gaps and areas for further exploration in the field of medicinal plant-agroforestry integration include:

- **Standardization of Agro-techniques:**
    - The sources highlight the need for standardizing agro-techniques for integrating medicinal plants in different agroforestry systems.

- **Commercialization and Value Chain Development:**
    - There is a need to explore strategies for ensuring a reliable supply of quality raw materials for the herbal products industry, as well as developing well-structured market linkages and value addition opportunities for medicinal plants produced in agroforestry systems (Sharma et al., 2017).

- **Policy and Institutional Support:**
    - Further research is required to understand the policy and regulatory barriers hindering the adoption of medicinal plant-agroforestry systems, and to develop effective policy frameworks and institutional mechanisms to promote their integration (Sharma et al., 2017).

- **Socioeconomic and Environmental Impacts:**
    - Comprehensive assessments of the socioeconomic and environmental impacts of medicinal plant-agroforestry systems are needed to quantify their benefits and inform decision-making (Sharma et al., 2017).

- **Collaborative Approaches:**
    - Exploring collaborative models, such as joint forest management with farmers and contract farming with drug companies, to facilitate the large-scale cultivation of medicinal plants in agroforestry systems (Jhariya et al., 2019).

- **Genetic Resources and Biodiversity Conservation:**
    - Research on the conservation and sustainable utilization of genetic resources of medicinal plants, as well as their integration with biodiversity conservation efforts in agroforestry, is an important area for further exploration (Jhariya et al., 2019).

15.6 Some potential future trends and opportunities for scaling up medicinal plant-agroforestry initiatives include

- **Increasing Demand for Medicinal Plants:**
  - The sources highlight the growing demand for medicinal plants and herbal products, both nationally and globally, which presents a significant opportunity to scale up medicinal plant cultivation through agroforestry (Rao et al., 2018).
- **Policy and Institutional Support:**
  - The establishment of initiatives like the National Mission on Medicinal Plants (NMMP) by the Government of India indicates the potential for policy and institutional support to promote the large-scale cultivation of medicinal plants in agroforestry systems (Rao et al., 2018).
- **Collaborative Approaches:**
  - The sources suggest that collaborative models, such as joint forest management with farmers and contract farming with drug companies, can facilitate the widespread adoption of medicinal plant-agroforestry systems (Rao et al., 2004).
- **Diversification of Income and Livelihood Opportunities:**
  - Integrating medicinal plants in agroforestry systems can provide farmers with diversified income streams and enhance their overall livelihood security, thereby incentivizing the scaling up of these practices (Rao et al., 2018).
- **Ecosystem Services and Environmental Sustainability:**
  - The ability of medicinal plant-agroforestry systems to contribute to biodiversity conservation, soil health, and climate change mitigation can drive their adoption as sustainable land-use practices (Patel-Weynand et al., 2017)

- **Technological Advancements:**
  - Innovations in agro-techniques, propagation methods, and value addition processes can improve the feasibility and profitability of medicinal plant cultivation in agroforestry systems, facilitating their scaling up (Rao et al., 2018).
- **Capacity Building and Knowledge Dissemination:**
  - Strengthening extension services, farmer training programs, and knowledge-sharing platforms can help overcome the knowledge barriers and promote the widespread adoption of medicinal plant-agroforestry systems (Rao et al., 2018).

15.7. Conclusion

Livelihood improvement through the integration of medicinal plants in agroforestry offers a multifaceted approach to enhancing farmer well-being and economic prosperity. The research underscores the significant benefits of cultivating medicinal plants within agroforestry systems, providing diversified income streams, food security, and ecological sustainability. Integrating medicinal plants in agroforestry not only boosts livelihood opportunities but also contributes to biodiversity conservation and climate resilience. Policymakers, practitioners, and researchers are urged to prioritize and support sustainable agroforestry practices that incorporate medicinal plants, recognizing their potential to drive economic growth, environmental stewardship, and community well-being.

References

Acheampong, E., Marfo, E., & Addo-Danso, S. 2014. Chainsaw operators, alternative livelihood options and climate change mitigation. *Kumasi: CSIR Forestry Research Institute of Ghana*.

Adhikari, P., Araya, H., Aruna, G., Balamatti, A., Banerjee, S., Baskaran, P., & Verma, A. 2018. System of crop intensification for more productive, resource-conserving, climate-resilient, and sustainable agriculture: Experience with diverse crops in varying agroecologies. *International journal of agricultural sustainability*, *16*(1), 1-28.

Ahvazi, M., Khalighi-Sigaroodi, F., Charkhchiyan, M. M., Mojab, F., Mozaffarian, V. A., & Zakeri, H. 2012. Introduction of medicinal plants species with the most traditional usage in Alamut region. *Iranian journal of pharmaceutical research: IJPR*, *11*(1), 185.

Ajayi, O. C., & Place, F. 2012. Policy support for large-scale adoption of agroforestry practices: experience from Africa and Asia. *Agroforestry-The Future of Global Land Use*, 175-201.

Akter, R., Hasan, M. K., Kabir, K. H., Darr, D., & Roshni, N. A. 2022. Agroforestry systems and their impact on livelihood improvement of tribal farmers in a tropical moist deciduous forest in Bangladesh. *Trees, Forests and People*, *9*, 100315.

Alavalapati, J., Nair, P. K. R., & Barkin, D. 2001. Socioeconomic and institutional perspectives of agroforestry. *World forests, markets and policies*, 71-83.

Ambu, G., Chaudhary, R. P., Mariotti, M., & Cornara, L. 2020. Traditional uses of medicinal plants by ethnic people in the Kavrepalanchok district, Central Nepal. *Plants*, *9*(6), 759.

Asare, R. 2004. Agroforestry initiatives in Ghana: a look at research and development.

Asigbaase, M., Adusu, D., Anaba, L., Abugre, S., Kang-Milung, S., Acheamfour, S. A., ... & Ackah, D. K. 2023. Conservation and economic benefits of medicinal plants: Insights from forest-fringe communities of Southwestern Ghana. *Trees, Forests and People*, *14*, 100462.

Bawa, K. S., & Seidler, R. 1998. Natural forest management and conservation of biodiversity in tropical forests. *Conservation biology*, *12*(1), 46-55.

Bera, S. K. 2010. Globalisation and sustainable exports of Indian medicinal and aromatic plants: A protection study.

Bijalwan, A., Verma, P., Dobriyal, M. J., Patil, A. K., Thakur, T. K., & Sharma, C. M. 2019. Trends and insights of agroforestry practices in Madhya Pradesh, India. *Current Science*, *117*(4), 597-605.

Bisht, S., Kumar, A., Kumar, N., Singh, H. and Kumar, P. 2022. Biofuel Production by Using Biomass and Its Application. In book: Renewable Energy and Green Technology in CRC Press.

Bosselmann, A. S., Boadi, S. A., Olwig, M. F., & Asare, R. 2023. Social Challenges and Opportunities in Agroforestry: Cocoa Farmers' Perspectives. *Agroforestry as Climate Change Adaptation: The Case of Cocoa Farming in Ghana*, 93-119.

Chandra, L. D. 2016. Bio-diversity and conservation of medicinal and aromatic plants. *Adv Plants Agric Res*, *5*(4), 00186.

Chavan, S. B., Keerthika, A., Dhyani, S. K., Handa, A. K., Newaj, R., & Rajarajan, K. 2015. National Agroforestry Policy in India: a low hanging fruit. *Current Science*, 1826-1834.

Chavan, S. B., Kumar, N., Uthappa, A. R., Keerthika, A., Handa, A. K., Sridhar, K. B., & Ram, N. 2017. Tree management practices in agroforestry. *Forests, Climate Change and Biodiversity; ICAR: New Delhi, India*, 87-101.

Chen, S. L., Yu, H., Luo, H. M., Wu, Q., Li, C. F., & Steinmetz, A. 2016. Conservation and sustainable use of medicinal plants: problems, progress, and prospects. *Chinese medicine*, *11*, 1-10.Sofowora, A., Ogunbodede, E., & Onayade, A. 2013. The role and place of medicinal plants in the strategies for disease prevention. *African journal of traditional, complementary and alternative medicines*, *10*(5), 210-229.

Current, D., Lutz, E., & Scherr, S. J. 1995. The costs and benefits of agroforestry to farmers. *The World Bank Research Observer*, *10*(2), 151-180.

Duffy, C., Toth, G. G., Hagan, R. P., McKeown, P. C., Rahman, S. A., Widyaningsih, Y., ... & Spillane, C. 2021. Agroforestry contributions to smallholder farmer food security in Indonesia. *Agroforestry Systems*, *95*(6), 1109-1124.

Endamana, D., Angu, K. A., Akwah, G. N., Shepherd, G., & Ntumwel, B. C. 2016. Contribution of non-timber forest products to cash and non-cash income of remote forest communities in Central Africa. *International Forestry Review*, *18*(3), 280-295.

Fahad, S., Chavan, S. B., Chichaghare, A. R., Uthappa, A. R., Kumar, M., Kakade, V., & Poczai, P. 2022. Agroforestry systems for soil health improvement and maintenance. *Sustainability*, *14*(22), 14877.

Hamilton, A. C. 2004. Medicinal plants, conservation and livelihoods. *Biodiversity & Conservation*, *13*, 1477-1517.

He, J., Ho, M. H., & Xu, J. 2015. Participatory selection of tree species for agroforestry on sloping land in North Korea. *Mountain Research and Development*, *35*(4), 318-327.

Hosier, R. H. 1989. The economics of smallholder agroforestry: two case studies. *World Development*, *17*(11), 1827-1839.

Ibrahim, A. O., Adedeji, A. S., & Meduna, P. N. 2019. Constraints facing agroforestry practices among farmers in New Bussa, Nigeria. *Journal of Research in Forestry, Wildlife and Environment*, *11*(3), 133-141.

Jemal, O. M., Callo-Concha, D., & van Noordwijk, M. 2021. Coffee agroforestry and the food and nutrition security of small farmers of south-western Ethiopia. *Frontiers in Sustainable Food Systems*, *5*, 608868.

Jewel, K. N. A., Ahmed, F., Saifullah, M., Wadud, M. A., Shahjahan, M., & Rahman, G. M. M. 2019. Existing agroforestry practices and tree species abandons in some selective char and hill ecological zones of Bangladesh. *J. Agrofor. Environ*, *13*, 67-72.

Jhariya, M. K., Banerjee, A., Yadav, D. K., & Raj, A. 2019. Agroforestry and climate change: issues, challenges, and the way forward. In *Agroforestry and Climate Change* (pp. 1-34). Apple Academic Press.

Joint, F. A. O. 2008. *Management of agroforestry systems for enhancing resource use efficiency and crop productivity* (No. IAEA-TECDOC--1606). Joint FAO/IAEA Division of Nuclear Techniques in Food and Agriculture.

Kadam, S. T., & Pawar, A. D. 2020. Conservation of medicinal plants: A review. *Int. Ayurvedic Med. J*, *8*, 3890-3895.

Kala, C. P., Dhyani, P. P., & Sajwan, B. S. 2006. Developing the medicinal plants sector in northern India: challenges and opportunities. *Journal of Ethnobiology and Ethnomedicine*, *2*, 1-15.

Kanyagui, M. K., Sajithkumar, K. J., Velankar, Y., Mohan, R., Viswanathan, P. K., & Magnani, N. 2023. Livelihood challenges faced by women in rural India: exploration of solutions using participatory action research. *Development in Practice*, 1-15.

Kassie, G. W. 2018. Agroforestry and farm income diversification: synergy or trade-off? The case of Ethiopia. *Environmental Systems Research*, *6*, 1-14.

Kaushal R, Tewari S., Thapliyal SD., Kumar, A., Roy, T., Islam, S., Lepcha, STS., Durai, J. 2021. Build-up of Labile, non-labile carbon fractions under fourteen-year-old bamboo plantations in the Himalayan foothills. Heliyon 7(3):e07850

Khakurel, D., Uprety, Y., Karki, S., Khadka, B., Poudel, B. D., Ahn, G., & Rajbhandary, S. (2024). Assessing the Risks to Valuable Medicinal Plants in Nepal from Human Activities and Environmental Factors. *Global Ecology and Conservation*, e02860.

Kraft, P., Rezaei, E. E., Breuer, L., Ewert, F., Große-Stoltenberg, A., Kleinebecker, T., & Nendel, C. 2021. Modelling agroforestry's contributions to people—a review of available models. *Agronomy*, *11*(11), 2106.

Krishnan, P. R., Kalia, R. K., Tewari, J. C., & Roy, M. M. 2014. Plant nursery management: principles and practices.

Kumar, A., Dwivedi, G. K., Tewari, S., Sah, V. K., Singh, H., Kumar, P., Kumar, N., Kaushal, R. 2020a. Soil organic carbon pools under

Terminalia chebula Retz. based agroforestry system in Himalayan foothills, India. Current Science (00113891), 118(7).

Kumar, A., Dwivedi, G.K., Tewari, S., Paul, J., Anand, R., Kumar, N., Kumar, P., Singh, H. and Kaushal, R. 2020b. Carbon Mineralization and Inorganic Nitrogen Pools under *Terminalia chebula* Retz.-Based Agroforestry System in Himalayan Foothills, India. Forest Science. 66(5), 634-643

Kumar, A., Kumar, P., Singh, H. and Kumar, N., 2021a. Impact of plant functional traits on near saturated hydraulic conductivity of soil under different forests of Kempty watershed in Garhwal Himalaya. Indian Journal of Soil Conservation, 49(1), pp.38-44.

Kumar, A., Kumar, P., Singh, H., Kumar, N., 2021b. Adaptation and mitigation potential of roadside trees with bio-extraction of heavy metals under vehicle emissions and their impact on physiological traits during seasonal regimes. Urban Forestry & Urban Greening, 58, p.126900.

Kumar, A., Kumar, P., Singh, H., Kumar, N., 2021c. Modulation of Plant Functional Traits under Essential Plant Nutrients during Seasonal Regime in Natural Forests of Garhwal Himalayas. Plant and Soil 465(1):1-16

Kumar, A., Kumar, P., Singh, H., Kumar, N., 2021d. Relationship of physiological plant functional traits with soil carbon stock in the temperate forest of Garhwal Himalaya. Current Science 120(7): 1064-1068.

Kumar, A., Singh, H., Kumari, G., Bisht, S., Malik, A., Kumar, N., Singh, M., Raturi, A., Barthwal, S., Thakur, A. and Kaushal, R., 2022. Adaptive resilience of roadside trees to vehicular emissions via leaf enzymatic, physiological, and anatomical trait modulations. Environmental Pollution, p.120191.

Kumar, A., Tewari, S., Singh, H., Kumar, P., Bisht, S., 2021e. Biomass accumulation and carbon stock in different agroforestry systems prevalent in the Himalayan foothills. Current Science 120(6), 1083-1088

Kumar, A., Tewari, S., Singh, Indra, Panday, R., Kumar, D. and Anand, R. 2018. Effect of nutrient sources on growth, yield and quality of turmeric under harad (*Terminalia chebula*) based agroforestry system Indian J. of Agroforestry, 20(2). 1-6

Kumar, N., Jeena, N., Kumar, A., Khwairakpam, R., Singh, H. 2021. Comparative response of rice cultivars to elevated air temperature in Bhabar region of Indian Himalaya: Status on Yield attributes. Heliyon 7(6):e07474

Kumar, N., Kumar, A., Jeena, N., Singh, R., and Singh, H. (2020). Factors Influencing Soil Ecosystem and Agricultural Productivity at Higher Altitudes. Microbiological Advancements for Higher Altitude Agro-Ecosystems & Sustainability. Springer. p. 55- 71.

Kumar, N., Singh, H., Giri, K., Kumar, A., Joshi, A., Yadav, S., Singh, R., Bisht, S., Kumari, R., Jeena, N., Khairakpam, R., and Mishra, G. 2024. Physiological and molecular insights into the allelopathic effects on agroecosystems under changing environmental conditions. Physiology and Molecular Biology of Plants. 248. 119255.

Kumar, N., Verma, S., Kumar, A., Dashila, H., Suyal, D.C., Kumari, G., Rawat, S., Jeena, N., Singh, M., Kumar, A. 2022. Molecular insights into stress-responsive genes in the mitigation of environmental stresses. Unravelling Plant-Microbe Synergy, Edition: 1, Publisher: Elsevier

Kumar, P. 2004. Valuation of medicinal plants for pharmaceutical uses. *Current Science*, 930-937.

Kumar, S., Chandrasekhar, C., Kumar, R., & Bhagavan, B. V. K. 2019. Performance of turmeric (*Curcuma longa* L.) genotypes for yield and yield attributing traits under high altitude conditions of Andhra Pradesh. *Journal of Pharmacognosy and Phytochemistry*, *8*(4), 1586-1589.

Kundra, D. N., & Prakashan, G. B. 2020. *A Text Book of Social Sciences for Class 6* (Vol. 6). Goyal Brothers Prakashan.

Leaman, D. J. 2007. Medicinal plant conservation and sustainable use: questions about an important global resource.

Lee, K. E., & Goulson, D. 2014. Medicinal plants, agroforestry and conservation. *Agroforestry Systems,* 88(4), 687-691.

Maccl, M., Lawson, C. S., Mortimer, S. R., Šmilauerova, M., Bischoff, A., Crémieux, L., ... & Steinger, T. 2007. Climate vs. soil factors in local adaptation of two common plant species. *Ecology*, *88*(2), 424-433.

Maitra, S., Hossain, A., Brestic, M., Skalicky, M., Ondrisik, P., Gitari, H., & Sairam, M. 2021. Intercropping—A low input agricultural strategy for food and environmental security. *Agronomy*, *11*(2), 343.

Majaura, M., Böhm, C., & Freese, D. 2024. The Influence of Trees on Crop Yields in Temperate Zone Alley Cropping Systems: A Review. *Sustainability*, *16*(8), 3301.

Mbuni, Y. M., Wang, S., Mwangi, B. N., Mbari, N. J., Musili, P. M., Walter, N. O., & Wang, Q. 2020. Medicinal plants and their traditional uses in local communities around Cherangani Hills, Western Kenya. *Plants*, *9*(3), 331.

Mercer, D. E., Frey, G. E., & Cubbage, F. W. 2014. Economics of agroforestry. In *Handbook of forest resource economics* (pp. 204-225). Routledge.

Mukhlis, I., Rizaludin, M. S., & Hidayah, I. 2022. Understanding socio-economic and environmental impacts of agroforestry on rural communities. *Forests*, *13*(4), 556.

Nair, P. R. 1985. Classification of agroforestry systems. *Agroforestry systems*, *3*, 97-128.

Ndah, N. R., Nanje Ekole, P., Agwa, M. H., Taku, J., Fonyikeh-Bomboh Lucha, C., & Agbor, D. T. 2023. Crop diversification and sustainability in a cocoa agroforestry system in Meme Division South West Region, Cameroon. *Asian journal of Research in Agriculture and Forestry*, *9*(2).

Palada, M. C., Mitchell, J. M., Becker, B. N., & Nair, P. K. R. 2003. The integration of medicinal plants and culinary herbs in agroforestry systems for the Caribbean: a study in the US Virgin Islands. In *III WOCMAP Congress on Medicinal and Aromatic Plants-Volume 2: Conservation, Cultivation and Sustainable Use of Medicinal and 676* (pp. 147-153).

Pandey R., Dubey S., Tewari L, Kumar A., Kumar D., Sharma N. and Tewari S.K. 2018. Performance of soybean under poplar-based agroforestry system in Terai region. Indian J. of Agroforestry, 20(1), 31-34.

Patel-Weynand, T., Bentrup, G., Schoeneberger, M., Karel, T. H., & Nair, P. K. R. 2017. Challenges and Opportunities. *Agroforestry Enhancing Resiliency in US Agricultural Landscapes under Changing Conditions*.

Paudel, S., Baral, H., Rojario, A., Bhatta, K. P., & Artati, Y. 2022. Agroforestry: opportunities and challenges in Timor-Leste. *Forests*, *13*(1), 41.

Pushpakumara, D. K. N. G., Marambe, B., Silva, G. L. L. P., Weerahewa, J., & Punyawardena, B. V. R. 2012. A review of research on homegardens in Sri Lanka: the status, importance and future perspective. *Tropical Agriculturist*, *160*, 55-125.

Rajeshwar Rao, G., Prabhakar, M., Venkatesh, G., Srinivas, I., & Sammi Reddy, K. 2018. Agroforestry opportunities for enhancing resilience to climate change in rainfed areas. *ICAR—Central Research Institute for Dryland Agriculture, Hyderabad*.

Ramanan, S. S., Arunachalam, A., & Singh, R. 2024. Horizons of medicinal and aromatic plants (MAPs) cultivation under agroforestry: Insights from India.

Rao, M. R., Palada, M. C., & Becker, B. N. 2004. Medicinal and aromatic plants in agroforestry systems. In *New Vistas in Agroforestry: A Compendium for 1st World Congress of Agroforestry, 2004* (pp. 107-122). Springer Netherlands.

Sánchez Toledano, B. I. 2017. Farmers' preferences and the factors affecting their decision to improve maize crops in Mexico.

Sharma, P., Singh, M. K., Tiwari, P., & Verma, K. 2017. Agroforestry systems: Opportunities and challenges in India. *Journal of Pharmacognosy and Phytochemistry*, *6*(6S), 953-957.

Sileshi, G. W., Mafongoya, P., Akinnifesi, F. K., Phiri, E., Chirwa, P., Beedy, T., & Jiri, O. 2014. Agroforestry: Fertilizer Trees.

Singh, V. S., & Pandey, D. N. 2011. *Multifunctional agroforestry systems in India: Science-based policy options* (pp. 2-35). Jaipur, India: Climate Change and CDM Cell, Rajasthan State Pollution Control Board.

Sistla, S. A., Roddy, A. B., Williams, N. E., Kramer, D. B., Stevens, K., & Allison, S. D. 2016. Agroforestry practices promote biodiversity and natural resource diversity in Atlantic Nicaragua. *PloS one*, *11*(9), e0162529.

Sobola, O. O., Amadi, D. C., & Jamala, G. Y. 2015. The role of agroforestry in environmental sustainability. *IOSR Journal of Agriculture and Veterinary Science*, *8*(5), 20-25.

Staton, T., Breeze, T. D., Walters, R. J., Smith, J., & Girling, R. D. 2022. Productivity, biodiversity trade-offs, and farm income in an agroforestry versus an arable system. *Ecological Economics*, *191*, 107214.

Suárez, A., Williams-Linera, G., Trejo, C., Valdez-Hernández, J. I., Cetina-Alcalá, V. M., & Vibrans, H. 2012. Local knowledge helps select species for forest restoration in a tropical dry forest of central Veracruz, Mexico. *Agroforestry systems*, *85*, 35-55.

Taleuzzaman, M., Ahmad, A., & Gilani, S. J. 2022. Socio-Economic Importance of Some Promising Edible Medicinal Plants from Rajasthan, India. In *Edible Plants in Health and Diseases: Volume 1: Cultural, Practical and Economic Value* (pp. 31-53). Singapore: Springer Nature Singapore.

Tewari, S., Tewari, L., Dubey, A., Kumar, A., Kumar, N., Kaushal, R. 2022. Use of the RAPD marker to determine the genetic diversity of various *Dalbergia sissoo* Roxb. (Shisham) genotypes and their evolutionary relationship in Indian subcontinents. Vegetos,

Tewari, S., Tewari, L., Kumar, A., Bisht, S, Kaushal, R. and Kumar, A. 2024. Germination Behaviour of Teak Genotypes and Effect of

Different Treatments on Seed Germination Characteristics in Himalayan Foothills, India. Indian Forestry, 150 (3): 244-254

Tewari, S., Tewari, L., Kumar, A., Kaushal, R. 2021. Performance of *Glycine max* L. varieties under *Tectona grandis* L. and *Dalbergia sissoo* Roxb. based agroforestry system Indian J. of Agroforestry, 20(2) 1-6

Thakur, P. S., & Dutt, V. 2007. Cultivation of medicinal and aromatic herbs in agroforestry for diversification under submontane conditions of western Himalayas. *Indian Journal of Agroforestry*, *9*(2).

Verma, T., Bhardwaj, D.R., Sharma, U., Sharma, P., Kumar, D., Kumar, A. and Kumar, A. 2023. Aroforestry systems in the mid-hills of the north-western Himalaya: A sustainable pathway to improved soil health and climate resilience. Journal of Environmental Management. 348, 119264.

Von Arx, G., Graf Pannatier, E., Thimonier, A., & Rebetez, M. 2013. Microclimate in forests with varying leaf area index and soil moisture: potential implications for seedling establishment in a changing climate. *Journal of ecology*, *101*(5), 1201-1213.

Wienhold, K., & Goulao, L. F. 2023. The embedded agroecology of coffee agroforestry: A contextualized review of smallholder farmers' adoption and resistance. *Sustainability*, *15*(8), 6827.

Yirga, S. A. 2019. Agroforestry for sustainable agriculture and climate change: a review. *International Journal of Environmental Sciences & Natural Resources*, *19*(5), 127-137.

Zerihun, M. F. 2021. Agroforestry practices in livelihood improvement in the Eastern Cape Province of South Africa. *Sustainability*, *13*(15), 8477.

Zhang, W., Su, K., Wang, Q., Yang, L., Sun, W., Ranjitkar, S., & El-Kassaby, Y. A. 2023. Agroforestry Species Selection for Forest Rehabilitation in the Asia-Pacific Region: A Meta-Analysis on High-Level Taxonomy. *Forests*, *14*(10), 2045.

CHAPTER SIXTEEN

AGROFORESTRY FOR THE FUTURE OF FOOD SECURITY AND CLIMATE RESILIENCE

ARADHANA[1], RAKESH NEGI[1], SARITA BISHT[2] AND MUKESH KUMAR[1]*

[1]DEPARTMENT OF BOTANY AND MICROBIOLOGY, GURUKULA KANGRI (DEEMED TO BE UNIVERSITY), HARIDWAR, INDIA
[2]FOREST RESEARCH INSTITUTE DEHRADUN

Abstract

In India, diverse agroforestry practices have emerged in response to varying ecological and socioeconomic backgrounds across different regions. The adoption of agroforestry by marginal farmers is seen as a pathway to strengthen resistance to the effects of climate change. This practice not only diversifies livelihood sources but also has a vital function in providing shade during heat waves and buffering soil moisture during dry spells. As agroforestry systems proliferate, they contribute to the establishment of new rural businesses, ranging from nurseries to processing units, and increase the demand for high-quality agricultural products and germplasm. The current climate variability poses challenges to ensure regular food production for the entire nation. Beyond its essential function in guaranteeing food security, agroforestry offers a wide range of applications and advantages that greatly enhance social cohesion, economic growth, and environmental sustainability. Present chapter delves into a detailed analysis of agroforestry, focusing on its potential to positively impact various aspects of environmental sustainability, including soil health, biodiversity preservation, water management, carbon sequestration, and microclimate regulation. Recognizing the ability of agroforestry interventions to provide both economic and environmental benefits, they are considered essential measures for communities to adapt and build resilience against the effects of climate change. The emphasis on sustainable agroforestry underscores its potential as a multifaceted solution to address food security issues and strengthen resistance to climate change, ultimately contributing to the promotion of sustainable agroforestry systems.

Keywords: Agroforestry, Carbon sequestration, Climate change, Climate resilience, Ecosystem Services, Food security.

16.1 Introduction

Agroforestry is a multifunctional, sustainable land use system in which trees thrive alongside crops and/or livestock on a single plot of land. Large, marginal, and small landholders in India have long engaged in agroforestry. Nevertheless, despite the efforts of several writers to gather data for their works, not much is known about these activities (Dhyani et al., 2011). Agroforestry emerges as a resilient choice amidst the challenges posed by climate change, providing adaptability against harsh weather events and temperature changes. The multifaceted benefits of agroforestry encompass economic, social, and environmental dimensions, addressing sustainability challenges. Agroforestry techniques also assist regenerative and sustainable agricultural methods by improving soil health, conserving biodiversity, and sequestering carbon. Agroforestry has significant effects on water management, community participation, and cultural values in addition to helping farmers to diversify their income sources. It is a tool for managing problems including risk reduction, water quality enhancement, and erosion management. It appears as a comprehensive strategy, tackling environmental, social, and economic aspects, thereby strengthening its position as a sustainable and resilient practice amid the challenges of a shifting climate (Kumar et al., 2020a&b: Kaushal et al., 2021). The implementation of agroforestry is increasingly acknowledged as a strategic measure to boost the resilience of agricultural communities against the impacts of climate change, consequently enhancing their overall well-being. Despite the challenges posed by changing climates and environmental variability, trees within and outside forests play an important part in supporting India's food security. These trees, integral to agricultural livelihoods, offer several benefits to society and the environment. Tree agroforestry is becoming more and more popular as a strategic intervention to enhance the resilience of farming communities to climate change, thereby improving their overall quality of life. This approach leverages community forestry, providing environmental and economic benefits, and has proven successful in bridging livelihood gaps and supporting forest landscape restoration. Extension professionals are essential in helping farmers design, establish, and maintain agroforestry systems, particularly given the escalating climate variability. While agroforestry is deemed more resilient than monocropping, economic incentives for farmers remain a challenge (Tewari et al., 2024; Kumar et al., 2021a&b; Kumar et al., 2022). The majority of agroforestry research has focused on technical aspects, and more extensionists are needed to promote agroforestry at the community level. Collaborative efforts between scientists and farmers, identifying desired species, developing indicators, and scaling up successful agroforestry systems, are crucial. Examples from different regions serve to illustrate how agroforestry projects can be optimized contextually to meet unique environmental and socioeconomic needs. Agroforestry technologies contribute to mitigate declining agricultural productivity, natural resource depletion, and ensuring food security. Agroforestry is gaining acknowledgment for its significance in addressing climate change and promoting agricultural sustainability (Tewari et al., 2022; Kumar et al., 2018; Kumar et al., 2021c&d).

The intersection of nature and food security is a critical concern for humanity, particularly in developing nations such as India, grappling with exponential population growth and heightened demand for natural resources. Despite significant increases in food production, challenges persist due to population growth and climatic variations. With over 75% of India's population reliant on agriculture, certain lands remain unsuitable for cultivation due to various factors. Trees, both inside and outside forests, play a crucial role in mitigating climate change and supporting food security by providing social and environmental benefits. In the face of population explosion and climate change, agroforestry practices vary across India, offering increased climate resilience and improved livelihoods. Despite soil deterioration, biotic pressure, degradation, and food insecurity in the country, agroforestry offers a realistic alternative for food security and climate resilience for the present and the future. Agroforestry proves essential in ensuring food safety amid unpredictable weather conditions, establishing a global link between livelihood and climate resilience. While agroforestry presents numerous direct benefits and ecosystem services, there exists a global gap between research and extension efforts in this field. The early development of agroforestry systems was shaped by the economic needs, cultural preferences, and environmental conditions specific to each region. Although these early systems demonstrated an understanding of ecological balance, it took time for such insights to be formally recognized in scientific literature. Embracing technological breakthroughs, bridging research gaps, and scaling up techniques are crucial steps to propel the profession forward. Interdisciplinary research, integrating ecological, economic, and social aspects, is identified as a pressing need (Verma et al., 2023; Kumar et al., 2020; Kumar et al., 2021e). Furthermore, the incorporation of technological advancements like machine learning and remote sensing may revolutionize the design, monitoring, and management of agroforestry systems. To scale up agroforestry, a comprehensive multi-stakeholder strategy is essential, encompassing supportive legislation, financial incentives, and effective information sharing. Agroforestry emerges not only as an environmentally sustainable practice but also as a pathway to social equality and economic feasibility, contributing to the broader goal of achieving sustainability amidst the complexities of climate change (Kumar et al., 2022; Pandey et al., 2018; Bisht et al., 2022). For agroforestry to be effective, practitioners must understand its fundamentals and their relation to community resilience, and addressing environmental challenges.

16.2 Benefits of Agroforestry

The fact that agroforestry benefits society, either directly or indirectly through ecosystem services, is not new. Trees in agroforestry systems give the rural community fuel wood, feed, fruits, and fibers to enhance people's quality of life by promoting livelihoods and lowering poverty. It makes sense that agroforestry fosters a worldwide synergy between livelihood and climate resilience. By producing food, fodder, and firewood all at once and reducing the effects of climate change, agroforestry systems hold the promise of improving livelihoods (Panday, 2007; DeFries et al., 2010). The agroforestry tree component enhances the soil's moisture regime by decreasing interception losses, cutting down on excess runoff, and raising the rate of infiltration and recharging. According to Sahu et al., (2015), trees can stop the groundwater table's rapid decline as well as reverse its upward trend. Agroforestry possesses numerous tangible and intangible benefits for society, yet numerous obstacles hinder its widespread adoption within the country. This chapter elucidates the various ways in which agroforestry could bolster livelihoods while also aiding in climate change mitigation efforts.

16.3 Agroforestry and food safety

Agroforestry is the practice of growing crops, animals, and/or other products with a tree component while simultaneously producing sustainable energy and a stable environment. Agroforestry has many physical and intangible advantages for society, but there are a lot of barriers preventing it from being widely adopted in the country. The ability of agroforestry to increase food security varies according to the circumstances, and many of these gains can be achieved both directly and indirectly, making a definite classification of them challenging (Ray et al., 2013; Challinor et al., 2014). Agroforestry is one such intervention that is being pushed. With proper planning and implementation, it may improve ecosystem services and have a good effect on income, food availability, and health (Nair and Toth, 2016; Rosenstock et al., 2016). Labor shortages, genetic diversity, and the fact that certain benefits take time to show are frequently the first barriers to agroforestry interventions. While agroforestry's environmental benefits are well known, its impacts and outcomes on food security are yet unknown. Agroforestry, offering a holistic and sustainable approach to agricultural practices, holds immense promise for ensuring future food security (Tewari et al., 2021; Kumar et al., 2021 Kumar et al., 2024).

Developing nations grapple with significant challenges such as food insecurity, extreme poverty, and environmental decline. Declining land holdings and per capita food production result from factors like reduced rainfall and soil fertility depletion. Hunger, conflicts, droughts, and food insecurity are common issues (Eyasu, 2002). Agroforestry, an evergreen agriculture approach, has been used for centuries to achieve agricultural sustainability and slow down negative effects like soil degradation and desertification. This approach holds considerable promise. To address food insecurity and climate change. Agroforestry shows promise as a long-term, holistic strategy for addressing future food security concerns. Because of its capacity to foster environmental sustainability, increase resilience, and support a range of livelihoods, agroforestry is an essential portion of the worldwide effort to guarantee a steady and secure food supply. The deliberate planting of shrubs and trees in agricultural landscapes diversifies production systems and promotes a more stable and resilient food source. In agroforestry, resilience mechanism for food security differs from person to person. (Meert et al., 2005). However, a lot of literature mentions food accessibility and availability as a component of food security (Schmidhuber and Tubiello, 2007). Agroforestry has significantly boosted household food security, supplying both cash and food for 72% of evaluated households. It reduces poverty, improves income and food security, and sustains environmental quality. Future interventions include home gardens, woodlots, and coffee farms. Despite less attention given to tree-based land use, agroforestry has reduced household vulnerability to shocks. Smallholder farmers are mainstreaming tree-based land use systems as resilience to social needs, as the poor are more exposed to change.

Agroforestry is a risk-averse option for overcoming food insecurity. Global warming is a real issue with serious consequences, particularly for Africa. The agricultural impacts are of greatest concern to developing countries, particularly in the tropics, because of their higher dependence on agriculture, subsistence level, low adaptive capacity, and limited institutional support. Most research has focused on major annual crops, with little on trees and perennial crop systems. Perennial trees face unique challenges due to their long duration and difficulty in changing varieties over short periods. Additional investigation is necessary to comprehend the resilience of these trees to the effects of climate change before advocating for their utilization.

The pivotal role of agroforestry in securing food supplies for the future is underscored by several crucial factors:

16.3.1 Agroforestry for food production

Tackling the obstacles related to food and nutritional security necessitates a variety of agricultural strategies, including boosting the yield of essential crops, fortifying them with essential nutrients, and broadening the cultivation of diverse edible plants like fruits, nuts, and vegetables to promote varied diets (Frison et al., 2011). The opportunity to broaden crop variety rests in the plentiful presence of underutilized native foods in forests and wooded regions, which frequently boast elevated levels of micronutrients, fiber, and protein in comparison to staple crops (Leakey, 1999; Malézieux, 2013). While these traditional foods typically originate from forests and wooded areas, their availability is declining as a result of deforestation and degradation of forested regions (FAO, 2010). Cultivation offers an alternative strategy in this situation. Moreover, by prioritizing genetic improvement and embracing effective agricultural techniques, both the quantity and quality of yields can be improved during cultivation, making planting an attractive choice. For many wild trees, such as indigenous fruits, it is possible to achieve a twofold or greater increase in yield through genetic selection (Jamnadass et al., 2011). Apart from directly supplying edible goods, trees integrated into agroforestry systems contribute to food production through various means, such as offering shade and structural support for crops, facilitating animal husbandry, and enhancing soil fertility.

16.3.2 Agroforestry for Fruits and Vegetables Production

Agroforestry plays a crucial role in boosting the yields of vegetables, which, along with fruits, contribute to diversify and nutritionally balanced diets rather than solely focusing on calorie intake (Susila et al., 2012). Trees can alter the microclimate for cultivated crops, particularly in challenging environmental conditions, and can also serve as support for climbing plants like yam (Maliki et al., 2012). In a notable initiative in East Africa, over 200,000 smallholder dairy farmers are cultivating fodder shrubs as supplementary feed. As a result of this approach, there has been a notable rise in milk production, allowing farmers to earn supplementary income through milk sales, exceeding USD 100 per cow annually. Additionally, it improves the effectiveness of milk distribution to urban consumers (Place et al., 2009). The notion of combining agriculture and trees is not new; in fact, the exact beginnings of agroforestry are unclear. Based on historical narratives and archeological finds, it seems that many ancient civilizations possessed agroforestry systems. The Mayan technique of combining multi-storied cropping systems made up of crops, trees, and bushes is known as a "forest garden" (Galhena et al., 2013). Like this, food crops like sorghum and millet were combined with perennial crops in ancient African agricultural systems (Rogé et al., 2017). The ancient agrosilvopastoral systems in the Mediterranean areas also point to the practice of agroforestry, which involved planting trees like olive, fig, and almond alongside grains and pastured animals (Calatrava and Sayadi, 2019).

16.3.3 Challenges in agroforestry for food and nutritional security

16.3.3.1 Policy constraints

The slow uptake of agroforestry practices is significantly hindered by policies governing seed access and exchange. Challenges arise due to the limited availability of tree seeds and the complex procedures required to obtain them, which obstruct farmers' efforts to incorporate trees into their farming systems. Moreover, the lack of acknowledgment of agroforestry as a viable investment sector by governmental bodies exacerbates the issue. In the absence of adequate support and incentives from policymakers, farmers may lack the motivation to adopt agroforestry practices on their land. Furthermore, there is a noticeable oversight in the collection of data on tree products and services, which leads to an underestimation of trees' importance to both livelihoods and ecosystems. This oversight further marginalizes agroforestry as a sustainable land-use option.

16.3.3.2 Constraints in delivering tree products to markets

Poorly structured markets present significant challenges to agroforestry initiatives. They lead to diminished returns for farmers and inflated prices for buyers, creating an imbalance in the agricultural economy. Factors such as the absence of collective bargaining among farmers, inadequate transport infrastructure, and the presence of numerous intermediaries further exacerbate this issue by driving down farm prices. Moreover, the high levels of wastage along the supply chain, particularly for perishable goods like fruit, pose additional obstacles to the success of agroforestry efforts. These challenges underscore the need for comprehensive reforms and support mechanisms to promote the sustainable development of agroforestry systems.

16.3.3.3 Underinvestment in research

Agroforestry faces several hurdles stemming from insufficient investment and attention in critical areas. Firstly, there's a lack of investment in developing high-yield tree lines and cultivars tailored for smallholder production, hindering productivity and profitability. Secondly, indigenous fruit trees, which hold the potential for diversifying and enriching agroforestry systems, receive limited focus in improvement efforts, constraining their contribution to farmers' livelihoods. Lastly, inadequate information on the nutritional value of indigenous food trees poses a challenge, as it limits the understanding of their potential health benefits and market value. Addressing these gaps in investment, research, and promotion is crucial for realizing the full potential of agroforestry in sustainable agriculture and food security initiatives. Additionally, there is a gap in understanding the effects of climate change on agroforestry systems, including interactions between pollinators and plants, which could impact fruit tree yields and exacerbate malnutrition in the face of climate change.

16.3.4 Options to enhance the role of agroforestry in supporting food and nutritional security

16.3.4.1 Policy opportunities

To enhance the role of agroforestry in sustainable development, policymakers should prioritize revising guidelines to support its growth.

This involves reforming regulations and legal restrictions to create a conducive environment for agroforestry practices. Clarity in land-use policy goals and regulations is crucial to recognizing the vital role of trees in development. Additionally, efforts should focus on strengthening farmers' access to markets for tree products, promoting economic viability. Countries like Brazil, China, and India could benefit from developing or updating agroforestry strategies to maximize their potential. Implementing minimum requirements for tree cover on farms, coupled with financial support for farmers, can facilitate the adoption of agroforestry. Moreover, embracing agroforestry as a climate-smart agricultural approach in national climate change adaptation plans will further elevate its significance in addressing environmental challenges. These initiatives collectively bolster the role of agroforestry in sustainable development efforts worldwide.

16.3.4.2 A three-step approach for agroforestry to support food and nutritional security

These interventions, including support for soil fertility replenishment technologies, participatory tree domestication, entrepreneurship, and value addition to tree products, as well as value chain analysis, aim to enhance the role of agroforestry by promoting sustainable practices, empowering local communities, and stimulating economic development in rural areas.

16.3.4.3 Planning for climate change

Adapting tree composition within agroforestry systems to evolving environmental conditions is essential for maximizing their effectiveness. This involves selecting "generalist" food trees that exhibit resilience and thrive in diverse climatic settings. Additionally, the development of vegetation maps and the exploration of changes under various climate scenarios are critical for informed decision-making regarding tree planting. Furthermore, conducting trials to assess genetic variation in climate-related traits of key fruit trees tailored to local communities enhances the capacity of agroforestry systems to withstand environmental challenges. These strategies collectively contribute to bolstering the role of agroforestry in sustainable land management and resilience-building efforts.

These options highlight various strategies and approaches to maximize the contribution of agroforestry to food and nutritional security amidst evolving challenges such as climate change and market dynamics.

16.4 Agroforestry and climate resilience

A wide range of industries and ecosystems are being impacted by climate change, which is presenting an unparalleled hazard to the world (Gössling et al., 2020). The agricultural sector is especially vulnerable as it depends so much on meteorological factors including temperature, precipitation, and seasonal patterns. According to research published by the Food and Agriculture Organization (FAO), yield losses in a variety of major crops might amount to as much as 30% by the middle of the century as a result of climate change. The world's food security is seriously threatened by this, particularly in developing countries where the majority of people get their living from agriculture (Mbow et al., 2020). Climate change has been associated with decreased yields and an increase in the occurrence of severe weather events such as storms, floods, and droughts, which have devastating effects on agricultural systems (Kumar et al., 2018). While protracted drought conditions may lead to desertification and soil degradation, which both have long-term implications on agriculture, high rains and flooding can cause erosion and the loss of arable land.

Agroforestry plays a crucial role in enhancing climate resilience, providing a sustainable and integrated approach to tackle the challenges posed by a climate change. These diverse roles showcase the adaptability of agroforestry in building resilience across ecological, economic, and social dimensions, making it a valuable strategy for sustainable and climate-resilient agriculture. In summary, agroforestry stands as a resilient and adaptive land-use strategy, contributing significantly to climate resilience in agricultural systems. Its multifaceted benefits extend beyond individual farms, positively impacting entire landscapes and communities facing the challenges of a changing climate. The role of agroforestry in climate resilience is multifaceted, encompassing various aspects that aid in the adaptability and sustainability of agricultural systems in the face of climate change. In summary, agroforestry plays a vital role in enhancing climate resilience by fostering diversified, adaptable, and sustainable agricultural systems that can withstand and respond to the challenges posed by a changing climate. The efficacy of global initiatives aimed at addressing climate change can be significantly improved by explicitly incorporating agroforestry and integrating forestry and farming agendas. Agroforestry practices, when implemented as per best practices, consider diverse ecological settings, market potential, and resource accessibility. Notably, agroforestry offers a comprehensive approach to support food access, income generation, health, and environmental stability while increasing climate change resilience.

Here are several key aspects that highlight the significance of agroforestry in climate resilience:

16.4.1 Agroforestry's resilient ecosystem services

Agroforestry systems play a pivotal role in fostering diverse ecosystem services, creating resilient environments through the intentional integration of trees with crops or livestock. Agroforestry has been shown to integrate production with various ecosystem services and commodities (McAdam et al., 2009). This deliberate combination results in a rich tapestry of flora and fauna, enhancing the system's capacity to withstand a multitude of climate stressors, including extreme weather events, pests, and diseases. Beyond mere biodiversity, these systems contribute significantly to microclimate regulation. The presence of trees provides shade, reduces temperature extremes, and moderates humidity levels, thereby creating more favorable conditions for the optimal growth of crops and the well-being of livestock, even in the face of unpredictable climate variability. Moreover, agroforestry practices contribute to efficient water management, addressing critical issues in agriculture. By preventing soil erosion, enhancing water infiltration, and reducing evaporation, trees in agroforestry act as natural buffers. This not only improves water retention in the soil but also mitigates the consequences of drought or erratic precipitation, ensuring sustainable availability of water for agricultural activities. Another vital contribution of agroforestry is its role in carbon sequestration. The integration of trees within these systems allows for the absorption and storage of carbon dioxide from the atmosphere, thereby reducing overall greenhouse gas concentrations and contributing to climate resilience.

16.4.2 Agroforestry's in biodiversity conservation and community resilience

Compared to traditional monoculture cropping systems, agroforestry (AF) offers multiple ecosystem benefits and enhances biodiversity while maintaining productivity and economic viability (Nguyen et al., 2013). Additionally, agroforestry systems (AFS) contribute to biodiversity by fostering genetic, species, and crop diversity, as well as by enhancing microhabitats both above and below ground, and supporting various wild, native, and non-native species critical for ecosystem services and environmental functions (Jose, 2012; Sistla et al., 2016; Sagastuy and Krause, 2019). Biodiversity conservation is a key outcome of agroforestry practices. The generation of complex ecosystems within these systems supports a diverse range of plant and animal species, enhancing ecological resilience. This diversity not only contributes to the health of the environment but also ensures that the agroforestry system remains adaptable to changing environmental conditions. Moreover, the range of plant species and components in agroforestry systems offers a means to diminish susceptibility to severe weather occurrences like storms, floods, or heat waves. This structural complexity enhances the overall resilience of the agricultural landscape, making agroforestry a sustainable and adaptive approach to modern farming practices. Agroforestry systems emerge as a crucial tool in enhancing the adaptive capacity of farmers in the face of changing climate conditions. The inherent flexibility of these systems, achieved through the intentional combination of trees, crops, and sometimes livestock, empowers farmers to navigate uncertainties. By cultivating a mix of products, agroforestry reduces the risks associated with a single crop failure, a vulnerability often exacerbated by climate-related factors. Beyond risk reduction, agroforestry significantly contributes to soil health improvement. The intricate root systems of trees within these systems play a pivotal role in enhancing soil structure, nutrient cycling, and moisture retention. This, in turn, fosters improved soil fertility, providing a robust foundation for sustained agricultural productivity, even in the midst of shifting climate patterns. The impact of agroforestry extends beyond individual farms to foster community resilience. The collaborative nature of agroforestry practices encourages community engagement and cooperation, establishing a shared responsibility for managing natural resources. This communal approach not only promotes sustainable land use practices but also enables collective responses to climate challenges. Economic diversification further reinforces climate resilience, as the integration of tree products like fruits, nuts, and timber diversifies income sources for farmers. This diversification is crucial, mitigating the susceptibility of communities to climate-related risks by reducing dependence on a single agricultural enterprise.

16.4.3 Agroforestry's pollination support

Additionally, agroforestry systems play a vital role in supporting pollination services (Maccagnani et al., 2020). The presence of diverse vegetation in these systems provides habitat and forage for pollinators such as bees and butterflies. This support enhances the resilience of crops that depend on pollinators for reproduction, contributing to overall agricultural sustainability. Agroforestry systems offer a multifaceted approach to enhancing agricultural resilience, with the diversity within these systems acting as a natural buffer against pests and diseases. By disrupting monocultures and introducing varied plant species, agroforestry mitigates the risk of widespread infestations, thereby contributing to the overall resilience of crops. Beyond pest and disease management, agroforestry provides a platform for livelihood diversification. The integration of income-generating activities related to tree products, such as fruits, nuts, and timber, offers farmers a diversified economic portfolio. This diversification not only helps buffer against economic shocks but also serves as a strategic response to climate-related uncertainties. Furthermore, agroforestry plays a vital role in cultural and social resilience. Many agroforestry practices align with local cultural traditions, and preserving or reintroducing these methods contributes to the cultural resilience of communities facing climate challenges. The integration of traditional knowledge and practices within agroforestry systems not only strengthens community ties but also fosters a sense of identity and continuity in the face of environmental changes. Agroforestry systems also contribute to risk reduction, particularly in regions prone to increased fire risk due to climate change. Well-designed agroforestry layouts can act as natural firebreaks, helping mitigate the risk of wildfires by reducing the spread of fires. This function becomes increasingly relevant in the context of climate-induced changes in fire patterns. Additionally, agroforestry systems serve as de facto seed banks, maintaining the genetic diversity of plant species. This genetic reservoir is a valuable resource for developing crops that are inherently more resilient to changing environmental conditions, ensuring the long-term sustainability of agricultural practices. In essence, agroforestry stands as a comprehensive and adaptable approach that not only addresses agricultural challenges but also contributes to the broader resilience of communities and ecosystems. Well-managed agroforestry landscapes present enticing opportunities for ecotourism, serving as a potential source of additional income for communities. The diversification of income streams achieved through ecotourism not only enhances economic resilience but also establishes a sustainable relationship between agriculture and tourism, offering a buffer against climate-related uncertainties. Agroforestry, particularly through practices like agro-silvopastoral systems, contributes to enhanced water use efficiency. Trees integrated into agricultural landscapes play a key role in regulating water flow, fostering judicious water use, and contributing to overall water conservation. Moreover, agroforestry systems, characterized by diverse vegetation, serve as natural buffers against natural disasters such as floods and landslides. This inherent quality significantly contributes to landscape stability and enhances community safety in the face of unpredictable natural calamities.

16.4.4 Agroforestry's multi-faceted contributions

Furthermore, agroforestry's role in promoting nutrient cycling, facilitated by the incorporation of nitrogen-fixing trees, stands as a cornerstone for improving soil fertility. This nutrient cycling resilience assumes paramount importance for sustaining agricultural productivity, especially amid the challenges posed by changing climate conditions. Agroforestry emerges as a dynamic contributor to sustainable development, offering benefits beyond traditional agricultural practices. One significant aspect is its potential to serve as an alternative energy source. By incorporating energy-yielding tree species, agroforestry contributes to renewable energy production, fostering resilience in regions susceptible to energy shortages or price fluctuations. This diversification of energy sources not only enhances energy security but also aligns with global efforts towards a more sustainable and climate-resilient future. Furthermore, agroforestry systems play a crucial role in enhancing water quality. By harnessing the inherent filtration abilities of trees and vegetation, agroforestry systems play a pivotal role in enhancing water quality. This is especially crucial for protecting water reservoirs from contamination and bolstering their ability to withstand the challenges posed by climate change. Beyond environmental benefits, agroforestry has direct implications for community health and nutrition. By providing a diverse array of tree fruits, nuts, and vegetables, agroforestry contributes to improved diets, enhancing the resilience of communities to health challenges exacerbated by climate stressors. The nutritional diversity offered by agroforestry systems not only addresses immediate food security concerns but also promotes long-term community well-being.

16.4.5 Agroforestry's impact on water management and quality

Agroforestry serves as more than just a farming practice; it becomes a living classroom for education and knowledge transfer. These systems offer valuable opportunities for learning about sustainable land management practices, empowering farmers and communities with the necessary tools to navigate and adapt to changing environmental conditions. Additionally, agroforestry allows for flexible land use planning, where the mix of trees and crops can be adjusted to respond to evolving climate conditions. This adaptability enhances the overall resilience of land management strategies, making agroforestry a cornerstone in building sustainable and climate-resilient communities. Agroforestry assumes a crucial role in environmental conservation, particularly in reducing pressure on natural forests. Agroforestry presents a sustainable and productive alternative to the depletion of natural forests by offering alternative sources of timber, fruits, and various other forest products. This conservation-focused method not only protects natural ecosystems but also enhances biodiversity resilience. It serves as a strategic tool in managing the delicate equilibrium between human requirements and environmental conservation, thereby ensuring the enduring health and sustainability of ecosystems. In floodplain and coastal areas, agroforestry becomes a key player in protection and resilience. The intentional planting of trees in these vulnerable zones serves as a natural defense mechanism. Trees absorb excess water, thereby preventing flooding and erosion, ultimately providing resilience to communities living in these areas. This dual functionality of agroforestry, both in conserving natural environments and protecting communities from environmental hazards, underscores its importance in fostering sustainable and climate-resilient landscapes.

16.5 Agroforestry as a catalyst for financial independence and gender equity

It's widely acknowledged that women face disadvantages in society compared to men. However, to tackle these disparities effectively, development experts need to grasp the social dynamics between men and women, as emphasized by Kiptot (2015). Agroforestry practices contribute to the economic empowerment of women. By involving the cultivation of non-timber forest products, agroforestry creates economic opportunities for women, fostering their financial independence. This economic empowerment not only enhances the resilience of women to climate-related challenges but also plays a pivotal role in promoting gender equity.

16.6. Conclusion

Agroforestry systems play a pivotal role in climate resilience through various interconnected mechanisms. Trees within these systems provide shade, offering crucial temperature regulation that mitigates extremes, particularly benefiting heat-sensitive crops in regions prone to high temperatures. Moreover, agroforestry practices embrace indigenous and local knowledge in land management, contributing to community resilience by adapting to changing climate conditions while preserving cultural heritage. The habitat created by agroforestry systems fosters beneficial soil microorganisms, promoting soil health and resilience. This, in turn, supports nutrient cycling and enhances plant resilience to climate stressors. Aligned with agroecological principles, agroforestry emphasizes ecological sustainability, integrating processes that promote long-term resilience in agricultural systems. Furthermore, these systems act as a buffer for food security during crises, ensuring sustained production and availability through the diversity of products and adaptable land use practices. Agroforestry goes beyond agricultural benefits, incorporating the cultivation of medicinal plants that enhance community health resilience by providing natural remedies for climate-related health challenges. Practices like contour planting and windbreaks contribute to erosion control, maintaining soil stability and preventing land degradation, ensuring the overall resilience of agricultural landscapes. Importantly, agroforestry fosters community networks focused on climate adaptation, creating shared experiences and collaborative management of agroforestry systems that contribute to collective resilience-building efforts within communities. In essence, agroforestry emerges as a multifaceted and integrative approach that addresses climate challenges while promoting sustainability, resilience, and community well-being. These additional roles showcase the breadth of agroforestry's contributions to climate resilience, emphasizing its capacity to address a wide range of environmental, social, and economic challenges associated with a changing climate. These roles underscore the comprehensive and interconnected contributions of agroforestry to climate resilience, spanning ecological, social, and economic dimensions. Research should support tree domestication to improve the yields of tree foods, and seek to enhance the complementarity and stability of food production in smallholders' agroforestry systems.

References

Bisht, S., Kumar, A., Kumar, N., Singh, H. and Kumar, P. 2022. Biofuel Production by Using Biomass and Its Application. In book: Renewable Energy and Green Technology in CRC Press.

Calatrava, J. and Sayadi, S. 2019. Evolution of farming systems in the Mediterranean high mountain: the case of the Alpujarra alta (Spain). *Sustainability*, *11*(3), 704.

Challinor, A.J., Watson, J., Lobell, D.B., Howden, S.M., Smith, D.R. and Chhetri, N. 2014. A meta-analysis of crop yield under climate change and adaptation. *Nature Climate Change*, *4*(4), 287-291.

Defries, R. and Rosenzweig, C. 2010. Toward a whole-landscape approach for sustainable land use in the tropics. *Proceedings of the National Academy of Sciences*, *107*(46), 19627-19632.

Dhyani, S.K., Kumar, R.V. and Ahlawat, S.P. 2011. *Jatropha curcas*: a potential biodiesel crop and its current R & D status. *Indian Journal of Agricultural Sciences*, *81*(4), 295-308.

Elias, E. 2002. Farmers' perceptions of oil fertility changes and management. Institute for Sustainable Development.

Food and Agriculture Organization. 2010. Global Forest Resources Assessment 2010, FAO Forestry Paper 163.

Frison, E.A., Cherfas, J. and Hodgkin, T. 2011. Agricultural biodiversity is essential for a sustainable improvement in food and nutrition security. *Sustainability*, *3*(1), 238-253.

Galhena, D.H., Freed, R. and Maredia, K.M. 2013. Home gardens: a promising approach to enhance household food security and wellbeing. *Agriculture* and *Food Security*, *2*, 1-13.

Gössling, S., Scott, D. and Hall, C.M. 2020. Pandemics, tourism and global change: a rapid assessment of COVID-19. *Journal of Sustainable Tourism*, *29*(1), 1-20.

Jamnadass, R., Dawson, I.K., Anegbeh, P., Asaah, E., Atangana, A., Cordeiro, N.J., & Simons, T. 2010. Allanblackia, a new tree crop in Africa for the global food industry: market development, smallholder cultivation and biodiversity management. *Forests, Trees and Livelihoods*, *19*(3), 251-268.

Jose, S. 2012. Agroforestry for conserving and enhancing biodiversity. *Agroforestry Systems*, *85*, 1-8.

Kaushal R, Tewari S., Thapliyal SD., Kumar, A., Roy, T., Islam, S., Lepcha, STS., Durai, J. 2021. Build-up of Labile, non-labile carbon fractions under fourteen-year-old bamboo plantations in the Himalayan foothills. Heliyon 7(3):e07850

Kiptot, E. 2015. Gender roles, responsibilities, and spaces: implications for agroforestry research and development in Africa. *International Forestry Review*, *17*(4), 11-21.

Kumar, A., Dwivedi, G. K., Tewari, S., Sah, V. K., Singh, H., Kumar, P., Kumar, N., Kaushal, R. 2020a. Soil organic carbon pools under *Terminalia chebula* Retz. based agroforestry system in Himalayan foothills, India. Current Science (00113891), 118(7).

Kumar, A., Dwivedi, G.K., Tewari, S., Paul, J., Anand, R., Kumar, N., Kumar, P., Singh, H. and Kaushal, R. 2020b. Carbon Mineralization and Inorganic Nitrogen Pools under *Terminalia chebula* Retz.-Based Agroforestry System in Himalayan Foothills, India. Forest Science. 66(5), 634-643

Kumar, A., Kumar, P., Singh, H. and Kumar, N., 2021a. Impact of plant functional traits on near saturated hydraulic conductivity of soil under different forests of Kempty watershed in Garhwal Himalaya. Indian Journal of Soil Conservation, 49(1), pp.38-44.

Kumar, A., Kumar, P., Singh, H., Kumar, N., 2021b. Adaptation and mitigation potential of roadside trees with bio-extraction of heavy metals under vehicle emissions and their impact on physiological traits during seasonal regimes. Urban Forestry & Urban Greening, 58, p.126900.

Kumar, A., Kumar, P., Singh, H., Kumar, N., 2021c. Modulation of Plant Functional Traits Under Essential Plant Nutrients During Seasonal Regime in Natural Forests of Garhwal Himalayas. Plant and Soil 465(1):1-16

Kumar, A., Kumar, P., Singh, H., Kumar, N., 2021d. Relationship of physiological plant functional traits with soil carbon stock in the temperate forest of Garhwal Himalaya. Current Science 120(7): 1064-1068.

Kumar, A., Singh, H., Kumari, G., Bisht, S., Malik, A., Kumar, N., Singh, M., Raturi, A., Barthwal, S., Thakur, A. and Kaushal, R., 2022. Adaptive resilience of roadside trees to vehicular emissions via leaf enzymatic, physiological, and anatomical trait modulations. Environmental Pollution, p.120191.

Kumar, A., Tewari, S., Singh, H., Kumar, P., Bisht, S., 2021e. Biomass accumulation and carbon stock in different agroforestry systems prevalent in the Himalayan foothills. Current Science 120(6), 1083-1088

Kumar, A., Tewari, S., Singh, Indra., Panday, R., Kumar, D. and Anand, R. 2018. Effect of nutrient sources on growth, yield and quality of turmeric under harad (*Terminalia chebula*) based agroforestry system Indian J. of Agroforestry, 20(2). 1-6

Kumar, N., Jeena, N., Kumar, A., Khwairakpam, R., Singh, H. 2021. Comparative response of rice cultivars to elevated air temperature in Bhabar region of Indian Himalaya: Status on Yield attributes. Heliyon 7(6):e07474

Kumar, N., Kumar, A., Jeena, N., Singh, R., and Singh, H. (2020). Factors Influencing Soil Ecosystem and Agricultural Productivity at Higher Altitudes. Microbiological Advancements for Higher Altitude Agro-Ecosystems & Sustainability. Springer. p. 55- 71.

Kumar, N., Singh, H., Giri, K., Kumar, A., Joshi, A., Yadav, S., Singh, R., Bisht, S., Kumari, R., Jeena, N., Khairakpam, R., and Mishra, G. (2024). Physiological and molecular insights into the allelopathic effects on agroecosystems under changing environmental conditions. Physiology and Molecular Biology of Plants. 248. 119255.

Kumar, N., Verma, S., Kumar, A., Dashila, H., Suyal, D.C., Kumari, G., Rawat, S., Jeena, N., Singh, M., Kumar, A., (2022). Molecular insights into stress-responsive genes in the mitigation of environmental stresses. Unravelling Plant-Microbe Synergy, Edition: 1, Publisher: Elsevier

Kumar, P., Tokas, J., Kumar, N., Lal, M. and Singal, H. R. 2018. Climate change consequences and its impact on agriculture and food security. *International Journal of Chemical Studies*, *6*(6), 124-133.

Leakey, R.R. 2010. Agroforestry: a delivery mechanism for multi-functional agriculture. *Handbook on Agroforestry: Management Practices and Environmental Impact*, 461-471.

Maccagnani, B., Veromann, E., Ferrari, R., Boriani, L. and Boecking, O. 2020. Agroecosystem design supports the activity of pollinator networks. *Entomovectoring for Precision Biocontrol and Enhanced Pollination of Crops*, 1-17.

Malézieux, E. 2013. Editorial. Underutilized fruit trees in Africa. Special issue. *Revue Fruits*.

Maliki, R., Cornet, D., Floquet, A. and Sinsin, B. 2012. Agronomic and economic performance of yam-based systems with shrubby and herbaceous legumes adapted by smallholders. *Outlook on AGRICULTURE*, *41*(3), 171-178.

Mbow, C., Rosenzweig, C.E., Barioni, L.G., Benton, T.G., Herrero, M., Krishnapillai, M. and Diouf, A.A. 2020. Food Security (No. GSFC-EDAA-TN78913).

McAdam, J.H., Burgess, P.J., Graves, A.R., Rigueiro-Rodríguez, A. and Mosquera-Losada, M.R. 2009. Classifications and functions of agroforestry systems in Europe. *Agroforestry in Europe: Current Status and Future Prospects*, 21-41.

Meert, H., Van Huylenbroeck, G., Vernimmen, T., Bourgeois, M. and Van Hecke, E. 2005. Farm household survival strategies and diversification on marginal farms. *Journal of Rural Studies*, *21*(1), 81-97.

Nair, P.R. and Toth, G.G. 2016. Measuring agricultural sustainability in agroforestry systems. *Climate Change* and *Multi-Dimensional Sustainability in African Agriculture: Climate Change* and *Sustainability in Agriculture*, 365-394.

Nguyen, Q., Hoang, M.H., Öborn, I. and van Noordwijk, M. 2013. Multipurpose agroforestry as a climate change resiliency option for farmers: an example of local adaptation in Vietnam. *Climatic Change*, *117*(1), 241-257.

Pandey R., Dubey S., Tewari L, Kumar A., Kumar D., Sharma N. and Tewari S.K. 2018. Performance of soybean under poplar based agroforestry system in Terai region. Indian J. of Agroforestry, 20(1), 31-34.

Pandey, D.N. 2007. Multifunctional agroforestry systems in India. *Current Science*, 455-463.

Place, F., Roothaert, R.L., Maina, L., Franzel, S., Sinja, J. and Wanjiku, J. 2009. The impact of fodder trees on milk production and income among smallholder dairy farmers in East Africa and the role of research. *World Agroforestry Centre Occasional Paper*.

Ray, D.K., Mueller, N.D., West, P.C. and Foley, J.A. 2013. Yield trends are insufficient to double global crop production by 2050. *PloS one*, *8*(6), e66428.

Rogé, P., Diarisso, T., Diallo, F., Boiré, Y., Goïta, D., Peter, B., and Snapp, S. 2017. Perennial grain crops in the West Soudanian Savanna of Mali: perspectives from agroecology and gendered spaces. *International Journal of Agricultural Sustainability*, *15*(5), 555-574.

Rosenstock, T.S., Lamanna, C., Chesterman, S., Bell, P., Arslan, A., Richards, M.B., and Zhou, W. 2016. The scientific basis of climate-smart agriculture: A systematic review protocol. *CCAFS Working Paper*.

Sagastuy, M. and Krause, T. 2019. Agroforestry as a biodiversity conservation tool in the atlantic forest? Motivations and limitations for small-scale farmers to implement agroforestry systems in north-eastern Brazil. *Sustainability*, *11*(24), 6932.

Sahu, M.L. and Yogesh, K. 2015. Exploring rainwater harvesting potential of *Terminalia belerica* Roxb. through rainfall partitioning and soil infiltration study. *Journal of Tropical Forestry*, *31*(3), 1-7.

Schmidhuber, J. and Tubiello, F.N. 2007. Global food security under climate change. *Proceedings of the National Academy of Sciences*, *104*(50), 19703-19708.

Sistla, S.A., Roddy, A.B., Williams, N.E., Kramer, D.B., Stevens, K. and Allison, S.D. 2016. Agroforestry practices promote biodiversity and natural resource diversity in Atlantic Nicaragua. *PloS one*, *11*(9), e0162529.

Susila, A.D., Purwoko, B.S., Reyes, M.R. and Palada, M.C. 2011. *Vegetable agroforestry systems in Indonesia* (No. Indonesian TMPEGS SANREM-CRSP 2009). Bogor Agricultural University.

Tewari, S., Tewari, L., Dubey, A., Kumar, A., Kumar, N., Kaushal, R. 2022. Use of the RAPD marker to determine the genetic diversity of various *Dalbergia sissoo* Roxb. (Shisham) genotypes and their evolutionary relationship in Indian subcontinents. Vegetos,

Tewari, S., Tewari, L., Kumar, A., Bisht, S, Kaushal, R. and Kumar, A. 2024. Germination Behaviour of Teak Genotypes and Effect of Different

Treatments on Seed Germination Characteristics in Himalayan Foothills, India. Indian Forestry, 150 (3): 244-254

Tewari, S., Tewari, L., Kumar, A., Kaushal, R. 2021. Performance of *Glycine max* L. varieties under *Tectona grandis* L. and *Dalbergia sissoo* Roxb. based agroforestry system Indian J. of Agroforestry, 20(2) 1-6

Verma, T., Bhardwaj, D.R., Sharma, U., Sharma, P., Kumar, D., Kumar, A. and Kumar, A. 2023. Aroforestry systems in the mid-hills of the north-western Himalaya: A sustainable pathway to improved soil health and climate resilience. Journal of Environmental Management. 348, 119264.

CHAPTER SEVENTEEN

MILLET-BASED AGROFORESTRY: A NATURE-POSITIVE FARMING TO ACHIEVE CLIMATE-RESILIENCE AND FOOD SECURITY

DINESHA S.[1*], AND SHRIDEVI B. TELI[2]

[1]DEPARTMENT OF FORESTRY,
NORTH-EASTERN HILL UNIVERSITY, TURA CAMPUS,
TURA-794002, MEGHALAYA, INDIA

[2]SRI KRISHNADEVARAYA COLLEGE OF HORTICULTURAL SCIENCES, ANANTAPUR-515002, ANDHRA PRADESH, INDIA

Abstract

Globally, population pressure, climate change, geopolitical tensions and inflation have pushed many countries into poverty, food and nutritional insecurity, forced displacement, etc. On the other hand, changing climate is causing unusual rainfall patterns, dry spells, and quality deterioration of crops, insects and pest outbreaks and ultimately reducing crop yield. To achieve food and nutritional security in this changing climate, it is necessary to use these limited resources scientifically and efficiently such as diverse agriculture, use of climate resilient crops, etc. Therefore, climate resilient crops like millets can be incorporated in the agroforestry system to promote efficient resource management and reduce malnutrition, food insecurity, and impact of climate change. Interestingly, empirical evidence showed that the government, farmers and institutions are making efforts for sustainable land use management practices like agroforestry, climate resilient agriculture, natural farming and organic farming. This is due to lower crop failure, higher crop diversity, and efficient site utilization through tree-crop multi strata arrangements. However, in the resource-poor arid and semi-arid regions, it is crucial to select compatible tree species, suitable crop combinations, and efficient competition management techniques to effectively manage soil and water resources. In this chapter, the decision to consider India and Africa was influenced by three key features: presence of some successful MbAF models in both the regions; millet production together accounts for 80 % of global production; and prevalence of malnutrition despite high production. Therefore, this chapter discusses the significance and potential of MbAF in these areas considering the "International Year of Millets-2023", highlighting region-specific successful MbAF models as well as their strengths, weaknesses, and providing recommendations to assist policymakers in taking the necessary actions. The chapter further recommends creating capable MbAF models to improve income and stabilize fragile arid and semi-arid farming conditions.

Keywords: Agroforestry, climate-resilient, food security, millets, nature-positive farming

17.1 Introduction

Millets are traditional grains that have been grown and consumed largely in both Africa and India. Millets are defined as small grains, annual, warm-season cereals that are members of the poaceae family. Due to their hardiness and ability to flourish in adverse weather conditions, millets are widely cultivated under dryland agriculture. These are mainly classified as major, minor and pseudo millets. Sorghum (*Sorghum bicolour*), Pearl millets (*Perisetum glocum*) and finger millet (*Eleusine coracana*) are major, foxtail (*Setaria italica*), little (*Panicumsumatrense*), kodo (*Paspalum scrobiculatum*), proso (*Panicum miliaceum*), barnyard (*Echinochloa esculenta*) and brown top (*Urochloa ramosa*) millets are minor and, amaranth (Amaranthus spp.) and buckwheat (*Fagopyrum esculentum*) are considered as pseudo millets. These are popularly known as "Nutri-cereals" as they contain more minerals, nutrients, and proteins than major staple food crops such as rice and wheat (Fig 1). These are mainly grown in arid and semi-arid climatic conditions of Asia, Africa and some parts of China, Europe, and United States, as they require a meager amount of water, soil fertility, pesticides, and management. India's annual production of millets accounts for around 80 % of all production in Asia and 37.5 % of global production. Among all the African nations, Niger, Sudan, Nigeria, and Mali are the major producers after India (Meena et al. 2021). Sorghum, pearl millet and finger millet are more prevalent in Africa and India. Though millets are nutri cereals, their production has been neglected in most parts of the world. The fall in the global area covered by millets may be attributed to several factors, including the green revolution, changed food habits, secured irrigation facilities, lack of government support and ensured returns from important commercial crops. India's millet production pattern showed a consistent growth with increased productivity in recent years. As a result, India has become the world's top producer and seventh-largest exporter of millets (Meena et al. 2021).

Agroforestry (AF) is a land use practice (sustainable and/or resilient) that integrates perennial large shrubs and trees with crops and livestock on farmlands and rural landscapes to enhance diversity, productivity, and agroecosystem sustainability (Dinesha et al. 2022). The combination of trees, millets, and livestock restores soil health and renews food security through closed nutrient cycling and effective resource utilization. Integration of trees along with crops can withstand adverse weather conditions (Ali et al. 2019; Bado et al. 2021). Currently, AF supplies 65% of India's timber requirements, and this is one of the ways to achieve the 33 % aim of increasing forest or tree cover (Bose 2015). Many studies have suggested AF as a solution to Africa's interconnected food, climate, energy, land, and water-related problems (Bayala et al. 2002; Payne, 2000; Rosenstock et al. 2019). To mitigate climate change, 71 % of African nations pledged to adopt AF and 50 % of the nations promised to plant trees to ecologically restore a total of 1.13 million km^2 of land (Rosenstock et al. 2019). It is advised to adopt eco-friendly farming methods such as agroforestry, climate-resilient agriculture, and organic farming in these areas. To increase the productivity of an agroforestry system (AFS), it is crucial to choose compatible tree species, congruous crop combination, and adopt proper management practices such as pruning, lopping, thinning, and other competition management strategies.

Climate change has caused significant challenges in recent years, reducing crop productivity, favoring the spread of weeds, pests, and other diseases, and altering the nutritional value and composition of crops (Sultan et al. 2013; Adhikari et al. 2015; Rao et al. 2022). Thus, there is a need for climate-resilient farming techniques that may increase agricultural output and withstand changing climatic conditions. Agroforestry and millets can both contribute significantly to the provision of numerous agroecosystem services. It can be scaled out even further by combining both crucial components and growing millets under various AFS. The MbAF can become a climate smart system and a workable option to address the current challenges of food, nutrition, energy, employment, and environmental security by regulating microclimate, protecting natural resources, and modifying hydrology, biogeochemistry, and agrobiodiversity.

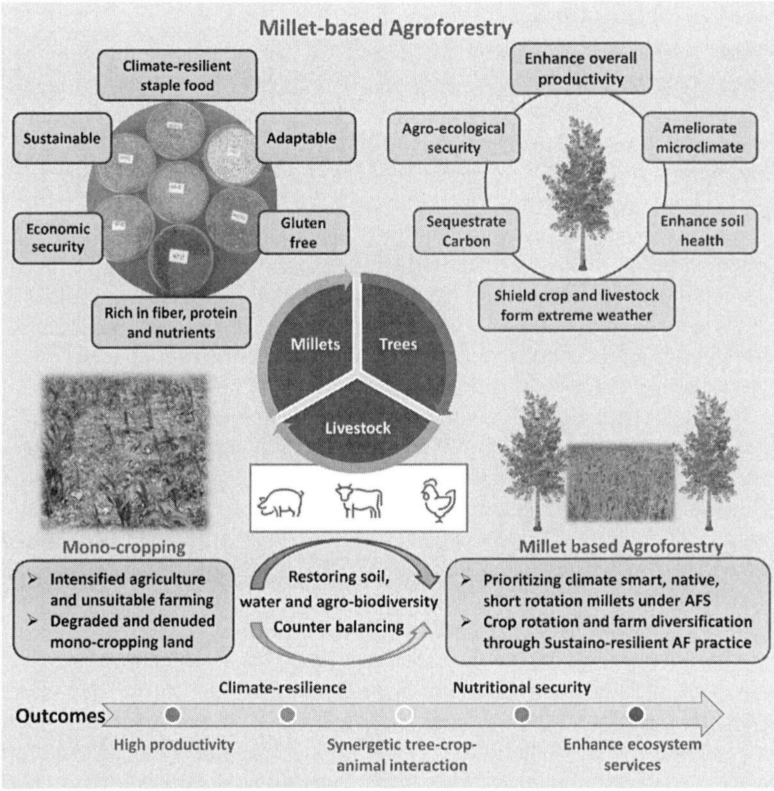

Fig 1: Millet-based Agroforestry system (MbAF) is a climate resilient and sustainable system that combines tree-crop-livestock on the same area of land. Along with its primary economic advantages, MbAF may also contribute to nutritional security and mitigation of climate change.

17.2 The role of MbAF in combating climate change

Millets have evolved special anatomical, physiological, biochemical, and life history traits to survive in adverse and dynamic environments (Pardo and Van Buren 2021). These species may serve as an untapped resource for developing the future generation of climate-resilient crops. According to Gaff and Oliver, evolutionary innovations of millets supported resilience such as C_4 photosynthesis with Kranz anatomy, altered stomata, salt glands, desiccation tolerance, and high water use efficiency, improved leaf water potential and deep root systems. Millets are considered as climate-resilient food crops as they thrive well in harsh climatic conditions, especially in fragile arid and semi-arid environments since their evolution and diversification were greatly influenced by exposure to arid environments. Millets have a less carbon (C) footprint, meager water requirement, and a shorter life cycle compared to the other food crops (Rao et al. 2019). They also produce enough output even in times of unusual rainfall, high temperature, and increased CO_2 levels (Saxena et al. 2018).

Many studies reported that farmers are shifting from conventional monoculture to eco-friendly cultivation techniques like agroforestry, climate resilient agriculture, and organic farming (Nayana and Veni 2020). This is due to their higher intermittent yield, increased production, lower crop failure, higher diversity of livestock and crop components, and efficient site utilization through tree-crop multi strata arrangements (Handa et al. 2019; Nayana and Veni 2020). Some perennial large shrubs and trees are an important component and provide various ecosystem services (Ashwath et al. 2023). For instance, *Gliricidia sepium*and *Pongamia pinnata* as manure species, *Leucaena leucocephala* and *Hardwickia binata* as a fodder yielding species, *Adansonia digitata*, *Zizypus*spp and *Vitellaria paradoxa* as fruit species, *Acacia senegal*and *Vachellia seyal* as gum producing species and *Aegle marmelos* and *A. indica* as medicinal species are providing various ecosystem services (Bayala et al. 2002; Rosenstock et al. 2019; Ashwath et al. 2023). These multipurpose species are more versatile because of their greater compatibility with agricultural crops in AFS and higher economic returns. In addition to these benefits, stressed soil also allows trees to pump nutrients and water up from deeper levels. Trees like *Balanites aegyptica, A. indica,* and *P. reticulatum* contribute to the natural regulation of crop pests (Abasse et al. 2013; Sow et al. 2020). AF practices are greatly recognized as a mitigation strategy for reduction of methane emission from ruminating livestock due to reduction of heat stress and increase in feed quality (Dinesha et al. 2022). A study in the Sahelian savannas of West Africa reported that rising temperatures have a worsening impact on agricultural yields regardless of variations in rainfall (Sultan et al. 2013). Additionally, traditional millet cultivars that are photoperiod-sensitive and have been grown by local farmers for centuries appear to be more resistant to future climate changes than modern cultivars developed for their high yield potential.

17.3 The role of MbAF in food and nutritional security

In arid and semi-arid regions of India, the states like Rajasthan, Maharashtra, and Karnataka produce two third of the country's millet production. In India, pearl millet contributes 61 % of total coarse grain production followed by 27 % from sorghum and 12 % from finger millet (Rao et al. 2019; Das et al. 2020). West Africa is home to large production of millets (Obilana 2003), mainly Nigeria (41%) followed by Niger (16%), Burkina Faso (7%), Mali (6.4%), Senegal and Sudan (4.8%). Among millets, sorghum, pearl millet and finger millet are distributed at global level. However, some minor millets such as fonio, black fonio and guinea, which are native and region specific to West Africa, are cultivated on a small scale. Vedas explained the importance of millets in the human diet as well as animal feed. Millets are a great source of carbohydrates, lipids, minerals, dietary fibers, vitamins, proteins, and other essential nutrients (Saxena et al. 2018; Sabuz et al. 2023). Millets are considered as 'superfoods', which are rich in macro and micronutrients (Fig. 2). High calcium is present in finger millet (364.0mg/100g), magnesium in foxtail millet (153.0mg/100g), iron in pearl millet (6.4mg/100g), phosphorus in sorghum (222.0mg/100g), fibers in foxtail millet (13.6g/100g), and protein (12.5g/100mg) in proso millet (Dhaka et al. 2021; Kadapa et al. 2023).

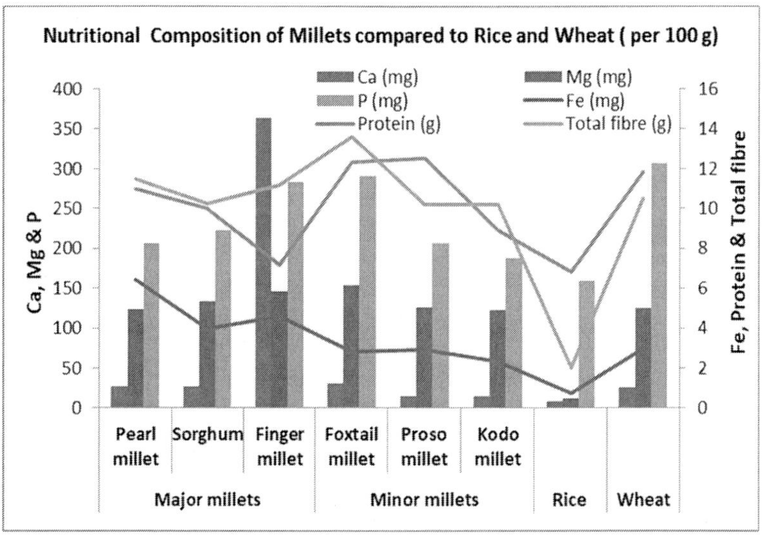

Fig. 2. Nutritional composition of millets compared to rice and wheat (per 100 g)

Millets are a rich source of vitamins like niacin, thiamine, riboflavin, and folic acid, high polyphenols (0.2-0.5%), and minerals (1-5%), compared to other staple foods (Kadapa et al. 2023). They also show various advantages over other cereals such as regulation of gut microbiota, anthropometric effects, antioxidant, anti-hyperglycemic, anti-cholesterol, and anti-hypertensive properties (Dhaka et al. 2021). Minor millets are considered as sacred grains in some rural parts India, where people consume them during fasting or celebration of festivals (Ravula et al. 2022). For instance, popped sorghum, foxtail millet bath, and amaranthus laddu are prevalent in certain parts of Karnataka and Telangana. Unfortunately, in many regions of Africa and India, in terms of area, production and consumption, minor millets have been gradually replaced by commercial staple food crops like rice and wheat. The limited demand for minor and pseudo millets is mostly because of unfamiliarity and a lack of public and scientific awareness, which is typically correlated with a dearth of research and development (Rocha et al. 2021). However, in many parts of the world, millets are used to prepare noodles, soups, hard drinks, pancakes, and cereal porridges due to its nutritional value (Sabuz et al. 2023). In India, millets mix a special drink prepared from multi-millets powder which is used by all aged people. Finger millet ball and sprouted powder is most preferred by diabetic patients as it is a rich source of Ca, low glycemic index, and high dietary fiber (Sarkar et al. 2015). As a result, millets are being given more attention and considered as a complete food.

17.3.1 Biofortification of millets

Globally, hunger and malnutrition have increased even more after a hit by COVID-19 pandemic and geopolitical tensions (Hendriks et al. 2022). The recent World Health Organization (WHO) report stated that 91.62 billion people are anemic, of which 30.2 % are women and 47 % are children under five. According to the global hunger index, two billion people are suffering from hidden hunger and nearly 50 % are from India. Biofortification is an effective tool to tackle hidden hunger as it is a cost effective permanent sustainable solution. Biofortification is defined as enrichment of nutrients in crops through agronomic, plant breeding and biotechnological intervention (De Valenca et al. 2017). Though millets are rich in some micro and macro nutrients, the presence of some antinutritional factors such as tannins and phytates reduce the bioavailability of minerals by chelating multivalent cations. Enhancement of nutrients can be done by two major strategies either by boosting nutritional level or by reducing antinutritional factors in millets. Biofortification has already been done with major millets such as sorghum, pearl millet and finger millet. They are also released under public domain by ICRISAT and Harvest Plus. These crops are biofortified for three common nutrients such as Fe, Zn and Vitamins. ICSR14001 is sorghum variety biofortified with Fe, whereas Dhanashakti hybrid is biofortified finger millet variety for Fe and Zn (Kadapa et al. 2023).

17.3.2 Biotechnological interventions in millet improvement

Biotechnological interventions such as genomics, proteomics, marker assisted selection (MAS), genome editing, genome sequencing, gene silencing, genome-wide association selection (GWAS), micropropagation, somaclonal variation and genetic engineering are playing vital role in millet improvement (Dhaka et al. 2021). Genetic engineering and MAS were used to improve nutrient content of millets, especially in pearl millet and sorghum. Mineral rich variety of finger millet 'Dapoli-2' was developed through somaclonal variation, which is rich in Fe and Ca and tolerant to blast (Sawardekar et al. 2017). Another study in finger millet reported that callus induction and shoot regeneration frequency was shown positive correlation with Ca content and negative correlation with protein and flavonoid content (Jamra et al. 2022). Genomic tools and MAS were used to identify Fe -Zn QTL location in pearl millet, and improved HHB-67 has been released. Draft genomic sequences of foxtail millet, pearl millet and finger millet can provide excellent sources for genetic improvement in these crops. Foxtail mosaic virus-based gene delivery system has been developed for functional characterization of genes using virus induced gene silencing approach which is used to silence expression of candidate genes in cereals (Dhaka et al. 2021). Genome wide association mapping for morphological characters was also available in finger millet and pearl millet (Sharma et al. 2018). Based on these observations, it is concluded that there is limited research on RNA interference, genome editing techniques such as clustered regularly interspaced short palindromic repeats (CRISPR), zinc-finger nuclease (ZFN) and transcription activation like effector nuclease (TALENS).

17.4 The role of MbAF in restoring agroecology

Millet based Agroforestry (MbAF) enhances microclimate and water table, increases agro-biodiversity, reduces soil and water erosion, improves soil flora and fauna, renews nitrogen and carbon, efficiently pumps water and nutrients at the time of scarcity (Singh 2003; Ceperley et al. 2012; Bright 2017). MbAF system uses resources efficiently and maintains a closed nutrient cycle, protects cattle and farmland from harsh weather, reduces CO_2, methane, and other greenhouse gas (GHG) emissions while providing various kinds of ecosystem services (Fig.2).

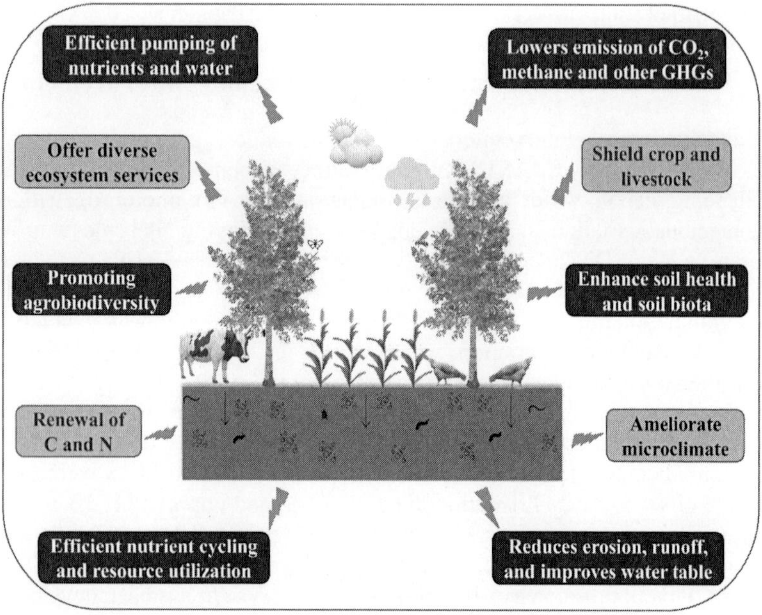

Fig. 2: Role of MbAF in restoring agroecology

The crop choice is dependent on soil quality, available irrigational facility, availability of seed and fertilizer, and rainfall rate (Bose 2015). The choice of a crop is often risky, but people in East Africa chose to grow maize because of the higher price and yield they could gain in a good year, even though the crop was more susceptible to drought and did not improve soil health (Dudley 2011). Some farmers are traditionally cultivating millets under trees due to their high nutritional value, drought tolerance, and ease of growth and to reclaim semi-arid or resource-poor land (Bose, 2015; Pardo and VanBuren 2021). Many exotic species such as Eucalyptus, Prosopis, Acacia, and Colophospermum have been tested for their abilities to improve degraded areas and increase production. *Z. mauritiana* and *P. cineraria* based AFS could be practiced for six years without significant reduction in yield. This model provides economic security and soil organic C was increased 8% in top layer and 13% in lower layers of AFS plots compared to mono-cropping (Ali et al. 2022). Traditional MbAF practices rely mainly on renewable resources, and they can be considered as a suitable way to increase productivity and minimize crop failure. For instance, *P. reticulatum* does not compete with millets for limited water, as evidenced by a study in West African Sahel (WAS) that found shrubs use water from the previous rainy season for the growth of the next year (Bright 2017). The MbAF is affordable and compatible with the low input farming systems, and it ensures sustainable management of soil, land resources, and ecosystem services (Bado et al. 2021).

17.5 Case Studies

Indian context

The Central Agroforestry Research Institute (CAFRI), Jhansi, and the South Asia Regional Program (New Delhi) of International Centre for Research in Agroforestry (ICRAF) collaboratively published successful AF models for several agro ecological zones in India (Handa et al. 2019). In this chapter successful MbAFs were recorded in five of India's 20 agro ecological zones (Table 1). MbAF such as foxtail millet under *M. azedarach* in Western Plain, Kutch and Part of Kathiawar Peninsula (AER-2), finger millet under *M.dubea* in Deccan Plateau-Hot arid (AER-3), sorghum and groundnut under teak and sorghum and cowpea under mango in Deccan Plateau-Hot semi-arid (AER-6), sorghum and pearl millet under poplar in Northern Plains (AER-9), and finger millet and proso millet under bamboo in Western Ghats and Coastal Plains (AER-19) were highly successful and more prevalent (Handa et al. 2019). Another report compiled by ICFRE recorded successful MbAF for instance sorghum and pearl millet under poplar in gangetic plain of north India, millet-based *C. mopane* AFS in Rajasthan and Gujarat, and finger millet-based *M. dubia* in Southern peninsula. In Karnataka state, several studies recommended *M. dubia* tree components in various AFS; this is due to its versatile characteristics such as higher compatibility with crops, sparse crown, nitrogen fixing, fast growing, enhanced soil fertility, high C sequestration and wider adaptability to diverse agro climatic conditions (Bhaskar et al. 2019; Kulkarni et al. 2020). Tamarind based brown top millet cultivation is prevalent in the dryland tracts of southern India (Maitra 2020). Trees like *P. cineraria* and *A. exelsa*, *A. senegal, Hardwickia binata and Colophospermum mopane are* widely cultivated with millet crops in Gujarat, Rajasthan, and Madhya Pradesh States (Dagar 2007). These trees are a boon to the arid and semi-arid regions, where they adapt to dry conditions and can be used as livestock fodder, a windbreak in the desert, and to improve nutritional security.In contrast, the productivity of pearl millet and cluster beans grown under five-year-old *A. indica* decreased because of the shading effect and bird damage caused by a dense tree canopy (Meena and Nagar 2018).

Table 1. Distribution of MbAF systems in India context

State	Tree-Crop components	Outcomes	Economic return	Reference
Karnataka and Maharashtra	Sorghum and Groundnut under Teak	Intercropping helped in soil moisture conservation	Higher economic return	Handa et al. 2019
Andhra Pradesh, Karnataka, and Telangana	Finger millet and *Meliadubia*at 8 m X 5 m spacing	Intercropping is advantageous in the initial four years	High net return (Rs. 4,44,169) and B: C ratio (18.64).	Bhaskar et al. 2019; Handa et al. 2019
Karnataka	Foxtail millet+*Melia dubia* at 8 m X 5 m	Enhanced soil health	Higher economic return	Kulkarni et al. 2020
Mizoram	Sorghum + *Meliadubia* at 2 m x 2 m	Minimized erosion in hills	Both ecologically and economically viable	Malik et al. 2009
Uttarakhand	Finger millet and barnyard millet under *G. oppositifolia*	Contributed more C sequestration in lower elevation	Improved return	Marital et al. 2022
Haryana	Pearl millet and *P. cinararia*	*P. cineraria* improves soil health	Increased yield	Bishnoi and Singh 2009
Rajasthan	Compared three treesfor pearl millet intercropping	Pearl millet under *Z. mauritiana*	Comparatively Ziziphus based AFS	Singh 2003

		was highly successful	recorded higher yield	
Karnataka	Finger millet and *G. sepium*	Increased microbial biomass in Gliricidia-based AFS	Increased yield of finger millet under Gliricidia.	Balakrishna et al. 2017
Haryana, Himachal Pradesh, and Uttarakhand	Poplar-based sorghum and pearl millet AFS	Performed better compared to sole cropping.	High economic return (2 lakh/ha/yr)	Handa et al. 2019
Uttarakhand	Poplar+wheat-pearl millet	Under poplar, wheat, and pearl millet more suitable	Economically viable system (B:C ratio is 2.54:1)	Charan 2017
Gujarat	Compared biomass production among bamboo, sorghum, and pearl millet	Bamboo has highest biomass production, and highest net energy balance.	Net income (bamboo> sorghum> pearl millet)	Patel et al. 2020
Maharashtra	Finger millet and proso millet under bamboo	Traditionally followed AFS	3.99 of B:C ratio	Handa et al. 2019
Gujrat, Madhya Pradesh,and Rajasthan	Foxtail millet under *Melia azedarach*	Improved income and livelihood	0.50 lakh/ha/yr	Vikrant et al. 2009; Handa et al. 2019

African context

In the West African Saheli (WAS), farmers adopted MbAF to diversify their agriculture and income along with minimizing risks, improving soil fertility, and reducing soil and wind erosion (Payne 2000; Bado et al. 2021). Integration of multipurpose trees like *Faidherbia albida*, *Adansonia digitata*, *Acacia senegal*, and *Ziziphus mauritiana* in MbAF has improved food and nutritional security and native, fast growing, highly economical and more tolerant of dry spells. In addition, these species have peculiar morphological and physiological traits that make them perfect for widespread MbAF throughout the region (Mokgolodi et al. 2011). For example, Baobab and Ziziphus-based AFS is widely practiced in WAS and was reported to have high demand and growing market for the Baobab and Ziziphus fruits (Bado et al. 2021). Integrating millets and an economic tree species through improved AFS could offer an opportunity to enhance the productivity of traditional farming systems in WAS (Payne 2000). The diversity of products from both crop and tree components would provide farmers with diversified sources of income round the year, for instance, the *Acacia senegal* provides both fodder and arabic gum while millets provide food and fodder (Ali et al. 2019).

In Burkina Faso, millet output was increased when intercropped with pruned *Vitellaria paradoxa* and *Parkia biglobosa* tree crowns in agroforestry (Bayala et al. 2002). This is due to crown pruning, which reduced transpiration rates and permitted more photosynthetically active radiation transmission below the crowns. However, long term effects may depend on the ability of the trees to maintain soil fertility and quick recovery from pruning. *F. albida* has various qualities that make it ideal for agroforestry. Therefore, farmers often retain it in their fields of sorghum and finger millet in countries like Mali, Senegal, Sudan, and Cameroon (Mokglodi et al., 2011; Payne, 2000).*Piliostigma reticulatum* and *Guiera senegalensis* based millet cultivation are prevalent in WAS. Studies recommended that millet intercropping with *P. reticulatum* is an ecological AFS for the Sahel that can remediate soils, increase crop yields, and buffer climate change (Bright 2017). The flowers of *P. reticulatum* have a repulsive effect on the flower beetles, which helped in keeping pests away and producing higher yield of millet in the field with *P. reticulatum* (Abasse et al. 2013).

Table 2. Distribution of MbAF systems in African context

Country	Tree-Crop components	Outcomes	Yield	Economics	Reference
Niger, Nigeria, Mali	Pearl millet + *Faidherbia albida*	Recommended MbAF for higher	Increase in growth and yield	Improved economic return	Payne 2000

and Senegal		millet production			
Senegal	Pearl millet-based*P. reticulatum*or *G. senegalensis*AFS	Millet grown under 1000-1500 shrubs ha^{-1}	Most prevalent MBAF in Senegal	Doubled millet yield for long term and high return	Bright 2017
Niger	Pearl millet + cowpea + Ziziphus	Low input AFS improved productivity and income	41% increased yield in ziziphus AFS compared to sole crop	Income rose by two to three times	Bado et al. 2021
Niger	Pearl millet + *A. senegal*	Increased C and overall soil fertility	Higher yield compared to monocrop	Improved economic return	Ali et al. 2019
Ethiopia, Zambia, and Malawi	Pearl millet + *F. albida*	Restored dry areas' soil fertility	Increased yield by 2.4 to 3.6-fold	Improved fertility and economic return	Mokgolodi et al. 2011
Niger	Pearl millet+ *P. reticulatum*	Flower feeding insects control	Significant increase in yield	Improved economic return	Abasse et al. 2013
Sudan	Parkland AFS + pearl millet +Sorghum	Under Parkland, advised to intercrop pearl millet, sorghum, and sesame	Highest yield achieved in Parkland based AFS than mono-cropping	Improved food security and high economic return	Fahmi et al. 2018

Burkin a Faso	*V. paradoxa* and *P. biglobosa* Parkland + Millet	Tree canopy pruning crucial for high production	Higher millet yield and biomass producti on	Improved soil health and other ecosystem services	Bayala et al. 2002
Burkin a Faso	Pearl millet + *Sclerocaryabirr ea*	High resource utilization	Prevalent MbAF	Improved yield and agrobiodivers ity	Ceperley et al. 2012

17.5 Comparison of MbAF in the context of India and Africa

Globally, many small and marginal farmers grow millets in the arid rainfed areas as a subsistence crop for local consumption. Commercial millet production is considered risky due to absence of large market outlets which create price fluctuations during bumper crops. Apart from grain production, millets are also cultivated for grazing, green fodder, and silage making. Some popular landraces in India are over three meters tall, produce higher fodder yield and less grain yield. Compared to India, African countries have limited application of improved technologies. Crops are usually grown without irrigation or chemical fertilizer, on light and well-drained soils that are poor in organic matter content. When irrigation is available, farmers prefer to cultivate more remunerative crops, especially maize (Kurukulasuriya and Mendelsohn 2008). There is seasonally high demand for pearl millet crop residues as fodder for milch animals in India. Short-duration millet cultivars are also grown under irrigation, before or after higher-value crops cultivation in the regions, where the season permits double cropping.

A study conducted in Zimbabwe on household decision-making regarding crop selection in dryland agriculture reported that millets are a low-risk crop while maize is a high-risk crop (Milner-Gulland et al. 1996). Various social, economic and biophysical factors are driving forces on changes in land use agropastoral systems in Saheli. A study conducted by Rasmussen et al reported that rapid events such as a drop in millet prices or a halt in circular migration had a higher impact on the system than other longer-term changes, such as increased rainfall variability. *A. Senegal* and *Ziziphus* spp. are more prevalent in both India and Africa. Some countries like Burundi, Ethiopia and Tanzania in sub-Saharan Africa are faced with acute food shortages. Therefore, Mokgolodi et al. Suggested that the development of environmentally and socially responsive agricultural systems, mainly MbAF, is vital to increase food and nutritional security in sub-Saharan Africa.

17.6 Economics of MbAF

The economic benefits of millets are obvious; these are cheaper to grow as they are hardy, drought resistant, less water intensive and there is no need for expensive fertilizers and pesticides. Their short duration helps farmers to take multiple crops in a year to get year-round income. The changing climate has increased crop failures due to erratic rainfall and drought. Hence, these millets act as an insurer to these adverse climatic conditions (Adhikari et al. 2015). India produces more than 50 million tons of millets, which accounts for 80 % of Asia's and 20 % of global production. India's millet productivity (1239 kg/ha) is higher than the global productivity (1229 kg/ha).

Though India is placed at a top position in millet production at global level, it exported only 1 % of its production to other countries in the financial year 2021 (APEDA). Canada, Russia, Ukraine, and USA are top importer of raw millets and exporter of value-added products. India is lacking its competition in value addition of millets. APEDA is leading the efforts to ramp up exports through special export strategy programs and policies with an export target of $100 million by 2025. MbAF is more economical compared to sole cropping (Table 1 and 2). Some studies reported that the MbAF system increased two- or three-fold higher income than income carned from a sole crop (Bright 2017; Bado et al. 2021). For instance, finger millet-based *M. dubia* showed a B:C ratio of 18.64 (Bhaskar et al. 2019), finger millet with bamboo resulted 3.99:1 (Handa et al. 2019) and higher B:C ratio (2.54:1) was also seen in pearl millet-based poplar AFS (Charan 2017). Hence, MbAF is a viable option to fulfill government ambition of doubling and diversifying farmers income.

17.7 Programs and policy support

Government of India promoting millets through increase in MSP of Millets, Initiative for Nutritional Security through Intensive Millet Promotion (INSIMP) and other programs and policies. It declared 2018 as the "National Year of Millets" and launched the Production Linked Incentive (PLI) scheme in 2021 to increase millet-based products. It also earmarked some budgetary outlay to promote millets as "Shri Anna" and encourage research and development through Indian Institute of Millet Research (IIMR). The government has introduced provision of seed kits and inputs to farmers, building value chains through Farmers Producer Organizations (FPO) and supporting the marketability of millets. IIMR has been recognized as the 'Centre of Excellence in millets' for research and development and it has also incubated 250 startups. NITI ayog, a government think-tank has launched four innovation challenges through Atal Innovation Mission such as processing efficiency, decentralized processing, marketing, and shelf-life augmentation to bridge the gap of supply chain, increase production and awareness.

India also introduced "National Agroforestry Strategy -2014" to promote AF at nation-wide. It is also promoting AF through the Sub-Mission on AF (SMAF) under National Mission for Sustainable Agriculture (NMSA). The Har Medh Par Pedh program encourages integrating trees in the bunds and boundaries of agricultural farms, canal banks and riverbanks (Manasa et al. 2022). In addition, the government initiative "Green India Mission" includes AF as a solution for different challenges in Indian agriculture (Dinesha et al. 2023). Improved germplasm and package of practices (POP) of successful Agroforestry models have been developed by CAFRI and ICFRE, which are being made available to the farmers, state forest departments and many other stakeholders. While in Africa, agroforestry is promoted through various initiatives at national level. For instance, "Kenya National Agroforestry Strategy" to scale up agroforestry in Kenya through policy mainstreaming and coordination, knowledge creation and management, and strengthening of value chains and incentives systems while embedding social inclusion. In addition, Harnessing Opportunities for Productivity Enhancement (HOPE) is an initiative supported by Bill and Melinda Gates Foundation, which is led by ICRISAT with 50 partner countries to increase millet production. The project provided improved crop varieties of millets, farmers' access to input markets and the ability to sell their produce at competitive market prices in some of the most resource-poor areas of sub-Saharan Africa and India.

17.8 MbAF: Strengths, weaknesses, and recommendations

Strengths, weaknesses, and recommendations are listed considering the earlier studies (Table 3). Millet based Agroforestry (MbAF) can be a sustainable farming practice in drier regions of Asia and Africa. As it is cost effective, integrates benefits of both millets and AF, and increases higher net returns. If we consider food security, MbAF is a viable option as millets and underutilized fruits are repositories of nutrients, vitamins, and proteins. It is also an instrument to reduce poverty in resource poor areas in India and Africa.

Table 3. Strengths, weaknesses, and recommendations for MbAF

Strengths	Weaknesses	Recommendations
Acts as an instrument to reduce poverty in resource-poor areas of Africa and India.	Lack of self-help organizations or businesses to institutionalize market operations at the local level	Encouraging value addition of millets and minor tree produce at local level with the help of SHGs, FPO's and cooperatives to boost local employment, production and alleviate poverty.
Steady rise in millet consumption due to their health benefits	Millet products are still seen by consumers differently from similar products derived from rice and wheat.	To increase consumption of millets, it is necessary to target millet's marketability with palatability based on taste preferences among people
Reduces cost of production, increases high net returns and per capita income.	Some tree-crop combinations are incompatible due to their competition for nutrients, water, and light. In addition, market fluctuations are a major concern. There is a lack of specific economic links and marketing strategies.	Providing MSP, subsidies for quality planting material, increasing market access and export, encouraging use of improved technologies, and connecting forward and backward linkages. Providing institutional credit and insurance cover for MbAF is needed.
It avoids crop failure. Increase C sequestration, maintain soil fertility and groundwater.	More emphasis on economy rather than agroecology and biodiversity in commercial millet production	AF needs to be covered in Ecosystem services accounting. Encouraging use of leguminous, fast growing, multipurpose compatible tree-crop combinations.

Gives nutritional and food security	Indian as well as African population, especially the women and children are facing malnutrition and other problems	Creating awareness, prioritizing both major and minor millets in public distribution systems, institutions, and mid-day meal schemes in schools on a large-scale basis to address malnutrition problem.
Acts as a source for resistant genes for various biotic and abiotic stresses. Hardy nature can withstand drought, pest, and diseases	Lack of targeted conservation cum commercialization intervention programs	Requires successful collaboration among expertise from diverse disciplines such as breeders, food chemists and food industry partners. Conservation of agrobiodiversity through ex situ and in situ conservation methods.
Traditional AFS already existed in some parts of Africa and India.	Lack of quality planting material, lack of proper infrastructure and market fluctuations are major drawbacks to grow in large-scale	Integration of livestock, research on best tree-crop combinations, educating farmers to adopt successful MbAF
Rising millets Production due to increased awareness	Unavailability of improved millet varieties, dominance of wheat and rice and still production is limited to dryland agriculture	Further research is needed on developing improved varieties and hybrids. Providing incentives to grow millets and educating farmers regarding food and nutritional security.
India and Africa are the major millets (nutri-cereals) producers	Lack of fully developed grades and standards for the selling of millets. Nutritional claims of millet haven't been fully validated.	Minimum standards and grades of millet need to be set by Food Safety Institutes. There is a need of extensive consumer surveys to evaluate how consumers perceive millets.
Many native communities engaged in small-	Following subsistence farming using age old practices	Mainstreaming of these communities under various government initiatives. Encouraging them to use

scale millet farming		improved techniques, conservation of native species and value addition of millet and tree products.
Trees like soapberry and neem contribute to the natural regulation of crop pests (Sow et al. 2020)	Baobab trees reduced millet yield by providing refuge for disease-carrying insects	Further research is needed to understand trees' role in managing habitat and preventing pests naturally.

Millets are hardy, less input intensive, resistant to pests and diseases and require minimal water, hence, reduce cost of production and act as insurance to crop failure. Therefore, integration of drought resistant trees and millets can also improve soil fertility, reduce land degradation, and increase C sequestration in soil. Some wild trees and millets are rich sources of resistant genes for biotic and abiotic stresses. Hence, they can be used as model plants or trees to study various stress resistant mechanisms. Both Africa and India have also established many traditional AF systems, and millets are also cultivated as sole and intercrop. It gives an opportunity to check compatibility of tree-crop combinations. Apart from ecological and food security benefits, MbAF can also generate employment opportunities in rural areas through awareness creation, value addition of millets and minor fruits or tree produce. In addition, some trees also regulate pests and diseases in crops.

Still there are many constraints that exist in promotion of MbAF. Millets are neglected in some parts of the world due to some socio-ecological conditions and they are considered a crop of poor farmers and marginal land. After the green revolution, consumer preference has shifted towards rice and wheat. Hence, there is more attention towards the economy rather than ecology. Lack of infrastructure, fair price, consumer preference, research on tree-crop combination and availability of quality planting material are the major concerns in promotion of MbAF. In addition, there is a lack of specific economic links and marketing strategies. Absence of value-added products, proper grades and standards are also a major concern in marketing. Apart from these, the African and Indian population especially women and children are also facing malnutrition and other challenges. These above challenges can be solved by taking some policy measures at global and national level.

There is a need to create awareness, prioritizing millets in PDS, institutions and mid-day meal schemes on a large-scale basis to address malnutrition problems. Millets' consumption can be increased by taking necessary steps to target millets' marketability with palatability based on taste preferences among people. Minimum standards and grades of millets need to be set by national and international related agencies. There is a need for coordinated millet research initiatives at the regional and state level like MIIRA (Millet International Initiative for Research and Awareness) at global level and IIMR at national level. Providing institutional credit and insurance cover for MbAF is needed. There is a need for a price support system for all millets including minor millets. Improving farmers' access to quality seed and planting material and further strengthening farmer access to markets for millets and tree products can enhance production. There is a need of synergizing national level policies, programs and institutions related to millets and AF to give boost to MbAF. Apart from this, encouraging value addition of millets and minor tree produce at local level with the help of SHGs, FPO's and cooperatives to boost local employment, production and alleviate poverty.

Conclusion

It is observed that tree-based millet production is more prevalent in India and African countries as compared to other countries. This is due to suitable agroecological conditions (arid and semi-arid), consumption and wider millet distribution. Among the millets, only pearl millet, finger millet and sorghum dominate in the food basket and there is less research on minor millets in the global context. India is far ahead in usage of improved technologies in millets production compared to Africa. Lack of quality seeds distribution, proper infrastructure and minimum support price are causing hindrance in productivity of millets. Even in India, the minimum support price is limited to major millets. Minor millets are not yet covered under the support price system. Hence, mainstreaming of millet's value chain is necessary in both India and Africa. More emphasis should be given to research and development to increase production and productivity.

In India, sole cropping of millet production is more prominent compared to agroforestry or tree-based cultivation practices. Hence, there is a need for further research and development in MbAF practices to explore their potential. Comparatively less yield reported in some AFS is due to accommodation of less plant population in comparison to sole crop. However, net returns (crop yield+tree yield) from MbAF will be higher than sole cropping. Tree crop interaction and architecture management is very important. Over the years the tree-crop interaction may turn negative if we don't address above ground and below ground competition. In some AFS crops can be grown in the initial period of up to tree canopy closure. Hence, knowledge and practice of agroforestry, agronomy and silviculture is very important to tackle these problems. In arid regions trees are mainly grown for fodder and fuel purposes. It would be better if they integrate fast growing, short rotation, multipurpose improved trees to meet nutritional security along with fuel, fodder, timber requirements. Ecosystem services accounting is crucial in agroforestry since many additional ecosystem services such as cultural, regulatory, and supporting services can be obtained in addition to provisional services. These MbAF practices also support various nations' efforts to achieve the land degradation neutrality (LDN) and Sustainable Development Goals (SDGs) of combating climate change, ending hunger and poverty, promoting decent employment and economic growth, and ensuring food security. MbAF is not only a solution for nutritional security and poverty alleviation in arid and semi-arid regions, but also an economically and ecologically viable option for climate-resilience. In conclusion, this India-led "International Year of Millets-2023" will shower some light on innovation and research in strengthening the value addition system and become a global hub in production as well as export.

References

Abasse T, Yayé A, Habou ZA, Adamou AI, Adam T. 2013. Influence of *Piliostigma reticulatum* on the infestation of millet plants by flower-feeding insects and *Coniestaigne fusalis* (Hmps) (*Lepidoptera: Pyralidae*) in agroforestry systems in the Aguié area in Niger. Journal of Applied Biosciences 66: 5140-5146

Adhikari U, Nejadhashemi AP, Woznicki SA. 2015. Climate change and eastern Africa: a review of impact on major crops. Food and Energy Security 4(2), 110-132

Ali, A, Aichatou, A, Maman, M. A, Issoufou, H. B. A, Djibo, E. S, Zoubeirou, M. A. 2019. Improvement of the productivity of millet (*Pennisetum glaucum* (L.) R. Br.) Intercropped with the Arabic gum tree (*Acacia senegal* (L.) Willd.) in agroforestry parkland in Niger. Advances in Agricultural Sciences 7: 74-84

Ashwath MN, Bodiga D, Wagmare B, Dechamma DNL, Hosur SR, Toushif PK, Dinesha S. 2023. Threatened and Near Threatened Underutilized

Edible Fruit Species of Southern India for Food Security and Diversifying Agroecology. Indian Journal of Ecology 50: 19-31

Bado BV, Whitbread A, Manzo MLS. 2021 Improving agricultural productivity using agroforestry systems: Performance of millet, cowpea, and ziziphus-based cropping systems in West Africa Sahel. Agriculture, Ecosystem and Environment 305: 107175

Balakrishna AN, Lakshmipathy R, Bagyaraj DJ, Ashwin R. 2017. Influence of alley copping system on AM fungi, microbial biomass C and yield of finger millet, peanut, and pigeon pea. Agroforestry Systems 91: 487-493

Bayala J, Teklehaimanot Z, Ouedraogo SJ. 2002. Millet production under pruned tree crowns in a parkland system in Burkina Faso. Agroforestry Systems 54: 203-214

Bhaskar V, Hanumanthappa DC, Bhavya NV, Somashekar KS. 2019. Growth, yield, and economics of finger millet (*Eleusine coracana*) in *Melia dubia*-based agroforestry system. International Journal of Current Microbiology and Applied Sciences 8: 1945-1950

Bose P. 2015. India's drylands agroforestry: a ten-year analysis of gender and social diversity, tenure, and climate variability. International Forestry Review 17: 85-98

Bright MBH. 2017. The Role of Shrub Agroforestry Systems in Increasing Food Security for the West African Sahel. Doctoral dissertation, The Ohio State University.

Ceperley N, Repetti A, Parlange M. 2012. Application of soil moisture model to Marula (*Sclerocaryabirrea*): Millet (*Pennisetum glaucum*) *agroforestry system in Burkina Faso.* In: Bolay JC, Schmid M, Tejada G, Hazboun E (eds) Technologies and Innovations for Development. Springer, Paris, France, p. 211

Charan S. 2017. Economics of poplar-based agroforestry models adopted by the farmers in Haridwar, Uttarakhand. Indian Journal of Ecology 44: 804-811

Dagar JC. 2007. Alternate Land Uses in Areas with Water Scarcity. In: On-farm land and water management, p. 160

Das B, Singh P, Sangeetha V, Bhowmik A, Ray P. 2020. Growth and Instability in Area, Production, Productivity and Consumption of Millets in India: An Analysis. Indian Journal of Extention Education 55: 158-161

Dhaka A, Singh RK, Muthamilarasan M, Prasad M. 2021. Genetics and genomics interventions for promoting millets as functional foods. Current Genomics 22: 154-163. https://doiorg/10.2174/1389202922666210225084212

Dinesha S, Hosur SR, Toushif PK, Bodiga D, Dechamma DNL, Ashwath MN, Pradhan D. 2023. Sustaino-Resilient Agroforestry for Climate resilience, Food security and Land degradation neutrality. In: Raj AJ, Jhariya MK, Banerjee A, Nema S, Bargali K (eds) Land and Environmental Management through Forestry. Wiley-Scrivener publishers, p. 217-245. https://doi.org/10.1002/9781119910527.ch9

Dinesha S, Raj A, Bhanusree MR, Balraju W, Rakesh S, Raj WG, Jha RK, Neeraj, Kumar K. 2022. Agroforestry Assisted Natural Farming in India: Challenges and Implications for Diversification and Restoration of Agroecosystem. International Journal of Environment and Climate Change 12: 1053-1069

De Valença AW, Bake A, Brouwer ID, Giller KE. 2017. Agronomic biofortification of crops to fight hidden hunger in sub-Saharan Africa. Global Food Security 12: 8-14

Fahad S, Chavan SB, Chichaghare AR, Uthappa AR, Kumar M, Kakade V, Poczai P. 2022. Agroforestry Systems for Soil Health Improvement and Maintenance. Sustainability 14: 14877

Fahmi MKM, Dafa-Alla DAM, Kanninen M, Luukkanen O. 2018. Impact of agroforestry parklands on crop yield and income generation: case study of rainfed farming in the semi-arid zone of Sudan. Agroforestry Systems 92: 785-800

Gaff DF, Oliver M. 2013. The evolution of desiccation tolerance in angiosperm plants: a rare yet common phenomenon. Functional Plant Biology 40: 315-328

Handa AK, Dev I, Rizvi RH, Kumar N, Ram A, Kumar D, Kumar A, Bhaskar S, Dhyani SK Rizvi J. 2019. Successful Agroforestry Models for Different Agro-Ecological Regions in India.

Hendriks SL, Montgomery H, Benton T, Badiane O, de la Mata GC, Fanzo J, Soussana JF. 2022. Global environmental climate change, covid-19, and conflict threaten food security and nutrition. bmj, p. 378. https://doi.org/10.1136/bmj-2022-071534

Jamra G, Shah P, Agarwal A, Sharma D, Kumar A. 2022. Endogenous phytonutrient, phytochemical, and phytohormone levels modulate in-vitro callus induction and plant regeneration in finger millet (*Eleusine coracana*) genotypes. Plant Biosystem 156: 700-709

Kadapa S, Gunturi A, Gundreddy R, Kalwala SR, Mogallapu UB. 2023. Agronomic Biofortification of Millets: New Way to Alleviate Malnutrition. In: Millets - Rediscover Ancient Grains. Intech Open. https://doi.org/10.5772/intechopen.110805

Kulkarni S, Rao S, Desai BK, Basavaneppa MA, Bhat SN, Yogeesh NL. 2020. Studies on status of soil microbial population in foxtail millet–

Melia dubia based agroforestry system under organic nutrient management practices. Journal of Pharmacognocy and Phytochemistry 9: 2064-2067

Kurukulasuriya P, Mendelsohn R. 2008. Crop switching as a strategy for adapting to climate change. African Journal of Agricultural and Resource Economics 2: 105-126. http://dx.doi.org/10.22004/ag.econ.56970

Maitra S. 2020. Potential horizon of brown-top millet cultivation in drylands: A review. Crop Research 55: 57-63

Manasa CPA, Salimath SK, Hedge R, Gooli M. 2022. Bund and Boundary Plantation: A Prominent Feature of Indian Agroforestry. Indian Journal of Agroforestry 24: 52-59

Meena RP, Joshi D, Bisht JK, Kant L. 2021. Global scenario of millets cultivation. In: Kumar A, Tripathi MK, Joshi D, Kumar V. (eds) Millets and Millet Technology. Springer, Singapore, p. 33

Meena SS, Nagar RP. 2018. Evaluation of neem strains and productivity of pearl millet and cluster bean under agroforestry system in semi-arid climate. Range Management and Agroforestry 39: 121-125

Milner-Gulland EJ, Mace R, Scoones I. 1996. A model of household decisions in dryland agropastoral systems. Agricultural Systems 51: 407-430

Mokgolodi NC, Setshogo MP, Shi LL, Liu YJ, Ma C. 2011. Achieving food and nutritional security through agroforestry: a case of *Faidherbiaalbida* in sub-Saharan Africa. Forest Studies in China 13: 123-131. https://doi.org/10.1007/s11632-011-0202-y

Nayana H Veni CP. 2020. SWOT analysis of Zero Budget Natural Farming. Biotica Research Today 3: 547-549

Obilana AB. 2003. Overview: importance of millets in Africa. In: Belton PS, Taylor JRN (eds.) Proceedings of the workshop on the proteins of sorghum and millets: enhancing nutritional and functional properties for Africa. Pretoria, South Africa, p. 28

Pardo J, VanBuren R. 2021. Evolutionary innovations driving abiotic stress tolerance in C4 grasses and cereals. The Plant Cell 33: 3391-3401.

Patel B, Patel A, Gami B, Patel P. 2020. Energy balance, GHG emission and economy for cultivation of high biomass varieties of bamboo, sorghum, and pearl millet as energy crops at marginal ecologies of Gujarat state in India. Renew. Ener. 148, 816-823

Payne WA. 2000. Optimizing crop water use in sparse stands of pearl millet. Agronomy Journal 92: 808-814.

Rao CS, Rakesh S, Kumar GR, Pilli K, Manasa R, Sahoo S, Swamy GN. 2022. Technologies, Programs, and Policies for Enhancing Soil Organic

Carbon in Rainfed Dryland Ecosystems of India. In: Plans and Policies for Soil Organic Carbon Management in Agriculture. Springer Nature, Singapore, p. 27

Rao ND, Poblete-Cazenave M, Bhalerao R, Davis KF, Parkinson S. 2019. Spatial analysis of energy use and GHG emissions from cereal production in India. Science of the Total Environment 654: 841-849

Rasmussen LV, Rasmussen K, Reenberg A, Proud S. 2012. A system dynamics approach to land use changes in agro-pastoral systems on the desert margins of Sahel. Agriculal Systems 107: 56-64

Ravula P, Kasala K, Chakraborty A. 2022. Farming, festivals, and food cultures among indigenous communities in Telangana, India. FolkLife 60: 115-134

Rocha V, Duarte MC, Catarino S, Duarte I, Romeiras MM. 2021. Cabo Verde's Poaceae flora: A reservoir of crop wild relatives' diversity for crop improvement. Front. Plant Science 12: 630217

Rosenstock TS, Dawson IK, Aynekulu E, Chomba S, Degrande A, Fornace K, Steward P. 2019. A planetary health perspective on agroforestry in Sub-Saharan Africa. One Earth 1: 330-344

Sabuz AA, Rana MR, Ahmed T, Molla MM, Islam N, Khan HH, Shen Q. 2023. Health-Promoting Potential of Millet: A Review. Separations 10: 80

Sarkar P, Kumar DHL, Dhumal C, Panigrahi SS, Choudhary R. 2015. Traditional and ayurvedic foods of Indian origin. Journal of Ethnic Food 2: 97-109

Sawardekar SV, Bhave SG, Sawant SS, Gokhale NB, Kelkar VG. 2017. Dapoli-2: A Mineral Rich Variety of Finger millet (*Eleusine coracana* L. Gaertn) Developed through Biotechnological Intervention. Journal of Agricultural Research and Technology 42: 8-14

Saxena R, Vanga SK, Wang J, Orsat V, Raghavan V. 2018. Millets for food security in the context of climate change: A review. Sustainability 10: 2228

Sharma D, Tiwari A, Sood S, Jamra G, Singh NK, Meher PK, Kumar A. 2018. Genome wide association mapping of agro-morphological traits among a diverse collection of finger millet (*Eleusine coracana* L.) genotypes using SNP markers. PloS one 13: e0199444

Singh G, Rathod TR. 2012. Resource use and crop productivity in a *Colophospermummopane* tree based agro-ecosystem in the Indian Desert. Applied Ecology and Environmental Research 10: 503-519

Singh R. 2003. Influence of Water Stress and Micro-Meteorological Changes on Pearl Millet Productivity under Arid Zone Agroforestry Systems. Journal of Agricultural Physics 3: 130-135

Soam SK, Venkatesan P, Sivaramane N, Sreekanth PD, Rakesh S, Tamak J, Rao CS. 2022. Evidence for the economic impacts of Eucalyptus and subabul-based agroforestry in Andhra Pradesh and Telangana. Indian Journal of Agroforestry 24: 52-60

Sow A, Seye D, Faye E, Benoit L, Galan M, Haran J, Brévault T. 2020. Birds and bats contribute to natural regulation of the millet head miner in tree-crop agroforestry systems. Crop protection 132: 105127

Sultan B, Roudier P, Quirion P, Alhassane A, Muller B, Dingkuhn M, Baron C. 2013. Assessing climate change impacts on sorghum and millet yields in the Sudanian and Sahelian savannas of West Africa. Environmental Research Letters 8: 014040

Vikrant M, Rakesh M, Munesh K. 2009. Effect of *Melia azedarach* L. intercropping on *Sorghum bicolor* (L.) Moench varieties in hills of Mizoram. Agricultura Scientia 6: 39-42

INDEX

Chapter 1

Chapter 2

Chapter 3

Chapter 4

Chapter 5

Chapter 6

Chapter 7

Chapter 8

Chapter 9

Chapter 10

Chapter 11

Chapter 12

Chapter 13

Chapter 14

Chapter 15

Chapter 16

Chapter 17